The Minerals, Metals & Materials Series

Takanari Ouchi · Gisele Azimi · Kerstin Forsberg ·
Hojong Kim · Shafiq Alam ·
Neale R. Neelameggham · Alafara Abdullahi Baba ·
Hong Peng
Editors

Rare Metal Technology 2022

TMS

Springer

Editors
Takanari Ouchi
University of Tokyo
Tokyo, Japan

Gisele Azimi
University of Toronto
Toronto, ON, Canada

Kerstin Forsberg
KTH Royal Institute of Technology
Stockholm, Sweden

Hojong Kim
Pennsylvania State University
University Park, PA, USA

Shafiq Alam
University of Saskatchewan
Saskatoon, SK, Canada

Neale R. Neelameggham
IND LLC
South Jordan, UT, USA

Alafara Abdullahi Baba
University of Ilorin
Ilorin, Nigeria

Hong Peng
University of Queensland
Brisbane, QLD, Australia

ISSN 2367-1181 ISSN 2367-1696 (electronic)
The Minerals, Metals & Materials Series
ISBN 978-3-030-92661-8 ISBN 978-3-030-92662-5 (eBook)
https://doi.org/10.1007/978-3-030-92662-5

This Springer imprint is published by the registered company Springer Nature Switzerland AG
The registered company address is: Gewerbestrasse 11, 6330 Cham, Switzerland

Preface

Rare Metal Technology 2022 is the proceedings volume of the eighth symposium on Rare Metal Extraction and Processing, TMS Annual Meeting & Exhibition 2022, held at the Anaheim Convention Center and Anaheim Marriott, Anaheim, California, USA from February 27 to March 3, 2022. This symposium is sponsored by the Hydrometallurgy and Electrometallurgy Committee of the TMS Extraction and Processing Division.

Rare metals have been attracting increasing attention over the past ten years; consequently, the number of participants in this symposium has been increasing. Recently, the efforts to achieve carbon neutrality have accelerated the development of clean technologies on an international level; therefore, the demand for rare metals has been increasing. However, the grade of rare metals in ores is declining, and the energy consumption and CO_2 emissions in mining and extraction processes are increasing. Therefore, the development of new process technologies has become an urgent requirement.

The Rare Metal Extraction and Processing Symposium has been organized to provide a forum for the discussion of the extraction process of rare metals from primary and secondary materials and residues as well as processing techniques used in rare metal production and mineral processing since 2014. The rare metals generally discussed in this symposium are antimony, bismuth, barium, beryllium, boron, calcium, chromium, gallium, germanium, hafnium, indium, manganese, molybdenum, platinum group metals, rare earth metals, rhenium, scandium, selenium, sodium, strontium, tantalum, tellurium, titanium, and tungsten, which are rare metals that are produced in low-tonnage compared to high-tonnage metals such as iron, aluminum, copper, lead, tin, zinc, and silicon.

The papers in this volume cover process development, process optimization, product characterization, environmental impacts, and plant engineering. Furthermore, the papers discuss various processing techniques for mineral beneficiation, separation, extraction, and purification of rare metals, which are based on biometallurgy, hydrometallurgy, pyrometallurgy, electrometallurgy, and other new extractive metallurgy techniques.

I would like to express my sincere gratitude to the co-organizers (co-editors) for their efforts to organize the symposium and edit the proceedings volume *Rare Metal Technology 2022*: Gisele Azimi, Kerstin M. Forsberg, Hojong Kim, Shafiq Alam, Neale R. Neelameggham, Alafara Abdullahi Baba, and Hong Peng. In addition, on behalf of the organizers (editors), I would like to acknowledge the TMS staff, Matt Baker and Patricia Warren, for their assistance in assembling and publishing this proceedings book. Finally, I would like to thank all the authors, speakers, and participants of this symposium and hope that we can continue to create a good forum for discussion to advance the field of rare metal extraction and processing in the next year and beyond.

Takanari Ouchi
Lead Organizer

Contents

Part I Processes for Rare Earth Elements

Rare-Earth and Critical Material Recovery from Peralkaline Volcanic Ores: Minerals Processing, Hydrometallurgy, and Solvent Extraction Separation 3
David Dreisinger

Extraction of Cerium, Lanthanum, and Neodymium from Alluvial Gold Mining Waste from the Bagre-Nechí Mining District in Colombia 17
Luver Echeverry-Vargas, Néstor Ricardo Rojas-Reyes, and Luz Marina Ocampo-Carmona

Rare-Earth Element Adsorption to Gypsum in Hydrometallurgical Processes 27
Farzaneh Sadri and Ahmad Ghahreman

Low-Cost Distillation Technology for Rare-Earth Recycling 41
Chinenye Chinwego, Hunter Wagner, Emily Giancola, Jonathan Jironvil, and Adam Powell

Extraction and Recovery of Rare-Earth Elements and Critical Materials from Coal Waste Using Low Cost Processing Methods: Acid Generation Using Acidithiobacillus Ferrooxidans Mediated Bio-Oxidation of Pyrite 51
Prasenjit Podder, Michael L. Free, and Prashant K. Sarswat

Extraction of Rare Earth Metals from Coal Ash Using Mild Lixiviants in a Single Step Process 63
Riya Banerjee, Saswati Chakladar, and Sanchita Chakravarty

Separation of Scandium by Phosphinic Acid Extractant from Sulfuric Acid Solution 71
Amilton Barbosa Botelho Junior, Denise Crocce Romano Espinosa, and Jorge Alberto Soares Tenório

Part II Solvent Extraction for Rare Metals

Physicochemistry in Service of Process Design: Case of Uranium Recovery by Liquid–Liquid Extraction 81
Alexandre Chagnes

Solvent Extraction Separation of Tungsten and Vanadium from Simulated Leaching Solution of Spent SCR Catalyst 93
Hanyu Wang, Guihong Han, Yanfang Huang, and Shengpeng Su

Extraction Behaviors of Vanadium(V) with Unacidified and Acidified N1923 from a Real Leachate of Vanadium-Titanomagnetite 101
Lei Liu, Zhaobo Liu, Nianwen Pu, Yunfeng Fu, Zhongyu Zhang, Shangchao Du, Guoshan Du, Ninglei Sun, Dehua Wang, and Xiaoyan Li

Novel Environmentally Friendly Leaching Process for Vanadium and Tungsten Recovery from Spent SCR Catalyst 111
Ana Belen Cueva Sola, Jong Hyuk Jeon, Rajesh Kumar Jyothi, and Jin-Young Lee

Effect of Antisolvent Type and Concentration on Morphology and Crystal Size of $(NH_4)_3ScF_6$ Obtained by Antisolvent Crystallization 119
Edward Michael Peters, Michael Svärd, and Kerstin Forsberg

Purification of a Low-Grade Molybdenite Ore for Industrial Steel Production 129
Alafara A. Baba, Christianah O. Adeyemi, Mamata Mohapatra, Mustapha A. Raji, Fausat T. Akanji, and Abdul G. F. Alabi

Hydrometallurgical Extraction of Rhenium from Mo-Bearing Leach Liquor of Molybdenite Flue-Dusts 135
Sadia Ilyas, Hyunjung Kim, Ijaz Ahmad Bhatti, Humma Akram Cheema, and Rajiv R. Srivastava

Extraction for Neodymium from NdFeB Magnet Using Supercritical Carbon Dioxide and Organophosphorus Ligands 147
Jiakai Zhang, Nattanai Kunanusont, Yusuke Shimoyama, and Gisele Azimi

Part III New Extractant and Biometallurgical Processing

Development of New Extractants for Platinum Group Metal Ions in Hydrochloric Acid Solutions 161
Hirokazu Narita

Development of Antagonistic Solvent Extraction Systems for Selective Separation of Copper, Cobalt, and Nickel in Ammoniacal Solution 169
Kurniawan Kurniawan, Jae-chun Lee, Jonghyun Kim, Ha Bich Trinh, and Sookyung Kim

Study on Pre-removal Antimony from Antimony–Gold Concentrate Using Slurry Electrolysis 179
Yonglu Zhang, Chengyan Wang, Xiaowu Jie, Wei Gao, and Shufeng Ruan

Bionanomining: A Revised Insight into Processing of South Africa's Complex Gold Ores 189
Daniel Ogochukwu Okanigbe, Abimbola Patricia Popoola, Nicholas Malatji, Tsietsi Lesufi, and Gift Sekgobela

Research Progress in Biohydrometallurgy of Rare Metals and Heavy Nonferrous Metals with an Emphasis on China 201
Jianzhi Sun, Bowei Chen, He Shang, Xiaolan Mo, and Jiankang Wen

Part IV Process Development and Optimization for Rare Metals

A Highly Selective Metal–Organic Complex Collector for Efficient Mineral Flotation 211
Zhao Wei, Haisheng Han, and Wei Sun

Tellurium Recovery—Development of a Novel Hydrometallurgical Process 225
Shijie Wang

CO_2-Optimized Recovery of Special Metals from Precipitation Residue by Selective Chlorination 237
S. Steinlechner and L. Höber

Comprehensive Processing of Mica Concentrate from Yaroslavsky GRK for Lithium Carbonate, Potassium Sulfate, Aluminum Sulfate, Rubidium, and Cesium 245
Aleksandr Suss, Nataliya Kuznetsova, Sergey Ostashin, and Evgeniy Isakov

Selective Sulfidation for Rare Earth Element Separation 259
Caspar Stinn and Antoine Allanore

Upgrading of Riruwai Cassiterite (Tin Ore) from Doguwa Local Government, North West Nigeria, Using Gravity Beneficiation Technique 279
Furqan Abdulfattah, Daniel Markus Bwala, Ibrahim Abdullah Rafukka, and Suleiman Bolaji Hassan

Method for Producing High-Purity $LiOH{\cdot}H_2O$ Using $Ba(OH)_2$ 289
H. Liu and G. Azimi

Part V Electrochemical Processing and High Temperature Processing

Direct Synthesis of Intermetallic Compounds Through Thermo-Reduction and Electrochemical Deposition 299
Shuhan Wang, Chao Du, Bao Zhang, Qi Wang, Xin Lu, Osamu Takeda, and Hongmin Zhu

Development of Molten Salt Electrolysis of MgO Using a Metal Cathode and Vacuum Distillation to Produce Ultra-High Purity Mg Metal 309
Jungshin Kang, Tae-Hyuk Lee, Hyeong-Jun Jeoung, Dong-Hee Lee, Young Min Kim, Kyung-Woo Yi, Toru H. Okabe, and Jin-Young Lee

Electrochemical Cell Design for Emf Measurements of Liquid Nd-Bi Alloys via Coulombic Titration in LiCl–KCl–$NdCl_3$ Electrolyte 317
Sanghyeok Im, Nathan D. Smith, Stephanie Castro Baldivieso, and Hojong Kim

Two-Phase Rare-Earth Alloys as Reference Electrodes in Molten Chlorides for Reliable Electrochemical Measurements 325
Nathan D. Smith, Stephanie Castro Baldivieso, Timothy Lichtenstein, Sanghyeok Im, and Hojong Kim

Fundamental Study of a Novel Electrolytic Process Using a Cu Cathode in MgF_2-LiF-KCl Molten Salt for Producing Mg Metal from MgO 333
Hyeong-Jun Jeoung, Tae-Hyuk Lee, Kyung-Woo Yi, Jin-Young Lee, Young Min Kim, Toru H. Okabe, and Jungshin Kang

Electrochemical Reduction of Iron Oxides in Aqueous NaOH Electrolyte Including Iron Residue from Nickel and Zinc Electrowinning Processes 341
Geir Martin Haarberg, Bo Qin, and Babak Khalaghi

Author Index 349

Subject Index 351

About the Editors

Takanari Ouchi is a lecturer at the Institute of Industrial Science, University of Tokyo. He received his Ph.D. in Nano-Science and Nano-Engineering from Waseda University in 2011. In this tenure, he developed electrochemical deposition processes to fabricate metal nanostructures with both well-controlled crystallinity and uniformity at a single nanometer scale and demonstrated the applicability of these processes to the fabrication of bit-patterned magnetic recording media for future hard disk drives. After completing his doctoral degree, Dr. Ouchi joined the Massachusetts Institute of Technology (MIT), where he developed liquid metal batteries, which is in principle a bi-directional electrolysis (electro-refining) cell, for application in grid-scale energy storage. As a research scientist, he led a research group to systematically investigate the electrochemical properties of liquid metal electrodes in molten salt electrolytes and developed novel lithium, calcium, and sodium liquid metal batteries. Since he began working as a research associate at the University of Tokyo in 2017, he has developed new recycling processes for rare and precious metals using pyrometallurgical and electrometallurgical methods. Since August 2021, he has been appointed as a lecturer in the Department of Materials and Environmental Science, Institute of Industrial Science, University of Tokyo. Dr. Ouchi has authored 27 peer-reviewed papers, including papers published in *Nature*, *Nature Energy*, and *Nature Communications*, seven conference proceedings, and six review papers; he has delivered over 100 talks at conferences. He has constantly contributed to creating a vibrant field for metal extraction. He is currently serving as the chairperson of the hydrometallurgy and electrometallurgy committee, a member of the Programming Committee, and an organizer of technical symposiums at The Minerals, Metals & Materials Society (TMS).

Gisele Azimi is an Associate Professor and Canada Research Chair in Urban Mining Innovations. She is jointly appointed by the Departments of Chemical Engineering & Applied Chemistry and Materials Science & Engineering at the University of Toronto. She is also a registered Professional Engineer in Ontario. Her research program is aligned well with the "Sustainability" and "Advanced Materials and Manufacturing" research themes. In her research program, she strives to achieve a sustainable future and mitigate the adverse effects of climate change through: (1) advanced recycling and urban mining of waste electrical and electronic equipment (WEEE), utilizing innovative recycling processes based on supercritical fluids; (2) industrial solid waste reduction through waste valorization to produce strategic materials like rare earth elements; (3) development of innovative materials with unique properties with far-reaching applications in structural and energy materials sectors; and (4) energy storage focusing on the development of a new generation of rechargeable batteries made of aluminum. She received her Ph.D. in 2010 from the department of Chemical Engineering and Applied Chemistry at the University of Toronto. Before returning to the University of Toronto as a faculty member in 2014, she completed two postdoctoral appointments at MIT in the departments of Materials Science and Engineering and Mechanical Engineering. She has received a number of awards including the McCharles Prize for Early Career Research Distinction, TMS EPD Young Leaders Award, Emerging Leaders Award in Chemical Engineering, Dean's Spark Professorship, Early Researcher Award, TMS Light Metals/Extraction and Processing Subject Award—Recycling, and Connaught New Researcher Award.

Kerstin Forsberg is Associate Professor in Chemical Engineering at KTH Royal Institute of Technology in Stockholm, Sweden. Her research is focused on separation processes, in particular crystallization. This knowledge is applied in collaborative projects concerning the recovery of resources from waste and to develop new innovative and more environmentally and economically sustainable processes. Forsberg is Head of the Division of Resource Recovery at KTH. She is also the Program Director of the master's program in Chemical Engineering for Energy and Environment and Deputy Director for the Research Platform for Industrial Transformation at KTH. As a member of the Hydrometallurgy and Electrometallurgy Committee at The Minerals, Metals & Materials Society (TMS), Dr. Forsberg has contributed to organizing technical symposia at TMS and as a guest editor for *JOM*.

Hojong Kim is an Associate Professor of Material Science and Engineering and Nuclear Engineering at Penn State University. He received a B.S. degree from Seoul National University and Ph.D. degree at MIT in the Uhlig Corrosion Laboratory. His current research focuses on electro-chemical processes for the separation of energy-critical elements and the development of corrosion-resistant materials. He is the recipient of the NSF CAREER award and the new doctoral new investigator award from the American Chemical Society. He served as chair (2017–2019) and vice-chair (2015–2017) of the Hydrometallurgy and Electrometallurgy Committee of The Minerals, Metals & Materials Society.

Shafiq Alam is an Associate Professor at the University of Saskatchewan, Canada. He is an expert in the area of mining and mineral processing with profound experience in industrial operations, management, engineering, design, consulting, teaching, research, and professional services. As a productive researcher, he has secured 2 patents and has produced over 170 publications. He is the lead/co-editor of 11 books, and one of the editorial board members of two mining and mineral processing journals named *Minerals* (an Open Access Journal by MDPI) and the *International Journal of Mining, Materials and Metallurgical Engineering*. He is the winner of the 2015 Technology Award from the Extraction & Processing Division of The Minerals, Metals & Materials Society (TMS), USA.

With extensive relevant industry experience as a registered professional engineer, Dr. Alam has worked on projects with many different mining industries. He is an Executive Committee Member of the Hydrometallurgy Section of the Canadian Institute of Mining, Metallurgy and Petroleum (CIM). During 2015–2017, he served as the Chair of the Hydrometallurgy and Electrometallurgy Committee of the Extraction & Processing Division (EPD) of TMS. Currently, he is serving on the TMS-EPD Awards Committee. He is a co-organizer of many symposia at international conferences through CIM and TMS. Dr. Alam is one of the founding organizers of the Rare Metal Extraction & Processing Symposium at TMS. He was involved in organizing the International Nickel-Cobalt 2013 Symposium and TMS 2017 Honorary Symposium on applications of Process Engineering Principles in Materials Processing, Energy and Environmental Technologies. Dr. Alam was a co-organizer of the 9th International Symposium on Lead and Zinc Processing (PbZn 2020). He is also involved in organizing the Rare Metal Extraction & Processing Symposium, co-located with the TMS 2022 Annual Meeting and Exhibition in Anaheim, California.

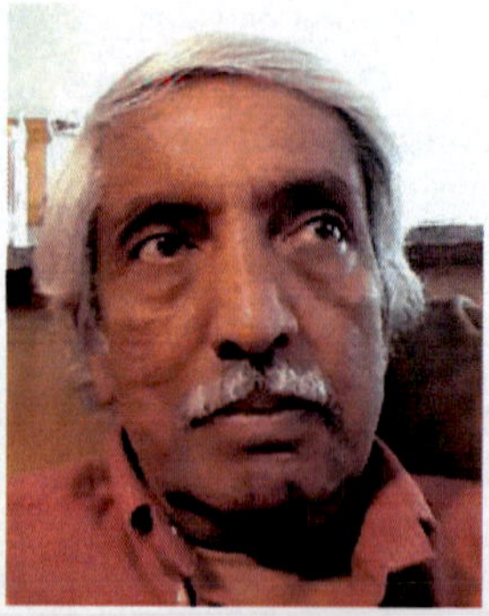

Neale R. Neelameggham IND LLC is involved in international technology and management consulting for metals and chemicals, thiometallurgy, energy technologies, soil reactor design with Agricoal, lithium-ion battery design, OrangeH2, rare earth oxides, and metal manufacturing.

He has more than 38 years of expertise in magnesium production and was involved in the process development of its startup company NL Magnesium to the present US Magnesium LLC, UT until 2011, during which he was instrumental in process development from the solar ponds to magnesium metal foundry. His expertise includes competitive magnesium processes worldwide and related trade cases.

In 2016, Dr. Neelameggham and Brian Davis authored the ICE-JNME award-winning paper "Twenty-First Century Global Anthropogenic Warming Convective Model." He is working on Agricoal®to greening arid soils, and at present energy efficient Orange hydrogen, and methane abatement. He authored the ebook *The Return of Manmade CO to Earth: Ecochemistry*, published through Smashwords in November 2018.

Dr. Neelameggham holds 16 patents and applications and has published several technical papers. He has served in the Magnesium Committee of the TMS Light Metals Division (LMD) since its inception in 2000, chaired in 2005, and since 2007 has been a permanent advisor for the Magnesium Technology Symposium. He has been a member of the Reactive Metals Committee, Recycling Committee, Titanium Committee, and Program Committee for LMD and LMD council.

Dr. Neelameggham was the Inaugural Chair, when in 2008, LMD and the TMS Extraction and Processing Division (EPD) created the Energy Committee and has been a Co-Editor of the Energy Technology Symposium through the present. He received the LMD Distinguished Service Award in 2010. As Chair of the Hydrometallurgy and Electrometallurgy Committee, he initiated the Rare Metal Technology Symposium in 2014 and has been a co-organizer to the present. He organized the 2018 TMS Symposium on Stored Renewable Energy in Coal.

Alafara Abdullahi Baba is a Professor of Analytical/Industrial and Materials Chemistry in the Faculty of Physical Sciences, University of Ilorin, Nigeria. He holds a Ph.D. in Chemistry from the University of Ilorin (2008). His dissertation entitled "Recovery of Zinc and Lead from Sphalerite, Galena and Waste Materials by Hydrometallurgical Treatments" was judged the best in the area of Physical Sciences at the University of Ilorin, Nigeria in 2010. His area of research covers analytical, industrial, and materials chemistry in general with special interests in solid minerals and materials processing through hydrometallurgical routes; reactions in solution for metallurgical applications; preparation of phyllosilicates, porous, and bio-ceramic materials; ore beneficiation including indigenous barite processing for oil and gas applications with developing eco-friendly-cum-low energy synthetic routes for industrial, technological utilities, among others.

Professor Alafara has been cited in many international acclaimed journals of high impact. As an academically inclined scholar with many distinct awards, several of his research breakthroughs of more than 135 publications have been presented at both national and international conferences and workshops for wider recognition. His Unilorin 198th Inaugural Lecture (2021) "Sustainable National Development: Mineral and Material Resources to the Rescue" affirmed the use of locally sourced minerals for industries, a condition for countries to save a significant of foreign earnings.

Professor Alafara has occupied various positions as a chairperson, secretary, and member of relevant academic and professional committees. He was Head, Department of Industrial Chemistry of the University of Ilorin between 2017–2020; Deputy Director, Central Research Laboratories (2014–2017); University of Ilorin Senate Representative for Students' Disciplinary Committee (2019–present); Secretary, Hydrometallurgy and Electrometallurgy Committee of the Extraction & Processing Division (2018–2021); Co-organizer of the Rare Metal Extraction & Processing Symposium and Lead Organizer, Energy Technologies & Carbondioxide Management Symposium (2021); Materials Characterization - Education Committee; EPD and Best Awards Committee Representatives and appointed EPD Council (Education Representative) of The Minerals, Metals & Materials Society (TMS), USA. He is also the current National Vice President of the Materials Science and Technology Society of Nigeria and Fellow of many professional societies including the Chemical Society of Nigeria and Materials Science and Technology Society of Nigeria.

He has supervised and is still supervising many undergraduate and postgraduate research works and is currently serving as an External Examiner to Postgraduate researchers in universities within and outside Nigeria.

Hong Peng is currently the Advance Qld Industry Research Fellow (mid-career) at the School of Chemical Engineering in the University of Queensland (UQ), Australia. He obtained a bachelor's degree in Minerals Engineering and a master's degree in Microbiology at Central South University, China followed by a Ph.D. in Chemical Engineering at UQ. Before joining UQ, Dr. Peng had experience as a chemical engineer in the industry working for the technology research center in Newcastle and Olympic Dam at BHP Billiton between 2006 and 2009. He was the recipient of the 2020 TMS Young Leaders Professional Development Award. Dr. Peng's research focuses on the fundamental aspects of mineral processing, interfacial colloid science, crystal kinetics, and precipitation as well as molecular dynamics simulation. These projects are of interest to the nanobubbles, mine tailings, zeolite, clay minerals, and base metals industries.

Part I
Processes for Rare Earth Elements

Rare-Earth and Critical Material Recovery from Peralkaline Volcanic Ores: Minerals Processing, Hydrometallurgy, and Solvent Extraction Separation

David Dreisinger

Abstract Rare-earth elements are found in many geological settings. Peralkaline volcanic mineralization is found in the Port Hope Simpson Critical Materials District in Labrador, Canada. Minerals such as allanite and fergusonite are found in abundance along with minor chevkinite, bastnasite, and monazite. In some deposits, zircon is present in significant concentrations. The complexity of these materials and the high intrinsic values contained require using every tool in the metallurgical toolkit in order to extract and recover the critical materials present. Conventional mineral processing, hydrometallurgical extraction, and solvent extraction separation are all required to maximize the potential values of the natural deposits.

Keywords Rare earths · Extraction · Mixed rare earth oxide · Foxtrot · Deep Fox

Introduction

The demand for critical materials is increasing dramatically due to climate-driven technological change. Canada is targeting net-zero carbon emissions by 2050 as is the European Union with over 10 countries phasing out internal combustion engines and aiming for 100% green energy by 2040. The United States has issued a number of executive orders to secure domestic critical material supply for the economy of the future. Concurrently, global automakers are making major investments in electric vehicle technologies. The impact of all these changes requires forming new supply chains for critical materials in the rare-earth element, battery materials, solar materials, and industrial metals sector.

Rare-Earth Elements (REEs) have a key role to play in technological change. REEs are used in many sectors including battery alloys, catalysts, ceramics, pigments and glazes, glass polishing powders and additives, metallurgical alloys, permanent magnets, and phospors [1]. The mix of uses and relative values of individual rare earths is constantly changing with technology and economic change. Currently, by

D. Dreisinger (✉)
Department of Materials Engineering, University of British Columbia, 309-6350 Stores Road, Vancouver V6T1Z4, Canada
e-mail: david.dreisinger@ubc.ca

T. Ouchi et al. (eds), *Rare Metal Technology 2022*, The Minerals, Metals & Materials Series, https://doi.org/10.1007/978-3-030-92662-5_1

value, over 90% of the rare earth market is made up of the permanent magnet rare-earth elements (Nd, Pr, Dy, and Tb). Adamas Intelligence estimates that the value of global magnet rare-earth oxide consumption will rise five-fold by 2030, from US$2.98B this year to US$15.65B at the end of the decade (2030). Adamas also predicts that a global shortage of neodymium, praseodymium, and didymium oxide will rise to 16,000 tons per annum in 2030.

All these changes in the market have focused government, industry, and academia on securing future supply to meet the climate challenges ahead of the global community. Future supply requires (1) economic deposits of rare-earth elements, (2) social license to develop these deposits, and (3) processing technology suited to the target deposit.

Historically, the source of rare earths has largely been confined to the treatment of carbonatite deposits (e.g., Mountain Pass, USA, and Bayan Obo, PRC), monazite from beach sands (e.g., India), or ion exchange clay deposits (PRC, Myanmar). Well-developed process flowsheets have been established for extraction from each of these REE feed sources [2].

Search Minerals Inc. (Canada) has been exploring and developing their Critical Rare Earth Element District in south east Labrador, for over a decade. The company is now poised for a "sprint to production" with strong support from the community, indigenous partners, and the Newfoundland/Labrador and Federal Governments of Canada.

Figure 1 shows the map of exploration targets advanced by the Labrador region. The five most advanced deposits are bolded (Deep Fox, Foxtrot, Silver Fox, Awesome Fox, and Fox Meadow). These deposits are all peralkaline volcanogenic in origin [3]. Peralkaline over-saturated deposits (quartz-bearing) are heavy rare-earth element

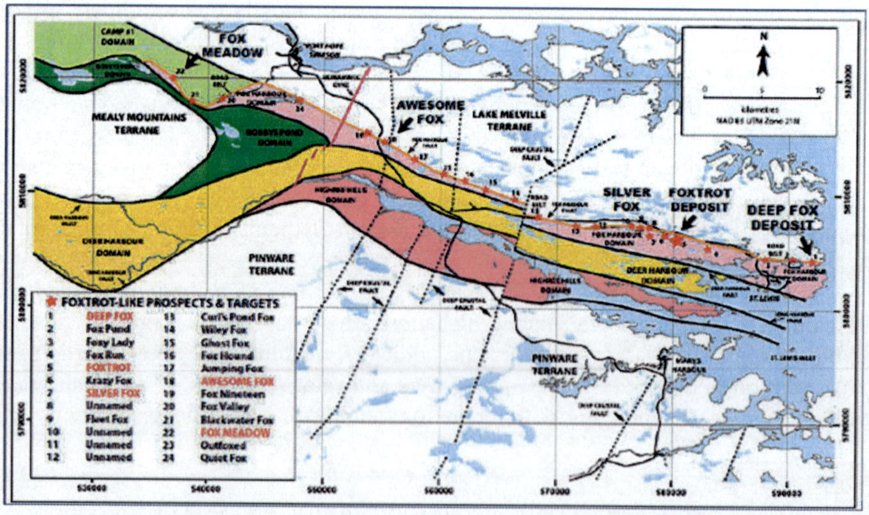

Fig. 1 Search minerals prospects and targets in Labrador, Canada

Table 1 NI 43-101 resources for Foxtrot and Deep Fox deposits [3, 4]

Deposit	Classification	Tonnes	Pr (g/t)	Nd (g/t)	Dy (g/t)
Foxtrot	Indicated	7,392,000	397	1485	191
	Inferred	1,958,000	423	1576	199
Deep Fox	Indicated	2,329,000	403	1486	206
	Inferred	3,902,000	357	1323	181

(HREE)-enriched in contrast to the carbonatite deposits which are light rare-earth element (LREE)-enriched (e.g., Mountain Pass and Bayan Obo). These deposits are formed by the concentration of the REE and other incompatible elements (e.g., Zr/Hf, Nb/Ta, F, U, and Th) in the upper portions of magma chambers. The magmas are either crystallized in place, or are transported to locations proximal to the magma chamber, or are transported to the surface and deposited as volcanic products.

Two of the deposits have had NI 43–101 resources declared (Table 1) [3, 4]. Search Minerals has completed an additional 7000 m of drilling for resource expansion and definition at Deep Fox in the summer of 2021 that will be incorporated in a further resource update in late 2021/early 2022. Channel samples of Fox Meadow and Silver Fox mineralization have also been obtained for mineralogy and testing.

Ore Processing

The processing of Search ores (mainly Foxtrot) has been reported in [5, 6]. The early work focused on conventional mineral beneficiation using flotation, gravity, and magnetic separation. The poor results from the early work led to the development of the Search Minerals Direct Extraction Process for the production of +99% pure mixed rare-earth oxide or equivalent mixed carbonate.

Recent beneficiation work that started on the Silver Fox deposit for Zr/Hf recovery has shown the potential of finer grinding and Wet High-Intensity Magnetic Separation to form a REE concentrate. This development is currently under further laboratory and engineering evaluation.

The use of solvent extraction and ion exchange to remove thorium and uranium from the processing solutions has also been studied.

These developments are outlined below.

Mineralogy Studies

A bulk sample was obtained from a Foxtrot channel sample and was stage crushed to K_{80} of 150 μm and then screened into two-size fractions: +38 and −38 μm for the mineralogical study at SGS Canada. Table 2 shows the minerals identified.

Table 2 Mineral list and formulas

Mineral	Mineral formula	Mineral	Mineral formula
Columbite (Fe)	(Fe, Mn)(Nb, Ta)$_2$O$_6$	Plagioclase	(NaSi, CaAl)AlSi$_2$O$_8$
Bastnasite	(Ce, La)CO$_3$F	K-Feldspar	KAlSi$_3$O$_8$
Synchysite	Ca(Ce, La)(CO$_3$)$_2$F	Biotite	K(Mg, Fe)$_3$(AlSi$_3$O$_{10}$)(OH)$_2$
Monazite	(Ce, La, Pr, Nd, Th, Y)PO$_4$	Quartz	SiO$_2$
Chevkinite	(Ce, La, Ca, Th)$_4$(Fe^{2+},Mg)(Fe^{2+},Ti,Fe^{3+})-(Ti, Fe^{3+})$_2$(Si$_2$O$_7$)$_2$O$_8$	Muscovites/Clays	KAl$_2$(AlSi$_3$O$_{10}$)(OH)$_2$
Fergusonite	(Y, Er, Ce, Fe)NbO$_4$	Amphibole/ Pyroxene	(Ca, Na)(Mg, Fe, Al, Ti)(Si, Al)$_2$O$_6$
Allanite	(Ca, Ce)$_2$(Fe^2,Fe^{3+})Al$_2$O-(SiO$_4$)(Si$_2$O$_7$)(OH)	Carbonates	CaCO$_3$
Zircon	ZrSiO$_4$	Fluorite	CaF$_2$
Apatite	(Ca, Ce, Y)$_5$(PO$_4$,SiO$_4$)$_3$(F, Cl, OH)	Hematite Ilmenite Magnetite	Fe$_2$O$_3$ FeTiO$_3$ Fe$_3$O$_4$

Mineral Abundance by QEMSCAN™

Figure 2 illustrates the normalized mass percent of the REE minerals (excluding zircon). It is apparent that allanite is the primary REE phase. The sample is dominated by quartz (35.8%) and K-feldspar (21.0%), moderate amounts of amphibole/pyroxene (13.7%), plagioclase (12.3%), minor Fe oxides (4.4%), biotite (3.9%) and muscovite/clays (1.6%), and trace amounts of other silicates, carbonates, fluorite, other oxides, and sulphides. REE-Zr minerals include mainly allanite (2.6%), zircon (2.5%), chevkinite (0.3%), fergusonite (0.2%), bastnasite/synchysite (0.1%), monazite (0.1%), and rare columbite. Most of the allanite (2.2%) occurs in the + 38 μm, but most of the zircon (1.5%) in the −38 μm fraction. Note that amphibole and pyroxene are grouped together due to their similar chemistries.

Grain Size Distribution

Figure 3 summarizes the D_{50} or 50% passing value from the cumulative grain size distribution of the fergusonite, bastnasite/synchysite, allanite, monazite, chevkinite, zircon, quartz/feldspars, muscovite, other silicates, oxides, and overall particle-size distribution for the Foxtrot sample. The grain size data indicates that it should be possible to liberate the REE minerals from the barren gangue minerals.

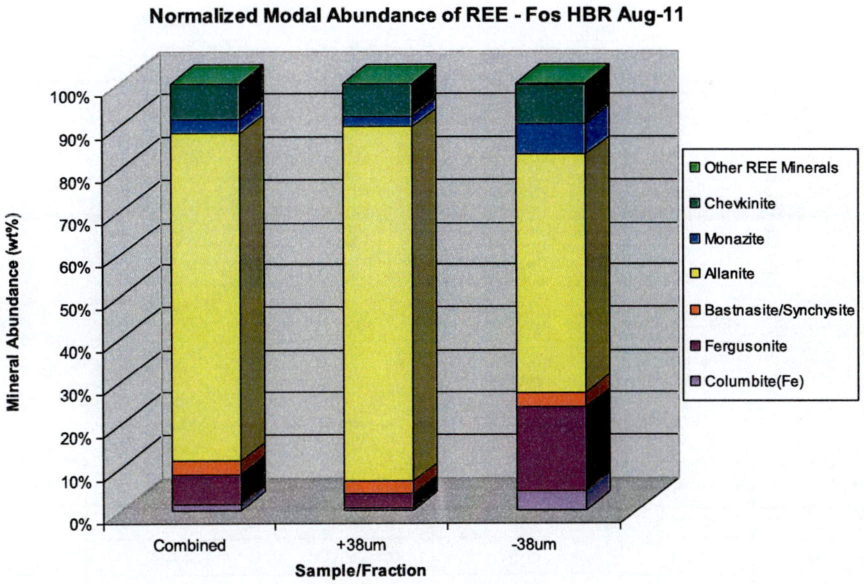

Fig. 2 Normalized mineral abundance of REE minerals

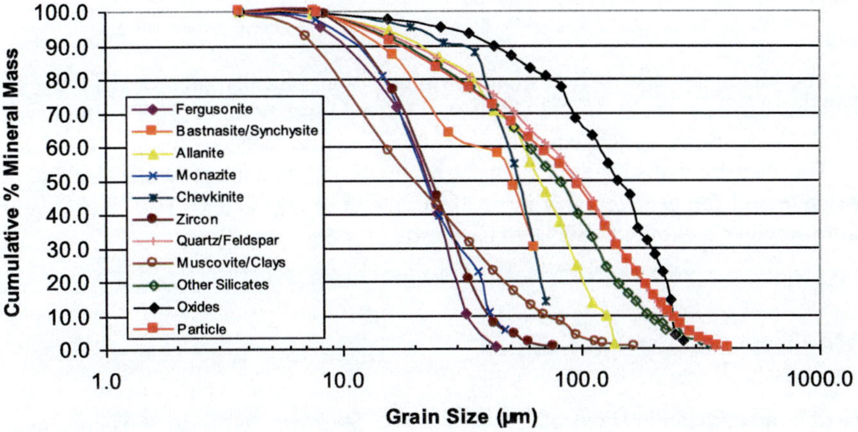

Fig. 3 Cumulative grain size distribution

Elemental Deportment

Elemental deportment was estimated for a number of key elements. The plots for two of the economic elements in the Foxtrot are shown below (Nd, Dy) representative of light and heavy REEs.

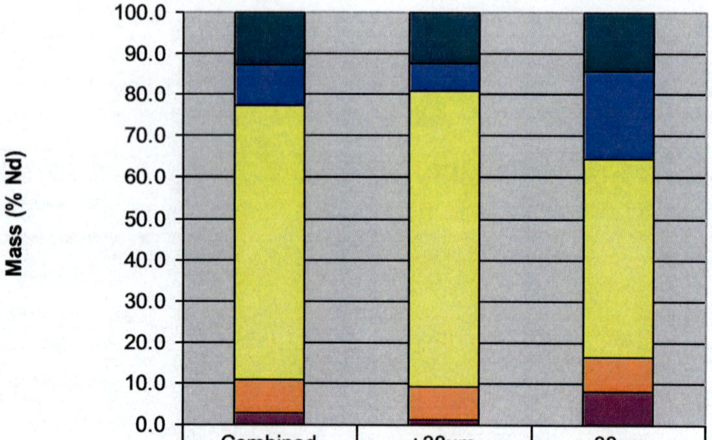

	Combined	+38um	-38um
Chevkinite	12.7	12.2	14.4
Monazite	10.1	6.8	21.3
Allanite	66.4	72.0	47.9
Bastnasite	7.9	7.7	8.5
Fergusonite	2.9	1.4	8.0

Fig. 4 Elemental deportment of neodymium

The elemental distribution of neodymium is graphically presented in Fig. 4. Allanite carries most of the Nd (66.4%), followed by chevkinite (12.7%), monazite (10.1%), bastnasite/synchysite (7.9%), and fergusonite (2.9%).

The elemental distribution of dysprosium is graphically presented in Fig. 5. Allanite and fergusonite carry most of the Dy at 49.3% and 40.5%, respectively, followed by chevkinite (8.8%) and bastnasite (1.4%).

Magnetic Separation Studies

Search mineralization (Foxtrot, Deep Fox, Fox Meadow, and Silver Fox) responds to magnetic separation (www.searchminerals.ca). The key elements of the new beneficiation work are as follows:

- Low-intensity magnetic separation (LIMS) removes ferro-magnetic minerals (e.g., magnetite).
- Wet high-intensity magnetic separation (WHIMS) recovers weakly magnetic materials from the samples (e.g., rare-earth minerals and iron-containing gangue minerals). This material forms the REE concentrate.
- Flotation is an option for recovery of a zircon concentrate

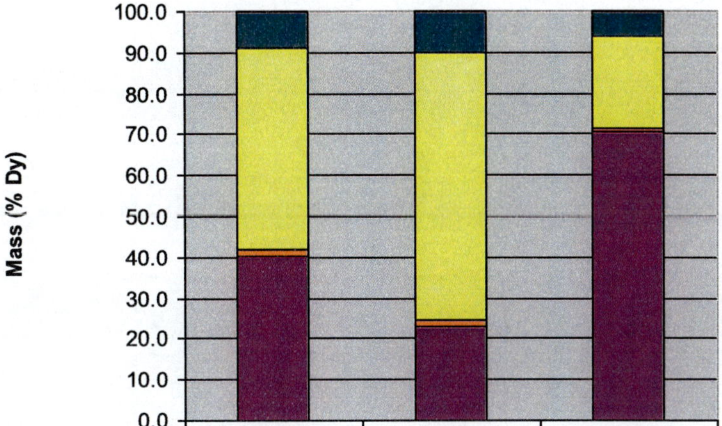

	Combined	+38um	-38um
Chevkinite	8.8	10.3	6.2
Allanite	49.3	65.0	22.3
Bastnasite	1.4	1.6	0.9
Fergusonite	40.5	23.1	70.6

Fig. 5 Elemental deportment of dysprosium

Table 3 shows typical results for LIMS and WHIMS recovery obtained at SGS [7] for a sample of Foxtrot mineralization at −53 μm. Zircon was not floated in this test.

The results of this test were excellent. The WHIMS Concentrate contains 27.7% of the original weight containing 4.52% TREO with 93.4% TREO recovery (TREO is total rare-earth oxide including Y). The LIMS concentrate contains 95.4% Fe_2O_3 (as magnetite) in 6.6% of original weight. The WHIMS tailing contains 65.7% of the original weight.

USE of SX and IX for Impurity Control in the Direct Extraction Process

The Search Minerals Direct Extraction Process was developed for the whole of ore treatment. The Direct Extraction process is now being adapted to the treatment of WHIMS magnetic concentrates. Figure 6 shows the new processing concept with the incorporation of LIMS, WHIMS, and Zr/Hf Flotation.

The Direct Extraction Process has been refined to enhance the purification of the final rare-earth oxide or carbonate product to control U and Th levels to less than 5 g/t. The simplified process flowsheet is shown in Fig. 7 and has been reported in [8]. The U IX process step is best applied in the releach solution loop. The original process

Table 3 LIMS and WHIMS magnetic separation (SGS [7]

		Assays (% of g/t)														
	Wt%	Nd	Pr	Tb g/t	Dy g/t	Y g/t	LREO	HREO	TREO	U g/t	Th g/t	Zn	ZrO_2	SiO_2	Fe_2O_3	CaO
WHIMS	27.7	0.62	0.18	139	843	4329	3.63	0.90	4.52	90	570	0.28	2.0	43.9	20.6	8.9
WHIMS Non-Mags	65.7	0.01	0.00	4	22	236	0.06	0.04	0.10	15.4	12	0.06	1.7	80.1	0.4	0.9
LIMS Mags	6.6	0.03	0.01	11	77	358	0.19	0.08	0.26	12.4	29	0.19	0.3	4.9	95.4	0.4
Head (dir.)	100.0	0.16	0.04	37	226	1116	0.91	0.24	1.14	32	135	0.13	1.3	66.3	11.8	2.8
		Distribution (%)														
	%	Nd	Pr	Tb	Dy	Y	LREO	HREO	TREO	U	Th	Zn	ZrO_2	SiO_2	Fe_2O_3	CaO
WHIMS	27.7	95.9	96.3	92.5	92.0	86.7	94.7	88.2	93.4	69	94.0	60.9	32.0	18.7	44.6	80.6
WHIMS Non-Mags	65.7	2.9	2.6	5.5	5.8	11.2	3.9	9.7	5.1	28.2	4.5	29.7	66.6	80.8	1.9	18.5
LIMS Mags	6.6	1.2	1.1	1.9	2.3	2.1	1.35	2.13	1.51	2.6	1.4	9.3	1.4	0.5	53.5	0.8
Head (dir.)	100	100	100	100	100	100	100	100	100	100	100	100	100	100	100	100

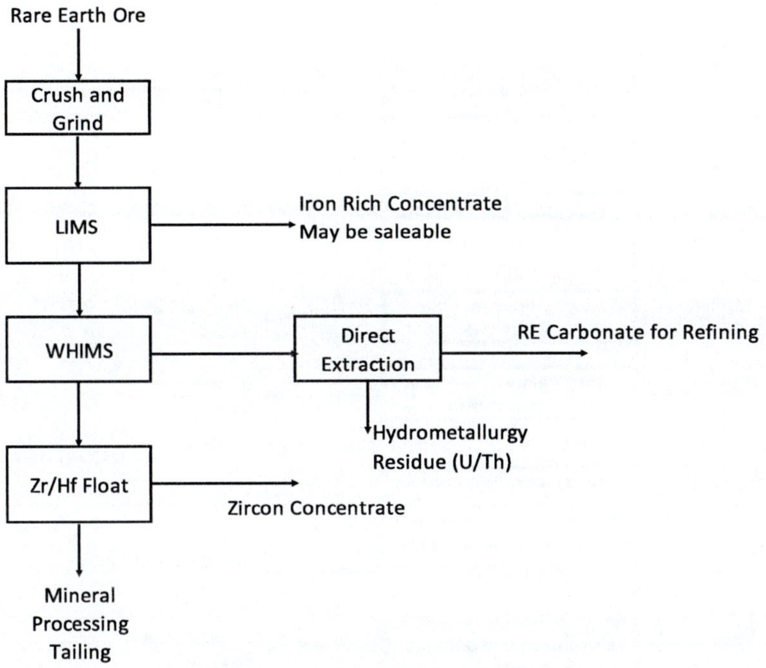

Fig. 6 Search minerals new simplified process flowsheet incorporating magnetic separation and direct extraction

for Th removal from releach solution involved pH adjustment to precipitate Th after releaching. This resulted in significant co-precipitation of REE and a circulating load of REE with increased costs for acid (H_2SO_4) and base (Na_2CO_3) with each cycle. Th removal by IX was tested (from the releach solution) but was unable to achieve suitably low residual Th levels. Accordingly, SX became the focus. Th SX from REE solutions has been reported by [9, 10]. A bench-scale and continuous pilot plant program was operated to test the process as part of an overall pilot plant operation for the application of the Direct Extraction Process. The details were

Organic Solution

- 1.0% Primene JMT (a primary amine), 2.5% Isodecanol (a modifier), and 96.5% Aromatic 150ND (a diluent).

SX Pilot plant:

- Aqueous Solution: Combined U-free releach solution adjusted to pH 1.5.
- Extraction: 2 stages at A/O advance ratio of 5:1, 45 °C, aqueous continuous.

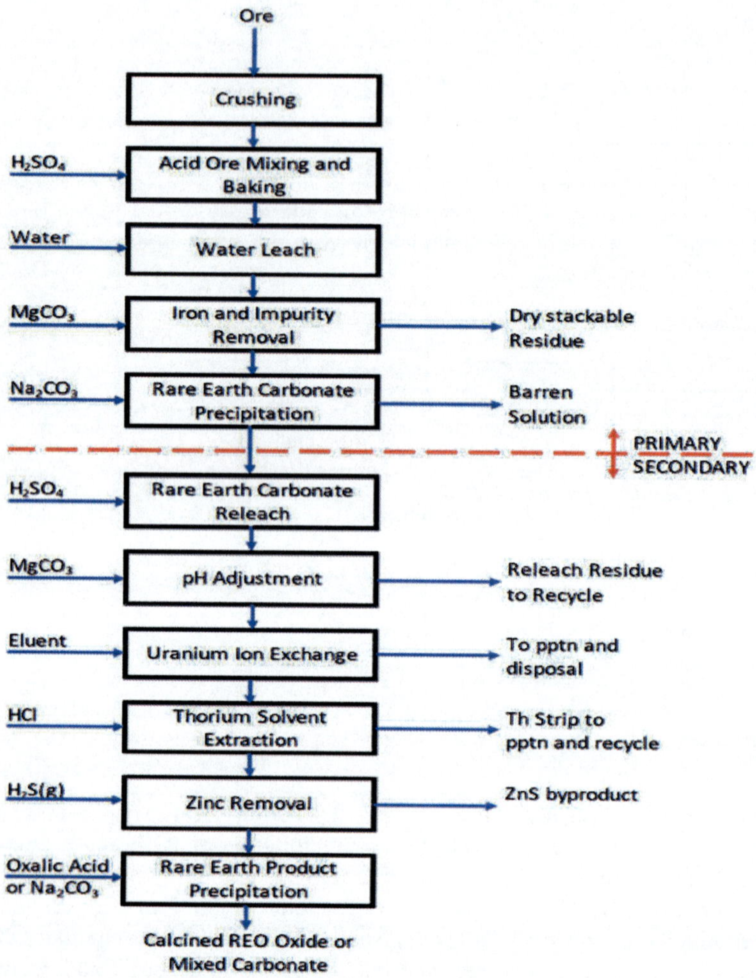

Fig. 7 The direct extraction process with U IX and Th SX

- Scrubbing: 2 stages at A/O advance ratio of 0.5:1 and A/O mixer-settler ratio of 1:1, ambient temperature, aqueous continuous. The scrub solution was 24 g/L H_2SO_4. The scrub solution was added to the extraction feed.
- Stripping: 2 stages at A/O advance ratio of 1.25:1 and A/O mixer-settler ratio of 1:1 at ambient temperature, aqueous continuous. The strip solution was 18 g/L HCl.

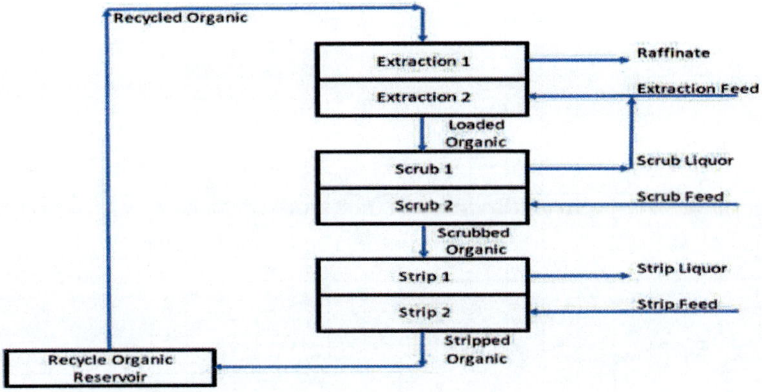

Fig. 8 The Th solvent extraction pilot plant

The pilot plant flowsheet is shown in Fig. 8. The pilot plant was operated for 5 days with consistent operation (Table 4). Th removal from the releach solution showed a drop from 4.72 mg/L Th to < 0.03 mg/L Th in purified raffinate versus <0.1 mg/L of Th target. The RE deportment to the strip solution was <1% (and will be recovered when recycled to the primary circuit).

Conclusions

The demand for critical materials is increasing dramatically, and this is likely to accelerate as the global response to climate change is mounted. The recovery of rare-earth elements is one sector that must explore new sources of supply to provide the magnet-making elements (Nd, Pr, Dy, and Tb) that are necessary for electric vehicle and wind power applications. The Search Minerals deposits in Labrador Canada have been investigated over many years. The new processing flowsheet being carried into the "sprint to production" utilizes beneficiation by LIMS and WHIMS followed by the application of the enhanced Direct Extraction Process on the REE containing WHIMS concentrate. The incorporation of a thorium solvent extraction process has allowed successful management of this critical impurity to the required levels in the mixed REE material directed to separation.

Table 4 Th SX pilot plant summary results

	Aqueous concentration (mg/L)							Organic solution (mg/L)						Fract
	Feed	Ext 1	Ext 2	Scr 1	Scr2	Str 1	Str 2	Ext 1	Ext 2	Scr 1	Scr 2	Str 1	Str 2	To Strip
La	3060	2670	2660	62.2	15.4	70.1	0.26	76	74	47	40	<3	<3	0.32%
Ce	6550	5670	5720	152	41.5	329	1.3	363	261	198	182	<3	<3	0.70%
Pr	758	658	663	18.5	5.12	43.2	0.16	35	34	26	24	<3	<3	0.79%
Nd	2730	2400	2420	68	19	135	0.48	119	117	88	80	<9	<9	0.69%
Sm	459	468	463	11	2.99	20.2	0.07	18	18	13	11	<4	<4	0.61%
Eu	27.2	23	23.3	0.52	0.14	0.74	<0.03	<3	<3	<3	<3	<3	<3	0.38%
Gd	464	420	426	8.8	2.17	7.4	<0.03	8	8	4	3	<3	<3	0.22%
Tb	71.1	64.6	64.9	1.36	0.33	1.08	<0.03	<3	<3	<3	<3	<3	<3	0.21%
Dy	412	371	376	7.47	1.77	4.96	<0.05	6	6	<4	<4	<4	<4	0.17%
Ho	79.3	71.5	72.1	1.36	0.3	0.62	<0.02	<2	<2	<2	<2	<2	<2	0.11%
Y	1750	1610	1540	27	5.3	6.48	0.03	18	18	6	3	<1	<1	0.05%
Er	214	191	193	3.53	0.74	1.28	<0.04	<4	<4	<4	<3	<4	<4	0.08%
Tm	29.6	26	25.9	0.49	0.1	0.16	<0.04	<4	<4	<4	<3	<4	<4	0.08%
Yb	164	149	147	2.85	0.58	0.91	<0.02	2	2	<2	<2	<2	<2	0.08%
Lu	21	18.8	18.9	0.34	0.06	0.08	<0.03	<3	<3	<3	<3	<3	<3	0.05%
Th	4.72	0.04	<0.03	<0.03	<0.07	34	0.8	<3	20	20	21	<3	<3	100%
U	0.03	0.02	0.02	<0.02	<0.02	<0.02	<0.02	<2	<2	<2	<2	<2	<2	
Al	2080	1910	1910	<0.8	<0.8	<0.8	<0.8	31	32	29	31	25	27	
Fe	9	8.2	8.3	0.3	<0.2	0.8	<0.2	2	<1	<1	<1	<1	<1	
Mg	10,500	9640	9640	2.23	0.72	0.26	0.29	<1	<1	<1	<1	<1	<1	

(continued)

Table 4 (continued)

	Aqueous concentration (mg/L)							Organic solution (mg/L)						Fract
Ca	1290	1240	1210	<9	<9	<9	<9	9	10	9	10	9	9	
Na	290	257	260	<2	<2	<2	<2							
Mn	319	289	290	0.12	<0.04	<0.04	<0.04	<0.4	<0.4	<0.4	<0.3	<0.4	<0.4	
Zn	2450	2280	2270	<0.7	<0.8	<0.9	<0.10	<6	<6	<6	<6	<6	<6	

Acknowledgements The author thanks the management of Search Minerals for permission to publish this paper. The financial support (to Search Minerals) of the Newfoundland Department of Industry, Energy and Technology, and from the Government of Canada Atlantic Canada Opportunities Agency (ACOA) is gratefully acknowledged.

References

1. Adamas Intelligence (2020) Rare earth magnet market outlook 2030. Adamas Intelligence, Ontario
2. Gupta CK, Krishnamurthy N (2005) Extractive metallurgy of rare earths. CRC Press, Boca Raton, USA
3. Roscoe Postle Associates (RPA) (2019) Technical report on the deep fox project in Labrador, Canada. NI 43–101 Report prepared by Masun KM. Search Minerals Inc., British Columbia
4. Roscoe Postle Associates (RPA) (2016) Technical report on the foxtrot project in Labrador, Canada. NI 43–101 Report prepared by Masun KM, Weir IC, Goode JR. Search Minerals Inc., British Columbia
5. Dreisinger DB, Clucas JD, Verbaan N, Grammatikopoulos T, Aghamirian M, Forstner C (2012) The processing of REEs from search minerals foxtrot resource. In: Goode JR, Moldoveanu G, Reyat MS, Metsoc CIM (eds) Proceedings of rare earths 2012. Montreal, pp 81–94
6. Dreisinger DB, Andrews G (2017) The economics of the search minerals direct extraction process for rare earth element recovery. In: Kim H et al (eds) Rare metal technology 2017. The minerals, metals & materials series, p 15. https://doi.org/10.1007/978-3-319-51085-9_1
7. SGS Canada (2021) Unpublished Results
8. Dreisinger D, Andrews G, Johnson MV (2021) Uranium and thorium removal from a rare earth sulfate solution by ion exchange and solvent extraction. In: Presentation at TMS Meeting 2021
9. Crouse DJ Jr, Brown KB (1959) The amex process. Ind Eng Chem 51(12):1461–1464
10. Amaral JCBS, Morais CA (2010) Thorium and uranium extraction from rare earth elements in monazite sulfuric acid liquor through solvent extraction. Miner Eng 23:498–503

Extraction of Cerium, Lanthanum, and Neodymium from Alluvial Gold Mining Waste from the Bagre-Nechí Mining District in Colombia

Luver Echeverry-Vargas, Néstor Ricardo Rojas-Reyes, and Luz Marina Ocampo-Carmona

Abstract The objective of this paper was to evaluate the extraction of cerium, lanthanum, and neodymium by exploring different leaching conditions with hydrochloric acid and sulfuric acid from a dephosphorized monazite concentrate obtained from alluvial gold mining waste from the Bagre-Nechí mining district in Colombia. The highest element extraction yields are achieved with H2SO4 as the leaching medium versus HCl at all evaluated concentration, time, and temperature conditions. The addition of 10% (v/v) H2O2 was found to increase rare-earth dissolution up to 93% in the H2SO4 medium. Thermodynamic analysis indicates that H2SO4 solutions show a higher capacity to solubilize REEs as opposed to HCl solutions, thus supporting the experimental results.

Keywords Rare-earth elements · Monazite · Hydrometallurgy · Mining waste

Introduction

Rare-earth elements (REEs) are a group of chemical elements comprising all lanthanides, yttrium, and scandium. They are often divided into two categories: light rare-earth elements (LREEs), comprising La, Ce, Nd, and Pr, and heavy rare-earth elements (HREEs), ranging from Sm to Lu, Sc, and Y [1]. Due to their particular spectroscopic and magnetic properties, REEs play a significant role in the global economy because of their importance in numerous fields of advanced materials science; therefore, the industrial demand for them has increased significantly [2].

L. Echeverry-Vargas (✉) · L. M. Ocampo-Carmona
Departamento de Materiales y Minerales, Universidad Nacional de Colombia, Calle 75 N. 79 A, 050034 Sede Medellín, Colombia
e-mail: ljecheverryv@unal.edu.co

L. M. Ocampo-Carmona
e-mail: lmocampo@unal.edu.co

L. Echeverry-Vargas · N. R. Rojas-Reyes
Instituto de Minerales CIMEX, Universidad Nacional de Colombia, Calle 75 N. 79 A, 050034 Sede Medellín, Colombia
e-mail: nrrojasr@unal.edu.co

T. Ouchi et al. (eds), *Rare Metal Technology 2022*, The Minerals, Metals & Materials Series, https://doi.org/10.1007/978-3-030-92662-5_2

In nature, REEs do not occur in their elemental state or as individual rare-earth compounds. REEs dispersed in the earth's crust occur as mixtures in many rock formations such as basalts, granites, gneiss, shales, and silicate rocks [3]. Their strong affinity for oxygen causes them to be found mainly as trivalent cations in carbonates, oxides, phosphates, and silicates [4]. Approximately 95% of all the world's rare-earth resources occur in only three minerals: monazite ((Ce, La, Nd) PO_4), bastnasite ((Ce, La, Y) CO_3F), and xenotime ((Y, Yb) PO_4). Therefore, these minerals are the main sources for REE extraction [5]. In monazite, the rare-earth oxide (REO) content ranges from 35 to 71%, while in xenotime and bastnasite it ranges from 52 to 54% and from 70 to 74%, respectively [6–8].

Research conducted in the region of the Bagre-Nechí mining district in Colombia indicates that this alluvial gold deposit contains mineral resources with the potential to be used to obtain REEs in the tailings of mining operations [9, 10]. The sustainability issue associated with mining activities in this area poses responsibility challenges to the mining areas, which should be geared towards the establishment of new technologies and processes that enable the recovery of resources from the waste produced, particularly the recovery of the monazite present in mining waste, in order to obtain REEs. In this regard, detailed studies of REE recovery from mine tailings produced at this site are needed. Although the REE content in tailings may be lower than in primary sources, their processing can be justified if environmental benefits are considered, e.g., mine site remediation and soil restoration. In addition, tailing mining does not rely on an effective collection, transportation, and sorting scheme; recyclable material is simply available on-site [11].

The objective of this paper is to study different conditions of REE dissolution with HCl and H_2SO_4 from a dephosphorized rare-earth concentrate obtained from alluvial gold mining waste from the Bagre-Nechí mining district in Colombia. The leaching conditions evaluated include acid concentration, process temperature, treatment time, and the effect of hydrogen peroxide on REE recovery. In addition, a thermodynamic analysis of the products formed in the leaching process was carried out by drawing species distribution diagrams.

Materials and Methods

Materials

In this paper, a dephosphorized rare-earth concentrate previously obtained by physically concentrating monazite from residual sands from alluvial gold mining was used as raw material. The monazite concentrate was subjected to an alkaline dephosphorization process followed by water rinsing, filtering, and drying. The reagents used in this work were analytical grade and their solutions were prepared with distilled water.

Table 1 Evaluated leaching conditions

Variable	Condition					
Acid concentration, M	0.5	1	2	3	4	
Time, min	15	30	60	90	120	
Temperature, °C	30	40	50	60	70	80

Leaching Tests

The rare-earth concentrate was subjected to leaching processes in HCl and H_2SO_4 solutions at different concentrations, temperatures, and treatment times, as shown in Table 1, in order to evaluate the effect of these variables on the leaching yield of rare-earth elements. The leachates were conducted in a three-hole Pyrex reactor equipped with a reflux condenser, constant agitation was maintained using a mechanical stirrer, and the slurry density was kept constant at 30 g/L. In all experiments, undissolved impurities remained as residue. Lanthanum, cerium, and neodymium contents present in the acid leach solutions were analyzed using a PerkinElmer Optima 8300 model ICP-OES (Inductively Coupled Plasma-Optical Emission Spectrometer).

Results and Discussion

Effect of Acid Concentration

The effect of HCl and H_2SO_4 concentration within the 0.5–4.0 M range on the extraction of La, Ce, and Nd was studied. Temperature was kept constant at 80 °C, leaching time at 90 min, and slurry density at 30 g/L, as shown in Fig. 1a, b. From these figures, it can be noted that increasing the concentration of HCl and H_2SO_4 enhances the elements' extraction efficiency, suggesting that acid is the limiting reagent in the leaching reaction. When the concentration of HCl changes from 0.5 to 4.0 M, the extraction of Ce increases from 1.5 to 30.4%, that of Nd from 23.3 to 48.1%, and that of La from 55.6 to 77.5%. On the other hand, when the concentration of H_2SO_4 is changed from 0.5 to 4.0 M, the extraction efficiency of Ce increases from 8.5 to 80.4%, that of Nd from 26.3 to 83.6%, and that of La rises from 62.3 to 91.2%. In all cases, it is evident that the extraction efficiency of La, Ce, and Nd is higher in H_2SO_4 than in HCl, and the extraction of La is higher, followed by that of Nd, with Ce in the last place.

The lower extraction yield for cerium is possibly due to the fact that Ce is oxidized from its trivalent to a tetravalent state (Ce^{3+} to Ce^{4+}) during the roasting process and exists as CeO_2 [12]. In this case, CeO_2 is poorly soluble at low concentrations of HCl and H_2SO_4 (0.5 M); a higher oxidation state in Ce favors the dissolution of La and Nd [13]. The results indicate that the oxidation capacity of CeO_2 improves with

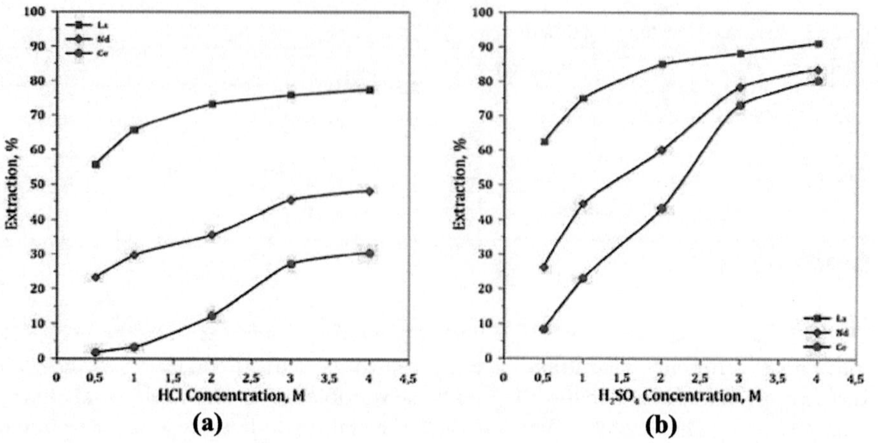

Fig. 1 Effect of **a** HCl and **b** H_2SO_4 concentration on leaching of La, Nd, and Ce from treated monazite

the increase of hydrogen ion concentration due to the increase in acid concentration. This leads to the formation of soluble compounds.

Effect of Temperature

In general, the dissolution of any metallic element is affected by the leaching temperature because the dissolution reaction is an endothermic process [14]. In this paper, experiments were conducted to evaluate the leaching behavior of La, Ce, and Nd in HCl and H_2SO_4 in a temperature range from 30 to 80 °C, keeping the leaching time constant at 90 min, the concentration of HCl and H_2SO_4 at 2.0 M, and the slurry density at 30 g/L. As seen in Fig. 2a, b, the extraction efficiency of La, Ce, and Nd increases when the leaching temperature is raised. The results indicate that increasing the temperature from 30 to 80 °C in HCl leaching increases Ce extraction efficiency from 1.0 to 9.3%, Nd extraction increases from 22.4 to 35.6%, and that of La increases from 51.2 to 73.2%. When the same evaluation is performed in H_2SO_4 leaching, it is observed that Ce extraction increases from 5.2 to 34.3%, Nd extraction efficiency increases from 22.7 to 61.1%, and that of La increases from 49.8 to 85.1%. The results show that, in all cases, the extraction of La, Ce, and Nd has a higher efficiency in sulfuric acid than in hydrochloric acid. Furthermore, in all scenarios, the order of extraction is Ce < Nd < La.

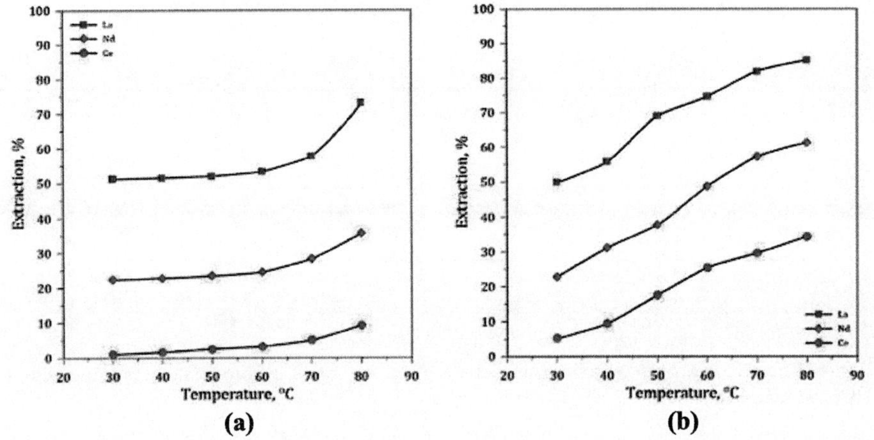

Fig. 2 Effect of temperature on leaching of La, Nd, and Ce from treated monazite with **a** HCl and **b** H_2SO_4

Effect of Time

In this paper, the effect of leaching time on the recovery of La, Ce, and Nd was evaluated, varying the time between 15 and 120 min. The concentration of HCl and H_2SO_4 was kept constant at 2.0 M, the leaching temperature at 80 °C, and the slurry density at 30 g/L. Figure 3a, b shows that increasing the leaching time in aqueous HCl solution increases the extraction of Ce (from 5.1 to 10.4%), Nd (from 28.1 to 37.3%), and La (from 59.8 to 75.4%), while increasing the leaching time in the

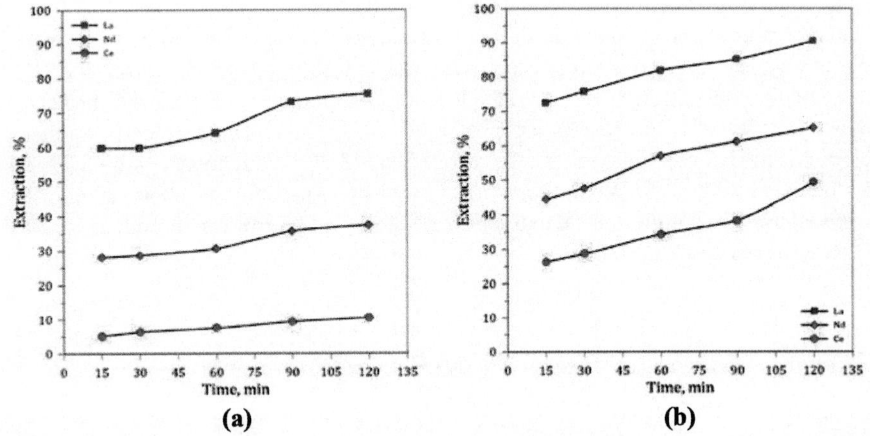

Fig. 3 Effect of leaching time on dissolution of La, Nd, and Ce from treated monazite with **a** HCl and **b** H_2SO_4

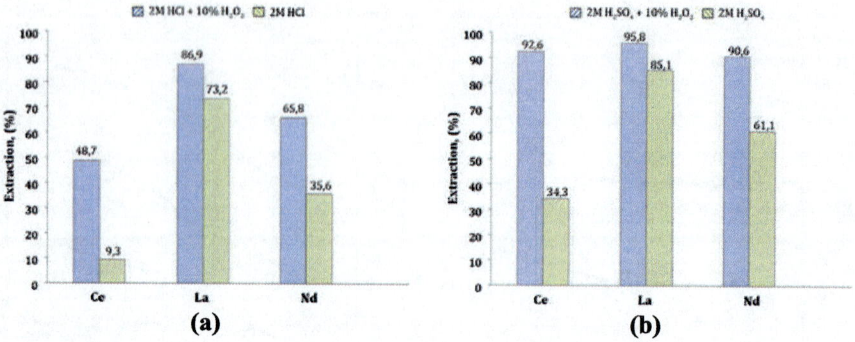

Fig. 4 Effect of hydrogen peroxide on leaching of La, Nd, and Ce from treated monazite with **a** HCl and **b** H_2SO_4

H_2SO_4 medium increases the recovery of Ce (from 26.3 to 49.4%), Nd (from 44.3 to 65.2%), and La (from 72.5 to 90.4%). Leaching times longer than 120 min in HCl and H_2SO_4 solutions had no significant effect on these elements' extraction efficiency.

Effect of the Addition of Hydrogen Peroxide (H_2O_2)

Hydrogen peroxide (H_2O_2) is widely used in hydrometallurgical processes due to its oxidizing characteristics [15–18]. In addition, it is considered one of the cleanest and most versatile chemicals, and is widely used in the industry [19, 20]. In acidic solutions, H_2O_2 is one of the strongest oxidants known. The oxidizing capacity of hydrogen peroxide is so strong that it is considered a highly reactive oxygen species [21].

The effect of H_2O_2 concentration on the extraction of La, Ce, and Nd was investigated by keeping the hydrogen peroxide ratio at 10% (v/v) in solution, the HCl and H_2SO_4 concentration at 2.0 M, the leaching temperature at 80 °C, and the slurry density at 30 g/L. The results are presented in Fig. 4, b. The extraction of these elements in HCl solution yields a Ce recovery of 48.7%, Nd recovery of 65.8%, and La recovery of 86.9%. When the action of hydrogen peroxide was evaluated in an aqueous H_2SO_4 solution, a Ce extraction efficiency of 92.6% was found, as well as 90.6% for Nd and 95.8% for La.

Thermodynamic Analysis for LREE Leaching

The monazite used in this study was subjected to a roasting process in conditions where it eliminated phosphate; therefore, when performing the thermodynamic analysis, PO_4^{3-} was excluded from the system. The LREEs present in the treated monazite, such as lanthanum, cerium, and neodymium in their oxidation state, exhibit a strong tendency to form complex ions with strong acids such as HCl and H_2SO_4, as reported in several papers [22–24]. The results indicate that, under the leaching conditions evaluated, the order of extraction of the elements of interest is as follows: Ce < Nd < La. It can also be seen that higher extractions of these elements are achieved with H_2SO_4 as opposed to HCl. To further investigate the leaching reactions of these elements in the dephosphorized monazite product, distribution diagrams of La, Ce, and Nd in HCl and H_2SO_4 were plotted, as shown in Figs. 5 and 6, respectively. The diagrams for these systems were made using MEDUSA® software and supplemented with the Hydra database, using the values of the critical stability constants reported in the NIST database.

The species distribution diagrams for La^{3+}, Ce^{3+}, and Nd^{3+} ions—the concentrations of these species were calculated from the chemical assay—in 2.0 M hydrochloric acid aqueous solution—characteristic concentration of the leaching experiments carried out in this work—are shown in Fig. 5. In this system, Ce is leached in its trivalent form (Ce^{3+}) and subsequently forms two soluble complexes with the chloride ion ($CeCl^{2+}$ and $CeCl_2^+$) at pH values below 4.0. A similar result occurs for La and Nd, where these species leach in their trivalent form (La^{3+} and Nd^{3+}) and then form species with Cl^- ($LaCl^{2+}$, LaC12 + , NdC12 + and $NdCl^{2+}$) at pH values below 8.0 for lanthanum and 7.0 for neodymium.

According to the species formation diagrams, the leaching capacity of Ce is comparatively higher than that of La and Nd at pH < 4. Under these conditions, cerium is found mostly as $CeCl_2^+$. However, experimental data indicate that the

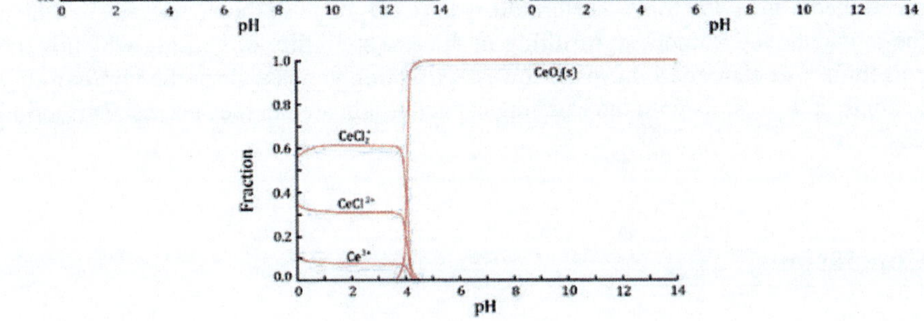

Fig. 5 Distribution diagrams of species for Ce^{3+}, La^{3+}, and Nd^{3+} as a function of pH in the Cl^- system at 25 °C with the following concentrations: 2.0 M Cl^-, 0.01 M La^{3+}, 0.007 M Ce^{3+}, and 0.008 M Nd^{3+}

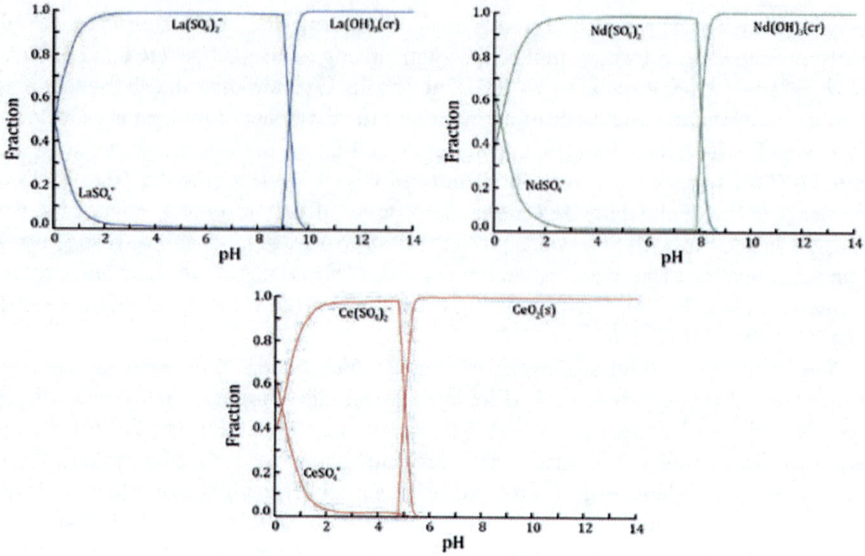

Fig. 6 Distribution diagrams of species for La^{3+}, Ce^{3+}, and Nd^{3+} as a function of pH in the $SO_4{}^{2-}$ system at 25 °C with the following concentrations: 2.0 M $SO_4{}^{2-}$, 0.02 M La^{3+}, 0.02 M Ce^{3+}, and 0.01 M Nd^{3+}

recovery of Ce is lower than that of the other elements. As noted above, the possible reason for cerium's low leachability is that it is oxidized to its tetravalent state (Ce^{4+}) by roasting, forming cerium dioxide (CeO_2), which is poorly soluble.

Figure 6 corresponds to the species distribution diagrams for the La-SO_4-H_2O, Nd-SO_4-H_2O, and Ce-SO_4-H_2O systems for 2.0 M sulfuric acid concentration. The concentration of the elements of interest was determined by chemical assays. These diagrams show that lanthanum tends to form $LaSO_4{}^+$ and $La(SO4)_2{}^-$ complexes at pH < 9. Neodymium forms $NdSO_4{}^+$ and $Nd(SO4)_2{}^-$ complexes up to pH 8. In the case of cerium, $CeSO_4{}^+$ and $Ce(SO4)_2{}^-$ complexes are formed at pH < 5. As mentioned above, Ce forms cerium dioxide (CeO_2) due to the conditions to which the monazite was subjected, resulting in the low solubility of cerium, which is the reason for this element to have the lowest extraction yields and for the impossibility to contrast the results with the leaching capacity indicated in the species distribution diagrams.

Conclusions

- A higher extraction yield of the elements Ce, La, and Nd was achieved with sulfuric acid as a leaching medium versus hydrochloric acid in all evaluated concentration, time, and temperature conditions.

- In all leaching scenarios evaluated, the order of extraction was Ce < Nd < La.
- The addition of 10% (v/v) hydrogen peroxide increases the dissolution of rare earths up to 93% in the H_2SO_4 medium.
- The thermodynamic analysis indicates that H_2SO_4 solutions exhibit a higher capacity to solubilize REEs as compared to HCl solutions, thus supporting the experimental results.
- Due to the dephosphorization conditions of monazite, Ce formed cerium dioxide (CeO_2), which is poorly soluble and, therefore, makes it impossible to contrast the extraction of Ce with the information provided by the species formation diagrams.

Acknowledgements The authors wish to thank the Colombian Ministry of Science for the financial support granted to the Project "Recovery of rare earth elements from minerals present in black sands, residue from alluvial gold mining in El Bagre-Antioquia," code 110180863804 within Call for Projects 808 of 2018 - CFP for Science, Technology and Innovation projects and their contribution to the country's challenges. And to the company Mineros Aluvial S.A.S BIC for the donation of the residual sands from alluvial gold mining.

References

1. Lucas J, Lucas P, Mercier TL, Rollat A, Davenport W (2014) Rare earths rare earths production and use. Elsevier
2. Martins TS, Isolani PC (2005) Terras raras: Aplicações industriais e biológicas. Quim Nova 28(1):111–117. https://doi.org/10.1590/S0100-40422005000100020
3. Krishnamurthy N, Gupta CK (2015) Extractive metallurgy of rare earths, 2nd edn. CRC press
4. Hedrick JB (1992) Rare earths, Mining Engineering (Colorado) (USA), 44(6):571–572
5. Cotton S (2006) Lanthanide and actinide chemistry. John Wiley & Sons
6. Kanazawa Y, Kamitani M (2005) Rare earth minerals and resources in the world. J Alloys Compd 408–412:1339–1343. https://doi.org/10.1016/j.jallcom.2005.04.033
7. Jordens A, Cheng YP, Waters KE (2013) A review of the beneficiation of rare earth element bearing minerals. Miner Eng 41:97–114. https://doi.org/10.1016/j.mineng.2012.10.017
8. Silva RG et al (2019) Processing of sulfuric liquor to obtain rare earth concentrates and selective cerium removal by thermal oxidation. In: XXVIII Encontro Nacional de Tratamento de Minérios e Metalurgia Extrativa, p 13
9. Kerguelen J (2016) Caracterización y aprovechamiento de recursos minerales en colas de terrazas aluviales del Distrito Bagre-Nechí. Universidad Nacional de Colombia
10. Lamus Molina AM (2005) Minerologia Aplicada Al Uso Y Aprovechamiento De Las Arenas Negras(El Bagre, Antioquia). Universidad Nacional de Colombia
11. Tunsu C, Menard Y, Eriksen DØ, Ekberg C, Petranikova M (2019) Recovery of critical materials from mine tailings: a comparative study of the solvent extraction of rare earths using acidic, solvating and mixed extractant systems. J Clean Prod 218:425–437. https://doi.org/10.1016/j.jclepro.2019.01.312
12. Kumari A et al (2019) Advanced process to dephosphorize monazite for effective leaching of rare earth metals (REMs). Hydrometallurgy 187:203–211. https://doi.org/10.1016/j.hydromet.2019.05.013
13. Xu Y, Liu H, Meng Z, Cui J, Zhao W, Li L (2012) Decomposition of bastnasite and monazite mixed rare earth minerals calcined by alkali liquid. J Rare Earths 30(2):155–158. https://doi.org/10.1016/S1002-0721(12)60014-3

14. Marshall WL, Jones EV (1966) Second dissociation constant of sulfuric acid from 25 to 350° evaluated from solubilities of calcium sulfate in sulfuric acid solutions. J Phys Chem 70(12):4028–4040. https://doi.org/10.1021/j100884a045
15. Ma Y, Stopic S, Gronen L, Friedrich B (2018) Recovery of Zr, Hf, Nb from eudialyte residue by sulfuric acid dry digestion and water leaching with H2O2 as a promoter. Hydrometallurgy 181:206–214. https://doi.org/10.1016/j.hydromet.2018.10.002
16. Pecina T, Franco T, Castillo P, Orrantia E (2008) Leaching of a zinc concentrate in H2SO4 solutions containing H2O2 and complexing agents. Miner Eng 21(1):23–30. https://doi.org/10.1016/j.mineng.2007.07.006
17. Birloaga I, De Michelis I, Ferella F, Buzatu M, Vegliò F (2013) Study on the influence of various factors in the hydrometallurgical processing of waste printed circuit boards for copper and gold recovery. Waste Manag 33(4):935–941. https://doi.org/10.1016/j.wasman.2013.01.003
18. Huang J, Chen M, Chen H, Chen S, Sun Q (2014) Leaching behavior of copper from waste printed circuit boards with Brønsted acidic ionic liquid. Waste Manag 34(2):483–488. https://doi.org/10.1016/j.wasman.2013.10.027
19. Wu L, Zhao S, Lin L, Fang X, Liu Y, He M (2016) In-depth understanding of acid catalysis of solvolysis of propene oxide over titanosilicates and titanosilicate/H2O2 systems. J Catal 337:248–259. https://doi.org/10.1016/j.jcat.2016.01.028
20. Kertalli E, va Rijnsoever LS, Paunovic V, d'Angelo MFN, Schouten JC, Nijhuis TA (2016) Propylene epoxidation with hydrogen peroxide in acidic conditions. Chem Eng Sci 156:36–43. https://doi.org/10.1016/j.ces.2016.09.008
21. Wang Z, Guo S, Ye C (2016) Leaching of copper from metal powders mechanically separated from waste printed circuit boards in chloride media using hydrogen peroxide as oxidant. Proc Environ Sci 31:917–924. https://doi.org/10.1016/j.proenv.2016.02.110
22. Shuai G, Zhao L, Wang L, Long Z, Cui D (2017) Aqueous stability of rare earth and thorium elements during hydrochloric acid leaching of roasted bastnaesite. J Rare Earths 35(12):1255–1260. https://doi.org/10.1016/j.jre.2017.06.007
23. Panda R et al (2014) Leaching of rare earth metals (REMs) from Korean monazite concentrate. J Ind Eng Chem 20(4):2035–2042. https://doi.org/10.1016/j.jiec.2013.09.028
24. Kim R, Cho H, Han KN, Kim K, Mun M (2016) Optimization of acid leaching of rare-earth elements from mongolian apatite-based ore. Minerals 6(3). https://doi.org/10.3390/min6030063

Rare-Earth Element Adsorption to Gypsum in Hydrometallurgical Processes

Farzaneh Sadri and Ahmad Ghahreman

Abstract There has been an increasing demand on the REE supply in the last few decades and special attention has been directed towards the extraction process of these elements. The impurity removal step is a key step in the processing circuit as it can determine the purity of the final product. Lime neutralization is deemed a preferred process due to its high efficiency in the removal of impurities and low reagent cost. However, some drawbacks such as co-precipitation of REE with calcium sulfate dihydrate ($CaSO_4$ $2H_2O$, CSD), which results in a significant REE loss, renders this process inefficient. In this study, the mechanism of REE co-precipitation with gypsum as well as the effect of two main impurities (Fe and Al), on the REE uptake from the pregnant leach solutions, during the impurity removal step, are discussed.

Keywords REE · Solution purification · Calcium sulfate dihydrate (CSD) · Crystallization

Introduction

The fast-growing demand for REE in recent years has attracted special attention to their extraction process. The purity of the final REE product is a valuable key factor to the market. Therefore, impurity removal is an important step in the REE process circuits. Irrespective of the source of REE and the processing route, the pregnant leach solution (PLS) always contains impurities such as iron, aluminum (Al), magnesium (Mg), calcium (Ca), manganese (Mn), thorium (Th), and uranium (U), among which Fe and Al or at least one of them is present in almost all resources [1]. Besides, the recovery of REE from their secondary sources has attracted special

F. Sadri (✉) · A. Ghahreman (✉)
Department of Mining Engineering, Queen's University, 25 Union Street, Kingston, ON K7L 3N6, Canada
e-mail: farzaneh.sadri@queensu.ca

A. Ghahreman
e-mail: ahmad.g@queensu.ca

T. Ouchi et al. (eds), *Rare Metal Technology 2022*, The Minerals, Metals & Materials Series, https://doi.org/10.1007/978-3-030-92662-5_3

attention in recent years, from a waste management and recycling point of view as well as significant REE contents in these sources [2]. The secondary sources of REE are usually more concentrated in either REE or impurities.

Precipitation technique is the conventional method used in solution purification. It can be used for the precipitation of metal hydroxides, sulfides, and basic salts. Co-precipitation of other metal ions is common during precipitation, which can affect the loss of valuable metals, purity of the solution, and, overall, the economics of the precipitation process.

Most processing costs in REE extraction circuits are incurred during the hydrometallurgical treatment of the ore/concentrate where REE are chemically extracted [3]. Therefore, a reduction in the operational costs, an increase in the grade of the concentrate, and the recovery of any REE lost in the hydrometallurgical processes (through co-precipitation, etc.) are worth considering. In most of the flowsheets for REE PLS primary purification, the impurities are removed by a simple neutralization step where pH is adjusted using alkaline reagents such as lime, limestone, magnesium oxide, magnesium, and sodium carbonate [3]. Among the reagents, lime and limestone are the cheapest options; however, using calcium-based reagents in a sulfate media causes gypsum (also known as calcium sulfate dihydrate or CSD) formation and consequently REE loss to CSD [4–6]. Despite some studies, the mechanism of REE loss during neutralization and impurity removal is still not well understood.

In this study, a pure synthetic solution of REE was used to investigate the mechanism associated with REE loss to CSD. By this, the impurity removal from REE liquor was simulated by synthesizing CSD crystals doped with four REE: Ce and Nd (light REE, or LREE), plus Er and Tb (heavy REE, or HREE), in the absence and presence of Fe and Al as impurities with the purpose of using lime as the reagent to cut down the costs of the process.

Experimental Procedure

Materials

The analytical grade reagents were used in the experiment, and all aqueous solutions were prepared with de-ionized water. The materials and reagents are presented in Table 1. Nitric acid was used to make a 5% HNO_3 solution for dilution of analysis samples and the standard solutions. The experiment was performed in duplicate as a check on the experimental technique and precision.

Table 1 Materials and reagents used in this study

Reagent	Purity/concentration (%)	Company
Ce (III) sulfate octahydrate	99.9	Alfa Aesar, USA
Nd (III) sulfate octahydrate	99.9	Alfa Aesar, USA
Tb (III) sulfate octahydrate	99.9	Alfa Aesar, USA
Er (III) sulfate octahydrate	99.9	Alfa Aesar, USA
Calcium sulfate dihydrate	98	Acros, Canada
Calcium hydroxide	>99	Fisher Scientific, Canada
Aluminum sulfate hydrate	98	Sigma Aldrich, USA
Iron (III) sulfate hydrate	97	Sigma Aldrich, USA
Sulfuric acid	95–98	Fisher Scientific, Canada
Nitric acid	68–70	BDH, Canada
Ce, Nd, Er, and Tb standard solutions	1000 ppm	Alfa Aesar, Canada

Experimental Process and Analysis

A calcium-saturated solution regulated at pH 1.0 using sulfuric acid was prepared by adding calcium sulfate dihydrate powder to the solution and agitating the resulting mixture for 5 h. After filtering the excess calcium sulfate dihydrate, octahydrate sulfate salts of four REE, cerium (Ce) and neodymium (Nd) representing the light REE, and terbium (Tb) and erbium (Er) representing the heavy REE were added to the solution to make a total of 4.0 g/L REE-containing sulfate solution (1.0 g/L of each REE). Aluminum and iron sulfate salts were used to introduce Al/Fe as impurities to the CSD-saturated solution containing 4.0 g/L REE. The salts were added to concentrations of 4.0 g/L Al and 4.0 g/L Fe, unless otherwise stated. The solution synthesized was used as the starting solution in the experiments. A 5.0% $Ca(OH)_2$ solution with a pH value of 13.0 was used as a precursor to crystallize CSD from the sulfate medium. $Ca(OH)_2$ was gradually added to the crystallization reactor until the pH of the solution reached 5.0 (and 3.2 in some experiments), and the crystals were collected for further analysis. The crystallization experiments were performed in a pyrex beaker on a magnetic stirrer (AREX digital, VELP Scientifica) and the pH of the solutions was continuously measured by a pH meter (Fisher Scientific accumet XL600). The crystal morphology of the collected CSD crystals was analyzed with a FEI FEG-Nova NanoSem scanning electron microscope (SEM), after being sputter-coated with a thin layer of Cr. X-ray powder diffraction analysis (XRD) was carried with a Bruker AXS powder diffractometer. CSD crystals were digested in aqua regia solution (HNO_3:HCl = 1:3). Chemical analysis of the solutions was conducted with an Agilent Technologies 4200 MP-AES apparatus for the REE content and ICP-OES from Agilent Technologies (Vista Pro-axial CCD spectrometer) for the Ca and S concentration. Chemi-STEM energy dispersive X-ray (EDS) elemental mapping was performed on a FEI Tecnai Osiris S/TEM, operated at 200 kV. EDS mapping data was acquired and processed with ESPRIT software. The depth profile for the

Table 2 Chemical analysis of the solids

Species	Ca^{2+}	SO_4^{2-}	Ce	Nd	Tb	Er	H_2O
Solid (Wt. %)	17.300	40.800	0.430	0.440	0.200	0.198	40.632
Molar ratio	0.432	0.435	0.003	0.003	0.001	0.001	2.25

distribution of elements Ce, Nd, Tb, and Er in CSD crystals was investigated by Time-of-Flight Secondary Ion Mass Spectrometry (ToF–SIMS) (ION-TOF SIMS IV, Muenster, Germany). The chamber is pumped continuously to keep an ultra-high vacuum, in which the normal pressure is inferior to 3×10^{-9} Torr. A pulsed 25 kV Bi^+ primary ion beam with a target current of 0.42 pA (19.1 ns DC) and a field of view of 200×200 μm was used. Analysis was performed in static mode with 128,128 × pixels and ion dose lower than 6×10^{11} ions cm^{-2} at each analysis cycle. For depth profiling, the O_2^+ sputter gun operated at 3 kV, 0.1 nm/s and the Cs^+ sputter gun operated at 3 kV, 0.029 nm/s were used in the positive and negative modes, respectively. In both cases, the crater size was 500×500 μm, while the region of analysis located in the middle of the crater was 200×200 μm. It has to be kept in mind that sputter rates can differ significantly depending on the sputtered material.

Results and Discussion

Composition

The chemical composition of the obtained crystals from the experiment at pH 5.0 is listed in Table 2. Note that ICP-OES was used to determine the concentration of Ca^{2+} and SO_4^{2-} species and MP-AES for the four REE.

It is clear from Table 2 that the uptake of light lanthanides by CSD is higher than that of heavy lanthanides due to the less ionic radius differences between the impurity ion (LREE) and Ca^{2+} compared to the HREE (six-coordinated ionic radii of Ce^{3+}, Nd^{3+}, Tb^{3+}, Er^{3+}, and Ca^{2+} have been reported to be 1.01, 0.983, 0.923, 0.89, and 1.00 Å, respectively [7]).

Crystal Structure

The XRD pattern of the crystals (Fig. 1), which were prepared by the addition of a 5% $Ca(OH)_2$ precursor from pH 1.0–5.0, illustrates that the crystals were composed of calcium sulfate dihydrate (CSD) phase by the strong peak at $2\theta = 13.5148$. Thus, the crystals were confirmed to be dihydrate, confirming that no complete or partial phase transformation occurred in these experiments. The XRD patterns of CSD crystals in the absence and presence of REE show that the position of the peaks

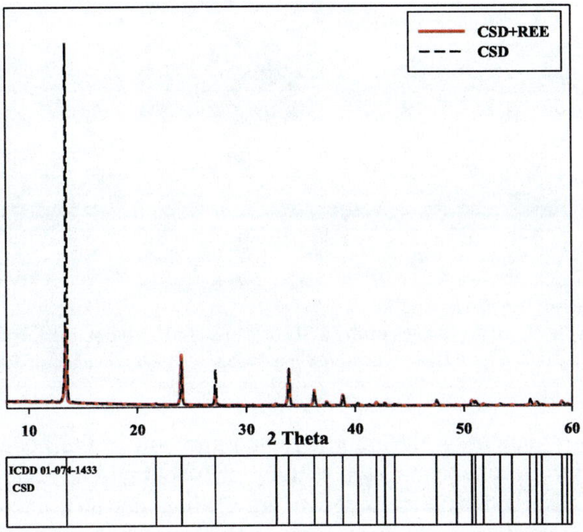

Fig. 1 XRD pattern of the crystals with 5.0% $Ca(OH)_2$ at pH 5.0

did not change with REE addition in the solution; This suggests that likely the substitution of REE ions with Ca^{2+} in the lattice during growth has not occurred, or it was a rare phenomenon occurring in very limited local areas and was hard to detect. Furthermore, the pattern remained unchanged, and no new peaks appeared, meaning that no distinct REE sulfate phase was detected in the XRD studies. These observations could likely suggest that the REE sulfates did not extensively form during the crystallization experiments or the uptake of REE has occurred via surface adsorption and not the incorporation into the crystal lattice, and this will be further explained in the next sections of the paper.

REE Capture by CSD Crystals

Morphological View

Figure 2 shows the surface morphology of CSD precipitates obtained from a crystallization experiment with a 5.0% $Ca(OH)_2$ from pH = 1–5 in the absence and the presence of REE ions in the solution. It can be observed that CSD has crystallized with a plate-like morphology which is the typical morphology for calcium sulfate dihydrate [8, 9]. Figure 2b represents the morphology of CSD crystals from the REE-containing calcium sulfate saturated solution. In the presence of REE in the solution, a needle-shaped morphology is formed that shows the growth is blocked in some directions and accomplished only along the c-axis. This is another evidence of the surface adsorption of REE to CSD.

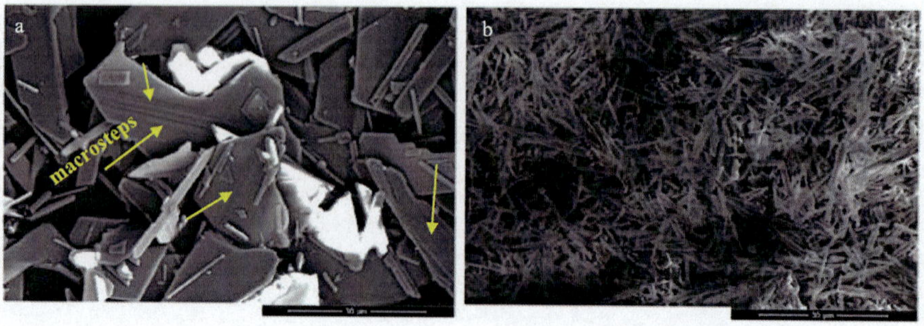

Fig. 2 SEM images of the CSD crystals at pH 5.0 obtained from a crystallization experiment with a 5.0% $Ca(OH)_2$ solution in the **a** absence and **b** presence of REE in the solution

Generally, the impurity incorporation in the host crystal lattice can occur via two mechanisms: step pinning or blocking that occurs when the impurity ions are adsorbed to the step edges or are accumulated on terraces ahead of migrating steps and slowed down the propagation velocity of the steps (impurity-step interactions [10]), and incorporation into the structure that occurs when the impurity ions are captured by the steps or kink sites along the step edges and become part of the growing crystal [11]. In Fig. 2a, the step bunches or macro-steps are visible on the surface of the crystals as dark lines, indicating that the growth is directed by advancing the steps. Figure 2b, on the other hand, shows that the presence of REE ions in the solution have caused a dramatic change in the step motion and therefore the final morphology of the crystals; the REE has slowed down the velocity of the steps or even have blocked their advance. The results confirm again that REE have a particular hindering effect on the dihydrate crystallization and as a result, more elongated and rod-like crystals are formed [12, 13].

It has been shown that lanthanides have a higher affinity towards the CSD lattice than other species such as Cd, Fe, Al, and Cr, due to a stronger complexation of lanthanides with sulfate ions (about 100 times larger affinity than Cd^{2+}) [14–17]. This happens because of the higher coordination number that can be obtained in the CSD crystal lattice. In addition, there is a great electrostatic binding force between lanthanide ions and SO_4^{2-} that creates a great attraction between the two ions. Besides, because of a very small difference in ionic radius of Ca^{2+} and lanthanide ions (1–6.2% for light REE and 7.7–13.9% for heavy REE), minimum lattice strain is induced by the uptake of an ion with an identical radius. Therefore, there could be preferential binding with sulfate ions at the surface of the CSD crystals and uptake of REE would occur [14, 15].

The EDS study was helpful towards the analysis of the REE distribution in the CSD crystals. The EDS elemental mapping confirmed the presence of REE in the structure of CSD or adsorbed on its surface. Figure 3 illustrates the TEM image of single particles of CSD and the REE distribution in CSD structure (Nd as a representative REE).

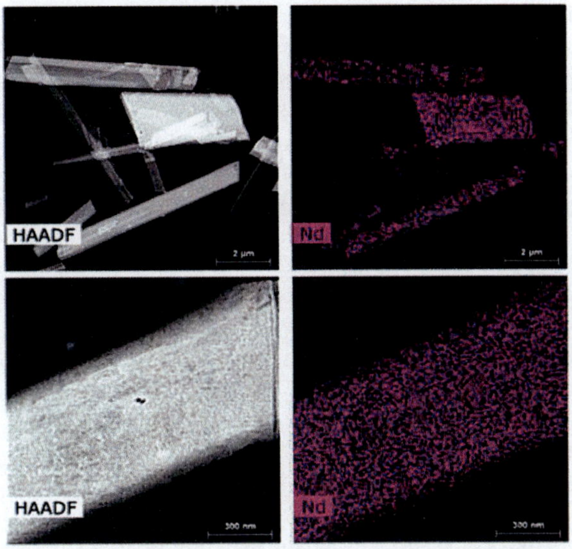

Fig. 3 (Left) TEM image of single particles of CSD, (right) the sites where REE ions are present

A Proof for the Suggested Mechanism

To prove the surface adsorption mechanism, a depth profile for the distribution of the four rare-earth elements (Ce, Nd, Tb, and Er) was obtained by ToF–SIMS analysis. From the results presented in Fig. 4, it is evident that a higher concentration of REE is detected on the surface, so it appears that the REE are mostly adsorbed on the surface. The fact that the REE intensity does not become null with depth is not indicative of much because the measurements are performed on particles, as opposed to a flat

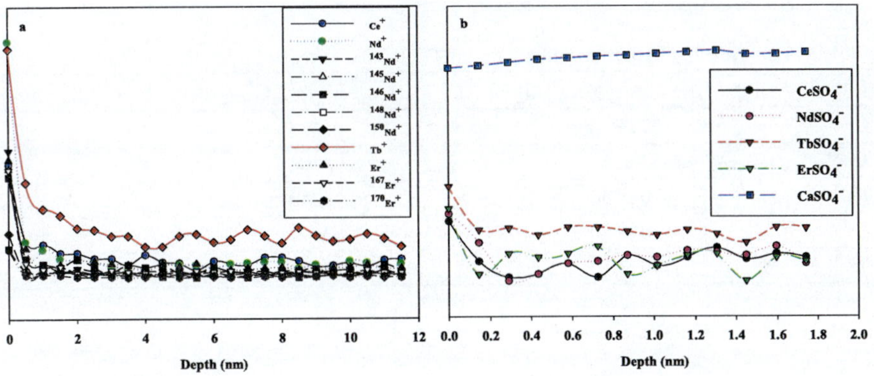

Fig. 4 ToF–SIMS depth profile of REE distribution in the structure of CSD crystals. **a** Positive ions and **b** Negative ions profile

surface, such that some fresh surface is always exposed. So, based on this alone, it cannot be concluded that the REE are present in the bulk and are substituting for Ca atoms. However, when performing negative ion analysis, we do observe ion species Ce-SO_4 (−), Nd-SO_4 (−), Tb-SO_4 (−), and Er-SO_4 (−) which is a fair indication that the REE could be substituting with Ca as part of the growth process. However, it is obvious that Ca has an opposing profile to REE, as its intensity has decreased from the depth to the surface of the crystals, suggesting that the REE have occupied its position on the surface and hence, blocked the growth. This is consistent with the EDS mapping of the REE in CSD crystals that was discussed before. Therefore, it can be concluded that surface adsorption likely is the main mechanism for the REE uptake by the CSD crystals and the isomorphous substitution is not a good mechanism to explain the co-precipitation of REE with CSD.

Thermodynamics View

Kinetic factors such as the growth rate of the host crystal, the rate of adsorption of the foreign ions upon the crystal surface, and the dehydration frequency of the dopant ion compared to the host ion may influence the uptake of dopant. From a thermodynamic point of view, a very important factor that affects the growth retardation process is the hydration enthalpy. It has been previously shown [18] that the dehydration of calcium ions on the crystal surface is the effective rate-controlling process in the crystallization and growth of calcium sulfate dihydrate. As is shown in Table 3, the hydration enthalpy of the REE ions present in the solution (which are in the range of 3360 $kJmol^{-1}$) is significantly larger than that of Ca^{2+} (1570 $kJmol^{-1}$), indicating a stronger water layer bonded with REE ions; therefore, the Ca^{2+} is not able to remove the water molecules to adsorb on the active growth sites, hence resulting in blocking the growth sites and growth process. It should be noted that the Gibbs free energy values are presented to confirm that the hydration process of all the presented elements is a thermodynamically spontaneous reaction.

Table 3 Experimental hydration enthalpies and Gibbs free energy of hydration for cations

Cation	$-\Delta H^0_{hyd(exp.)}$ ($kJmol^{-1}$)	$-\Delta_{hyd}G^*_{exp.}$ ($kJmol^{-1}$)
Ca^{2+}	1577 ± 6	1505
Ce^{3+}	3337 ± 10	3200
Nd^{3+}	3420 ± 50	3280
Er^{3+}	3550 ± 50	3495
Tb^{3+}	3540 ± 10	3400

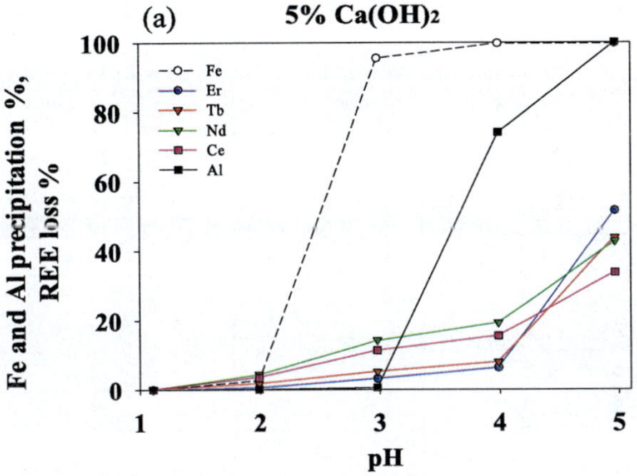

Fig. 5 Precipitation percentage of Fe, Al, and REE when using 5.0% $Ca(OH)_2$ as a reagent to elevate the pH of the solution containing 1.0 g/L of each REE, 4.0 g/L Fe, and 4.0 g/L Al from pH 1.0–5.0

REE Precipitation Behavior in the Presence of Two Main Impurities: Fe and Al

Figure 5 illustrates the precipitation curve for REE, Al, and Fe when adjusting the pH of the solution from 1.0 to 5.0 using 5.0% $Ca(OH)_2$. Iron and aluminum virtually completely precipitate at pH 3.1 and 5.0, respectively; these values are in agreement with the values reported in the literature [19–21].

It is clear from Fig. 5 that REE losses of 51.6% Er, 42.9% Tb, 42.7% Nd, and 33.7% Ce were obtained when using 5.0% $Ca(OH)_2$, which are significantly higher than the losses achieved in the absence of impurities in the solution. The higher REE loss when Fe and Al are present in the solution can be associated with the possibility of adsorption of REE ions to the (Fe, Al)OH_3 precipitates in the residue in a similar principle of CSD adsorption, which will be further discussed in sect. 3.5.

REE Precipitation Behavior at Different Impurity Contents of the Solution

To study the effect of impurities on REE loss in the impurity removal process, different concentrations of 4.0, 2.0, 1.0, and 0.5 g/L for Al and Fe were added to the starting solution while maintaining REE concentration constant at 4.0 g/L (1.0 g/L for each REE). The pH was increased to 3.2 by adding lime and the final solution was filtered and analyzed for the REE, Fe, and Al content. Table 4 shows the REE loss percentage at pH 3.2. As shown, a larger volume of lime is consumed when the initial Fe concentration of the solution is the highest (4.0 g/L) compared with 2.0, 1.0, and 0.5 g/L. That is because more iron ions are available in the solution which requires more neutralization reagent. Equations 1 and 2 depict the neutralization reactions used to precipitate dissolved Al and Fe impurities present in solution:

Table 4 Precipitation percentage of REE in different Fe and Al concentrations using 5.0 wt% lime slurry

Test no	Fe—initial solution (g/L)	Al—initial solution (g/L)	Volume of consumed $Ca(OH)_2$ (mL)	Precipitated solids (g)	% precipitation at pH 3.2			
					Ce	Nd	Tb	Er
1	4.0	4.0	18.3	2.57	16.3	17.6	5.8	4.5
2	4.0	2.0	18.2	2.46	16.5	17.4	5.2	4.4
3	4.0	1.0	18.0	2.42	16.3	17.3	4.9	4.9
4	4.0	0.5	18.0	2.35	15.4	17.3	5.3	4.3
5	2.0	4.0	14.5	1.71	11.9	10.8	2.8	1.7
6	2.0	2.0	14.0	1.60	11.8	11.4	2.2	1.7
7	2.0	1.0	14.0	1.61	11.4	11.7	1.5	1.4
8	2.0	0.5	14.0	1.75	11.2	11.3	1.9	1.8
9	1.0	4.0	11.8	1.21	9.5	10.1	1.5	1.2
10	1.0	2.0	12.0	1.20	9.8	9.8	1.5	1.0
11	1.0	1.0	11.2	1.10	9.2	9.6	1.9	1.2
12	1.0	0.5	10.8	1.00	9.3	9.5	1.3	1.0
13	0.5	4.0	10.3	1.11	4.7	5.2	0.7	0.7
14	0.5	2.0	10.0	0.91	4.9	4.8	0.8	0.6
15	0.5	1.0	9.0	0.78	4.2	5.00	0.2	0.2
16	0.5	0.5	8.4	0.72	4.0	4.3	0.7	0.6

$$Fe_2(SO_4)_{3(aq)} + 3Ca(OH)_{2(aq)} + 6H_2O_{(l)} \rightarrow 2Fe(OH)_{3(s)} + 3CaSO_4.2H_2O_{(s)} \quad (1)$$

$$Al_2(SO_4)_{3(aq)} + 3Ca(OH)_{2(aq)} + 6H_2O_{(l)} \rightarrow 2Al(OH)_{3(s)} + 3CaSO_4.2H_2O_{(s)} \quad (2)$$

As shown in Table 4, increasing the iron concertation from 0.5 to 4.0 g/L resulted in higher REE loss. This is caused by the larger amount of CSD precipitates that are formed when 4.0 g/L Fe is available in the solution which co-precipitates a higher proportion of the REE in the system by surface adsorption or by incorporation into the crystal lattice [6, 22]. Another reason for the increased REE loss likely is the iron hydroxide precipitates in this pH range (1.0–3.2) which can adsorb REE species from the PLS [22]. In addition, $Fe(OH)_3$ has a lattice site in the crystal structure that can physically trap elements inside the crystals as they grow [20, 22, 23]. Other investigators have reported that the hydroxides of Fe and Al display enrichment in REE and play a significant role in defining the REE distribution in bauxite, together with the other REE-bearing phases (i.e., monazite and xenotime) [22, 24]. In fact, REE are likely associated with the Al- and Fe-(oxy)-hydroxides as adsorbed cations [24]. In addition, several studies have used co-precipitation of Fe hydroxide and REE to concentrate REE. This method has a great potential for the determination of ultra-trace levels of REE in natural samples [25, 26]. The above phenomena can explain

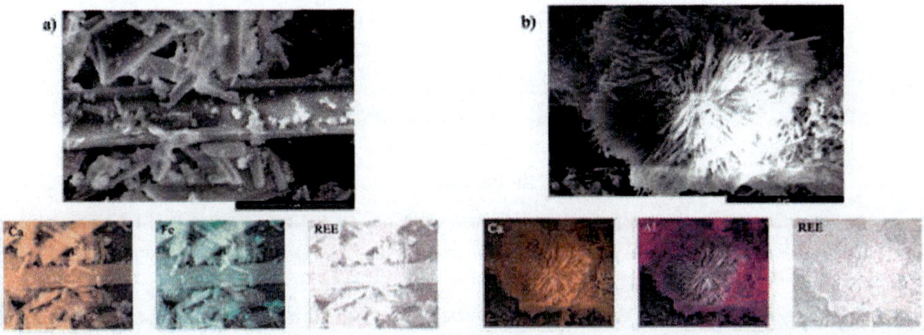

Fig. 6 SEM micrographs and EDS mapping of precipitates obtained from filtering the neutralized solutions at pH **a** 3.2 and **b** 5.0, using 5.0% $Ca(OH)_2$

the increased REE loss to precipitates when there is a higher Fe and Al content in the PLS. For instance, the SEM analysis was performed on the precipitates obtained from neutralization of the REE-, Fe- and Al-containing solution using a 5.0% $Ca(OH)_2$ slurry. Figure 6 illustrates the SEM micrograph and EDS mapping of the solid samples precipitated by neutralizing the solution from pH 1.0–3.2 and then from 3.2 to 5.0. As is shown in Fig. 6, REE were detected on all solid surfaces of CSD, Fe, and Al hydroxides, proving that REE can adsorb on Fe- and Al hydroxides as well as CSD.

Conclusion

Under a regular titration condition, the presence of REE ions was demonstrated to have a significant effect on the morphology of calcium sulfate dihydrate crystals and change the aspect ratio of the crystals. Dihydrate crystals formed in the absence of REE were of thick elongated plate shape. In the presence of REE, dihydrate crystals were in the form of needles. It proves that REE ions interact with the side surfaces of calcium sulfate dihydrate via adsorption and possibly some isomorphous substitution. REE also can be considered as growth inhibitors as they retard the growth of dihydrate crystals in certain directions and hence change the morphology.

Based on the characterization results, it was concluded that the main mechanism of the co-precipitation of REE ions with CSD from the solution at low pH values is the surface adsorption of these ions to the CSD surface. The XRD spectra of the CSD crystals showed the exact same patterns for CSD crystals obtained in the absence and presence of REE dopants in the solution. No new peaks or a peak shift was observed, indicating that the REE did not incorporate into the crystal structure of CSD (i.e. REE were not part of the CSD growth process) or at least the REE concentration was very low to be detected by XRD. This implies that the REE uptake by the CSD crystal is mostly governed by surface adsorption. Furthermore, the depth profile of the distribution of REE in CSD crystals by the ToF–SIMS technique clearly showed that more REE is available on the surface of the crystals rather than the CSD lattice. Hence, it can be concluded that the REE mostly exert their action upon the crystal surface.

Higher REE loss values were obtained in the presence of Fe and Al in the system (33.7% Ce, 42.7% Nd, 42.9% Tb, and 51.6% Er at 4.0 g/L initial concentrations of Fe and Al, each). Effects of different Fe and Al concentrations of 4.0, 2.0, 1.0, and 0.5 g/L in the initial REE liquor were also investigated. It was observed that REE loss to the residue is greatly affected by the impurity contents of the solution. For instance, the REE loss from a solution containing 4.0 g/L Fe were 16.3% Ce, 17.6% Nd, 5.8% Tb, and 4.5% Er while losses using 0.5 g/L Fe in the starting solution were 4.0% Ce, 4.3% Nd, 0.7% Tb, and 0.6% Er. This is due to a higher concentration of the impurity (i.e. Fe) requiring more reagent to neutralize generating more CSD precipitate that adsorbs REE. Furthermore, $Fe(OH)_3$ can physically adsorb REE on the surface of precipitates in addition to trapping them in some lattice sites inside the crystal structure.

References

1. Judge WD, Azimi G (2020) Recent progress in impurity removal during rare earth element processing: a review. Hydrometallurgy 196:105435. https://doi.org/10.1016/j.hydromet.2020.105435
2. Jyothi RK, Thenepalli T, Ahn JW, Parhi PK, Chung KW, Lee J-Y (2020) Review of rare earth elements recovery from secondary resources for clean energy technologies: grand opportunities to create wealth from waste. J Clean Prod 267:122048. https://doi.org/10.1016/j.jclepro.2020.122048
3. Verbaan N, Bradley K, Brown J, Mackie S (2015) A review of hydrometallurgical flowsheets considered in current REE projects. In: Symposium on critical and strategic materials. Victoria, British Columbia, British Columbia Ministry of Energy and Mines, vol 3, pp. 147–162
4. Kim R, Cho H, Han KN, Kim K, Mun M (2016) Optimization of acid leaching of rare-earth elements from mongolian apatite-based ore. Minerals 6:63. https://doi.org/10.3390/min6030063
5. Sadri F, Kim R, Yang Z, Ghahreman A (2018) The effect of calcium sulfate crystallization and the crystal modification on aqueous REE stability in Ca saturated REE-Ca-SO4-H2O systems. Hydrometallurgy 182:82–96. https://doi.org/10.1016/j.hydromet.2018.10.010
6. Sadri F, Kim R, Ghahreman A (2019) Substitution of calcium with Ce, Nd, Er, and Tb in the structure of microcrystals of calcium sulfates with controlled hydration water: a proposed mechanism. Cryst Growth Des 19:2621–2631. https://doi.org/10.1021/acs.cgd.8b01744
7. Shannon RD (1976) Revised effective ionic radii and systematic studies of interatomic distances in halides and chalcogenides. Acta Crystallogr Sect A Cryst Phys Diffr Theor Gen Crystallogr 32:751–767. https://doi.org/10.1107/S0567739476001551
8. Al-Othman A, Demopoulos GP (2009) Gypsum crystallization and hydrochloric acid regeneration by reaction of calcium chloride solution with sulfuric acid. Hydrometallurgy 96:95–102. https://doi.org/10.1016/j.hydromet.2008.08.010
9. Feldmann T, Demopoulos GP (2014) Effects of crystal habit modifiers on the morphology of calcium sulfate dihydrate grown in strong CaCl2-HCl solutions. J Chem Technol Biotechnol 89:1523–1533. https://doi.org/10.1002/jctb.4231
10. Land TA, Martin TL, Potapenko S, Palmore GT, Yoreo JJD (1999) Recovery of surfaces from impurity poisoning during crystal growth. Nature 399:442–445. https://doi.org/10.1038/20886

11. Davis KJ, Dove PM, De Yoreo JJ (2000) The role of Mg2+ as an impurity in calcite growth. Science 290:1134–1137
12. Dang L, Wei H, Wang J (2007) Effects of ionic impurities (Fe2+ and SO42-) on the crystal growth and morphology of phosphoric acid hemihydrate during batch crystallization. Ind Eng Chem Res 46:3341–3347. https://doi.org/10.1021/ie0614881
13. Gao X, Huo W, Zhong Y, Luo Z, Cen K, Ni M, Chen L (2008) Effects of magnesium and ferric ions on crystallization of calcium sulfate dihydrate under the simulated conditions of wet flue-gas desulfurization. Chem Res Chin Univ 24:688–693. https://doi.org/10.1016/S1005-9040(09)60007-6
14. Hamdona SK, Al Hadad UA (2007) Crystallization of calcium sulfate dihydrate in the presence of some metal ions. J Cryst Growth 299:146–151. https://doi.org/10.1016/j.jcrysgro.2006.11.139
15. de Vreugd CH, Witkamp GJ, van Rosmalen GM (1994) Growth of gypsum III. Influence and incorporation of lanthanide and chromium ions. J Cryst Growth 144:70–78. https://doi.org/10.1016/0022-0248(94)90012-4
16. Feldmann T, Demopoulos GP (2013) Influence of impurities on crystallization kinetics of calcium sulfate dihydrate and hemihydrate in strong HCl-CaCl2 solutions. Ind Eng Chem Res 52:6540–6549. https://doi.org/10.1021/ie302933v
17. Deng L, Zhang Y, Chen F, Cao S, You S, Liu Y, Zhang Y (2013) Reactive crystallization of calcium sulfate dihydrate from acidic wastewater and lime. Chin J Chem Eng 21:1303–1312. https://doi.org/10.1016/S1004-9541(13)60626-6
18. Smith BR, Sweett F (1971) The crystallization of calcium sulfate dihydrate. J Colloid Interface Sci 37:612–618. https://doi.org/10.1016/0021-9797(71)90339-0
19. Akcil A, Koldas S (2006) Acid Mine Drainage (AMD): causes, treatment and case studies. J Clean Prod 14:1139–1145. https://doi.org/10.1016/j.jclepro.2004.09.006
20. Dutrizac JE (1983) Jarosite-type compounds and their application in the metallurgical industry. Mineral Sciences Laboratories CANMET), Society of AIME, New York
21. Fernando WAM, Ilankoon IMSK, Syed TH, Yellishetty M (2018) Challenges and opportunities in the removal of sulphate ions in contaminated mine water: a review. Miner Eng 117:74–90. https://doi.org/10.1016/j.mineng.2017.12.004
22. da Silva RG, de Morais CA, Teixeira LV, de Oliveira ÉD (2018) Selective removal of impurities from rare earth sulphuric liquor using different reagents. Miner Eng 127:238–246. https://doi.org/10.1016/j.mineng.2018.08.007
23. Jeffery GH, Basset J, Mendham J, Denney RC (1989) Textbook of quantitative chemical analysis. Longman Scientific and Technical, London, copublished in the US with John Wiley & Sons Inc., New York
24. Mondillo N, Balassone G, Boni M, Chelle-Michou C, Cretella S, Mormone A, Putzolu F, Santoro L, Scognamiglio G, Tarallo M (2019) Rare Earth Elements (REE) in Al- and Fe-(Oxy)-hydroxides in bauxites of provence and languedoc (Southern France): implications for the potential recovery of REEs as By-products of bauxite mining. Minerals 9:504. https://doi.org/10.3390/min9090504
25. Li Y, Guo W, Wu Z, Jin L, Ke Y, Guo Q, Hu S (2016) Determination of ultra-trace rare earth elements in high-salt groundwater using aerosol dilution inductively coupled plasma-mass spectrometry (ICP-MS) after iron hydroxide co-precipitation. Microchem J 126:194–199. https://doi.org/10.1016/j.microc.2015.12.006
26. Qi L, Zhou M-F, Malpas J, Sun M (2005) Determination of rare earth elements and Y in ultramafic rocks by ICP-MS after preconcentration using Fe(OH)3 and Mg(OH)2 coprecipitation. Geostand Geoanaly Res 29:131–141. https://doi.org/10.1111/j.1751-908X.2005.tb00660.x

Low-Cost Distillation Technology for Rare-Earth Recycling

Chinenye Chinwego, Hunter Wagner, Emily Giancola, Jonathan Jironvil, and Adam Powell

Abstract Recycling has been proposed as a promising potential source of supply to meet some of the U.S. rare-earth demands for use in permanent magnets. The high growth rates of products that make use of rare-earth magnets—particularly wind turbines and electric and hybrid vehicles—show that their stock in use is on the rise and in the near term will become available as scrap feed for recycling. This study presents an overview of magnet recycling technologies and focuses on the techno-economic analysis of liquid metal leaching and distillation, including the effect of a new continuous gravity-driven multiple effect thermal system (G-METS) metal distillation technology on energy use and overall cost. The G-METS system can potentially reduce the energy consumption of the overall process to 64 kWh/kg, which is about 30% less than metal production from ore, and 61–67% less than the process using conventional distillation.

Keywords Rare earths · Permanent magnets · Recycling · Distillation

Introduction

The development of permanent magnets has evolved from lodestone in the 1920s to modern neodymium-iron-boron (NdFeB or "neo") magnets because of the need to obtain a good combination of maximum energy product and high Curie temperature. The high energy product value from rare-earth magnets and other permanent magnets enables the use of smaller magnets and higher magnetic strength applications. Although there are generally two rare-earth magnets, neo magnets and samarium cobalt (SmCo) magnets, neo magnets are the strongest and most affordable. Dysprosium (Dy) and cobalt addition improve the magnet's high-temperature characteristics [1]. Their applications are seen in technologies that require potent magnets such as wind turbine generators, vehicle traction motors in hybrid and electric vehicles (HEVs/EVs), speakers and headphones, magnetic resonance imaging

C. Chinwego (✉) · H. Wagner · E. Giancola · J. Jironvil · A. Powell
Worcester Polytechnic Institute, Worcester, MA, USA
e-mail: chachinwego@wpi.edu

T. Ouchi et al. (eds), *Rare Metal Technology 2022*, The Minerals, Metals & Materials Series, https://doi.org/10.1007/978-3-030-92662-5_4

(MRI) scanners, computer hard disk drives, and high-performance alternating current (AC) servo motors, among others. According to a forecast by Adamas Intelligence [2], an independent research and advisory company on strategic metals and minerals, there will be an increase in the global demand for NdFeB alloys and powders at a compound annual growth rate (CAGR) of 9.7% from 2020 through 2030, and thus an increase in the need for rare-earth elements (i.e., neodymium, praseodymium, dysprosium, and terbium). Meanwhile, global production will increase at a slower CAGR of 7.1%, leading to a supply market struggle to keep up with growing demand. NdFeB alloy global shortage is predicted to amount to 48,000 tonnes annually by 2030, estimated as the amount needed to produce 25–30 million electric vehicle traction motors. Didymium oxide (NdPr oxide) will also experience a shortage rise to 16,000 tonnes in 2030, equivalent to roughly three times the total annual output of Lynas Corporation and MP Materials [2]. Therefore, there is a need to ensure a sustainable supply of these elements by methods that include recycling. In achieving this, it is essential to ascertain the availability of the raw materials for recycling and the economic viability of the methods employed.

Several studies have reported various methods for recycling rare earths from end-of-life (EoL) products. Still, there are limited detailed techno-economic analysis (TEA) studies on these methods. This study presents a review of various processes for recycling rare earths from EoL magnet products and techno-economic analysis of a proposed low-cost alternative using a new distillation process.

Recycling Processes for Rare-Earth Permanent Magnet (REPM)

There are multiple processes for producing new REPMs from old ones. Figure 1 shows the rough cumulative energy and cost of the value chain for virgin REPMs from mining to magnet production. In this graph, one can group recycling processes in terms of how many "steps back" they take in the value chain to recover rare earths. The most efficient, which loses the least energy and value, is the magnet-to-magnet approach. But recycled magnets must have the same composition as scrap input, and scrap with mixtures of magnet alloys can be difficult to use. And introducing a decrepitation agent, generally, hydrogen can degrade the properties somewhat, requiring small additions of unalloyed rare-earth metals to recover properties like virgin magnets. On the other end, solvent extraction and sulfuric acid baking recover a mixed rare-earth oxide which can be inserted into the separation process, recovering pure oxides which can be made into any magnet alloy with properties indistinguishable from virgin magnets. But it requires the most energy, as it must re-do the separation, reduction to metal, and alloy production steps. Liquid metal leaching is between these two: it recovers a rare-earth magnet mischmetal, preserving the energy of metal reduction, but makes it difficult to tailor the REPM alloy.

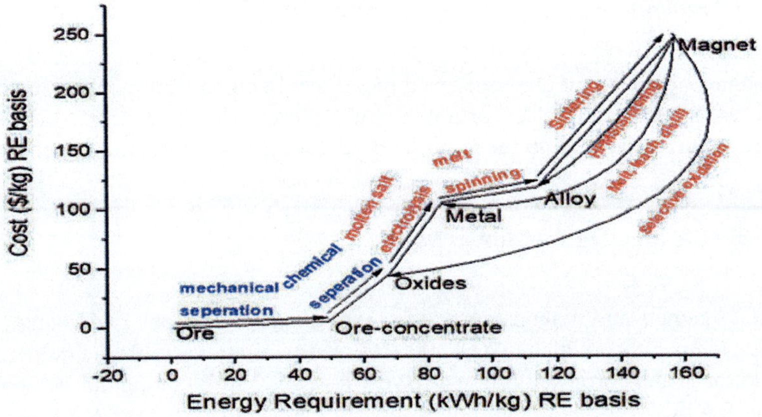

Fig. 1 Cumulative energy and cost [3–6] of various stages of NdFeB magnet production, with alloy production energy use roughly estimated; "RE basis" indicates per kilogram of contained rare-earth metal

The magnet-to-magnet approach has recently seen commercial deployment, as Urban Mining Company has set up operations in Texas. But for a pure recycling industry, it will require one of the other two to supply the unalloyed rare-earth metals required for good properties.

Liquid Metal Leaching and Distillation

This liquid metal extraction process uses molten magnesium or other group IIA metals such as Ca or Ba for rare-earth magnet recycling. Previous studies have shown that this technology extracts rare-earth alloys (Nd, Dy) from removed magnets at a temperature of 1273 K and an Mg/magnet ratio of 10 [7]. The Nd and Dy extraction efficiencies were 100% and 60%, respectively. Furthermore, the addition of Calcium suppressed the oxidation of Dy, thus increasing the extraction efficiency [8]. The use of bismuth (Bi) instead of Mg can increase the Dy recovery above 90% and using both Mg and Bi enables some separation of light and heavy rare earths [9].

Proposed Leaching and Distillation Process

The proposed process is similar to the two-stage leaching using Mg and Bi as described by Ott and McCallum [9], but with a new efficient distillation step called Gravity-driven Multiple Effect Thermal System (G-METS) [10]. It consists of the following steps.

Crushing and Sorting

The magnet scrap is first processed to remove non-magnetic components, including nickel or other coatings, for example, by preferential degradation [11], and if necessary, further ground to the required particle size for the leaching operation.

Leaching RE Metals Into Molten Mg

Crushed magnets and magnesium are fed into a leaching vessel, and heated to the leaching temperature, which is 700–1000 °C. Neodymium is soluble in liquid magnesium, but iron and boron solubilities are very low, and Mg-B compounds which form remain as solids. Various studies have used Mg:magnet mass ratios from 1:1 to 10:1; here we consider 3:1 to 10:1 ratios.

The diffusion coefficient of Nd in liquid Mg was measured $4.61 \times 10^{-8} \pm 1.32$ cm^2/s at a leaching temperature of 700 °C, and the coefficient was found to increase with leaching temperature to $8.98 \times 10^{-8} \pm 3.50 \times 10^{-8}$ cm^2/s at 800 °C [12]. Here, we assume a leaching temperature of 1000 °C, and a time of 6 h, following the work of Akahori et al. [7]. At that temperature, Nd recovery rate is close to 100%, and Dy recovery rate is 60%.

Separating the Melt from Fe-B Alloy Particles

After leaching, liquid Mg-RE and scrap particles are transported to the separation unit. The denser Fe-B alloy particles settle on the crucible bottom; thus, the Mg-Re solution can be easily separated from the melt. A filtration or centrifugal separation unit completes the separation of liquid Mg-RE from Fe-B alloy particles. We nonetheless assume a small amount of Mg remains with the Fe-B alloy particles. Note that due to the limited leaching recovery rate of heavy rare-earth elements such as dysprosium, this alloy particle stream contains a large fraction of those elements from the scrap.

G-METS and Vacuum Distillation of the Mg-RE Melt

The Mg-Re melt is distilled to recover the Mg and RE separately, based on the difference in vapor pressure. This proposed method distills this melt in two steps. The first uses the new G-METS process for distilling most of the Mg. Studies have shown that this process can reduce the energy required to distill magnesium by as much as 90% versus today's batch distillation process [10]. This process flows liquid Mg-RE alloy through a steel distiller, so maximum rare-earth content is limited to 20–30 wt% to prevent corrosion of the steel. That said, if the material comes in with 90–97 wt% Mg, G-METS can efficiently remove 65–85% of the magnesium. The second step is less-efficient conventional vacuum distillation using Mg sublimation, leaving behind solid metal consisting almost entirely of rare earths. As a result, the rare-earth extraction efficiency mainly depended on the Mg:magnet ratio [7, 8, 12, 13].

Liquid Bismuth Leaching of Fe-B Particles

Leaching in liquid bismuth is used to recover more dysprosium from the Fe-B particle stream after the magnesium leaching process, following [14]. This again selectively removes rare earths, with low iron but its 85% recovery of Magnesium and 80% recovery of Bismuth will be targeted and further recovered using both G-METS and the traditional distillation process.

Techno-Economic Analysis

Studies carried out on the techno-economic analysis (TEA) of rare-earth recycling have focused on secondary sources like clay waste [15], coal ash [16], and fluidized catalytic cracking waste [17]. Another study on the TEA of recycling HDD magnets considered other precious materials including Au, Ag, and Pd alongside REE recovery [18]. This study focused on magnet scrap recycling from EoL NdFeB magnets to obtain RE alloy as a precursor for magnet production.

A process flow diagram was developed to estimate the economic feasibility of the magnesium and bismuth leaching process. The TEA was carried out using the ARPA-E METALS tool spreadsheet v1.0 from the U.S. Department of Energy [19]. A complete material and energy balance was developed and used in the ARPA-E METALS tool. Information was pulled from these to determine the variable cost, including raw materials, utilities, and miscellaneous materials. The mass balance contained a total of 12 elements from the magnet scrap and leaching metal.

Various adjustable parameters such as Mg/scrap feed ratio, Bi/leaching vessel waste ratio, and Mg and Bi losses to waste streams were modeled to determine their effects on the overall target recovery and energy use. Due to high uncertainty in scrap acquisition, the cost of scrap was omitted from this analysis, the cost can be added to the analysis separately. The Streams and Compositions sheets in the ARPA-E METALS Tool calculated the value flows for each input and output stream based on their compositions and verified perfect closure of the mass balance for all 12 elements. Various online vendors were used to determine the values for each compound/element, and their best- and worst-case scenarios were found.

The energy requirement for the leaching vessel was calculated from the equation:

$$\sum \dot{m} c_p \Delta T = \dot{m}_{sc} c_{psc} (T_f - T_0) + \dot{m}_{Mg} c_{pMg} (T_f - T_0) + \dot{m}_r c_{pr} (T_f - T_r)$$

Table 1 Assumptions used to determine the cost calculations for the best, reference, and worst-case scenarios. FTE is full-time equivalent, FCI fixed capital investment, VC variable costs, and FC fixed costs

	Best case	Reference case	Worst case
Mg/magnet ratio	5:1	10:1	10:1
Mg metal	\$3000 per ton	\$3000 per ton	\$7000 per ton
Bi metal	\$7000 per ton	\$7000 per ton	\$14,000 per ton
Electricity	0.06 per kWh	0.09 per kWh	0.18 per kWh
Operating labor	8FTEs	12FTEs	16FTEs
Other labor	75% OL	85% OL	90% OL
Maintenance	2% FCI	7.5% FCI	10% FCI
Indirect cost	10% of VC and FC	15% of VC and FC	25% of VC and FC
Insurance	1% FCI	1% FCI	1% FCI
Royalties	1% FCI	1% FCI	1% FCI

In this equation, $\dot{m}_{sc}$ is the mass flow rate of the scrap, $\dot{m}_{Mg}$ is the mass flow rate of the Mg refill,and$\dot{m}_r$ is the mass flow rate of the liquid Mg recycle from the distillation operation. Moreover, c_{psc},c_{pMg}, and c_{pr} are the specific heat capacities of the scrap, magnesium, and Mg recycle, respectively. T_0 is the room temperature at which the scrap will be fed into the leaching vessel, T_r is the weighted average temperature of Mg recycle exiting the G-METS and conventional distillation processes, and T_f is the temperature of extraction of the leaching vessel. This was similarly used to obtain the energy required for the second leaching process with bismuth.

Assumptions used for best, worst, and reference cases are shown in Table 1.

Model Results

Figure 2 shows the proportions of each material flow compared to the product flow. Thus, for a plant with a target of 205 metric tons of RE alloy per year and an additional 11 tons of Dy metal, about 745 metric tons of scrap feed, ~149 metric tons of Mg, and 104 tons of bismuth will be required. Magnesium and bismuth are recycled after distillation. The raw material costs can be calculated based on these flows as shown in Table 2.

The energy requirement for the process can be seen in Fig. 3, with a total energy usage of 64.28 kWh/kg of RE (mixed rare-earth metals obtained). These are given for an Mg/scrap feed mass ratio and Bi/waste mass ratio of 10:1. This increases to 248.1 kWh/kg without G-METs. A decrease in these ratios will lead to an overall decrease in the energy used by the equipment. The energy consumption for the first G-METS distiller was estimated to be 15.3 kWh/kg of RE product and 6.58 kWh/kg of Dy metal based on the thermal model described in the G-METS distillation patent

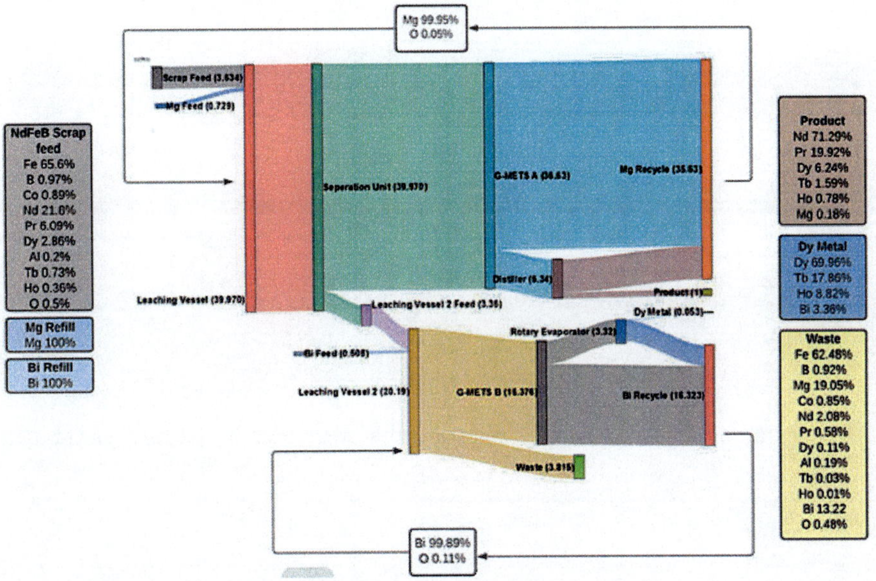

Fig. 2 Material balance flow ratios for Mg/Bi leaching and distillation process; mass flow rates in parentheses are ratios to the main product output stream

Table 2 Raw materials, flow rates, and their values for a 205 ton/a RE alloy recovery and an additional 11 ton/a of Dy metal (using bismuth) with an 85% plant capacity factor

Streams	Flow rate (kg/h)	Cost (USD/kg)	Annual cost (USD)
Scrap feed	100	–	–
Mg feed	20.05	3	447,866
Bi feed	13.93	7	726,000
Total	133.98	10	1,173,866

application [20]. This is estimated for an 85% Mg distillation and 80% Bi distillation, respectively. For the conventional distillation and the rotary evaporation, the energy usage was estimated at 25.36 kWh/kg of RE product and 15.75 kWh/kg of Dy, respectively, based on the fraction of Mg and Bi distilled.

Also, assuming a best- and worst-case energy cost of \$0.06 USD/kWh and \$18 USD/kWh, the annual electricity consumption is expected to be \$900,993 and \$2,702,980, respectively.

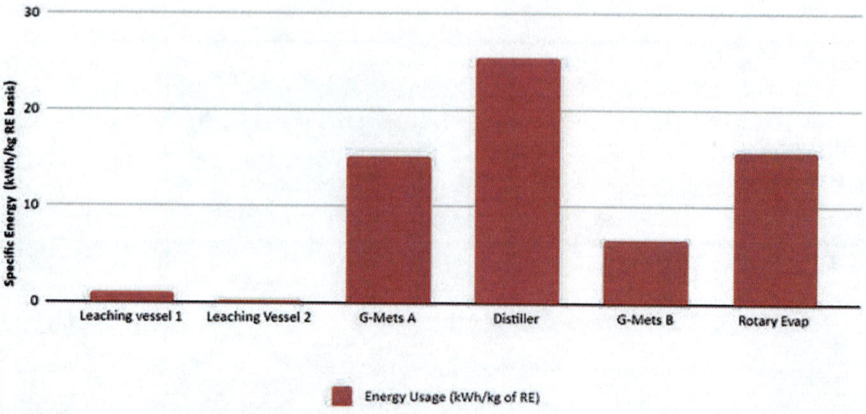

Fig. 3 Energy requirement of various equipment for the Mg/Bi leaching process with G-METS distillation

Discussion

The introduction of the G-METS system has the potential to reduce the energy required and thus the cost of utilities for the process by 67% for the reference case as shown in Table 3. This thereby reduces the total annual cost and production cost by 40% compared to traditional distillation. This cost can be minimized further by reducing the Mg/scrap ratio, as can be observed by adjusting this parameter in the material balance spreadsheet on the ARPA-E METAL tool. Figure 4 shows three annual operating cost stack scenarios for an RE magnesium/bismuth leaching plant. The utilities cost is 14% of the total annual cost, while raw materials, which include the cost of the magnesium and bismuth metals, are ~15% of the total annual cost. The bulk of the cost is seen in the operating labor and other labor, i.e., the laboratory costs, plant overheads, and supervision which amount to 56% of the total annual cost.

Table 3 Cost of utilities, total annual cost, and cost of RE/Dy per kg with and without G-METs for Best-, Reference-, and Worst-case scenarios

	With G-METS			Without G-METS		
	Cost of utilities	Total annual cost	Cost per kg of RE alloy plus Dy metal	Cost of utilities	Total annual cost	Cost per kg of RE alloy plus dy metal
Best case	$586,996	$4,817,432	$22.32	$1,511,814	$7,408,612	$34.32
Reference case	$1,148,766	$8,045,319	$37.27	$3,505,969	$13,365,268	$61.91
Worst case	$2,747,656	$14,062,271	$65.14	$8,294,015	$24,878,820	$115.25

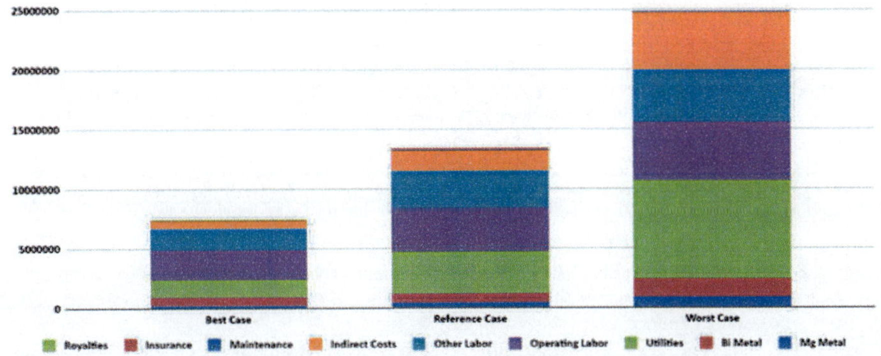

Fig. 4 Three annual operating cost stack scenarios for a 205ton/a RE alloy recovery and further 10 ton/a of Dy (using Bismuth) using G-METS as well as conventional distillation

Conclusion

Techno-Economic Analysis of recovering REEs from NdFeB scrap using a novel G-METS distillation system was performed for a magnesium and bismuth leaching process. The G-METS system can potentially reduce the energy consumption of the overall process to 64 kWh/kg, which is about 30% less than metal production from ore, and 61–67% less than the process using conventional distillation. The processing cost is also shown to be lowered from 34–115$/kg for the best-case and worst-case scenarios without G-METS distillation to 22–65$/kg with it. The largest operating cost factor is labor, whose cost per kg of product would likely decline with increasing scale. These costs compare with today's NdPr metal prices of $100–120/kg, which would make operating margins much larger with G-METS than without it. The G-METS distillation process can therefore be considered a potential enabling technology for liquid metal leaching and distillation.

References

1. Ulewicz R, Wyslocka E (2015) 24th Int Conf Met Mat Proc, 1680. https://www.researchgate.net/publication/304676716_MAGNETS_HISTORY_THE_CURRENT_STATE_AND_THE_FUTURE
2. Adamas Intelligence (2021) UBS rare earth forecasts only tell half the story (Adamas Intelligence, 2021). https://www.adamasintel.com/ubs-ndpr-forecasts-miss-mark/. Accessed 27 Mar 2021
3. Fernandes IB, Abadías Llamas A, Reuter MA (2020) JOM 72(7):2754–2769
4. Peiró LT, Méndez GV (2013) JOM 65(10):1327–1340
5. B. Inc, Rare earth elements to make a strong comeback after 2020 setback, says Beroe Inc. https://www.prnewswire.com/news-releases/rare-earth-elements-to-make-a-strong-comeback-after-2020-setback-says-beroe-inc-301271188.html. Accessed 01 Oct 2021

6. Statistica (2021) REO price by type 2020. Statista. https://www.statista.com/statistics/617249/price-range-of-selected-rare-earth-oxides/. Accessed 01 Oct 2021
7. Akahori T, Miyamoto Y, Saeki T, Okamoto M, Okabe T (2014) Magnesium Technol 2014:35
8. Ellis TW, Schmidt FA (1995) Recycling of rare earth metals from rare earth-transition metal alloy scrap by liquid metal extraction. U.S. Patents 5,437,709. https://patents.google.com/patent/US5437709A/en Accessed 17 May 2021
9. Ott RT, McCallum RW (2021) Recovering rare earth metals from magnet scrap, U.S. Patent 10,323,299, 18 June 2019. https://patents.google.com/patent/US10323299B2/en. Accessed 17 May 2021
10. Telgerafchi AE, Espinosa G, Rutherford M, Powell A, Dussault D (2021) Magnesium 2021:145
11. Ott B, Spiller DE, Taylor PR (2019) Rare earth magnet recovery from hard drives by preferential degradation. Springer. https://link.springer.com/content/pdf/10.1007%2F978-3-030-10386-6.pdf Accessed 24 June 2021
12. Na HW, Kim YH, Taek Son H, Ho Jung I, Shin Choi H, Bum Kim T (2014) Curr Nanosci 10:128
13. Xu Y, Chumbley LS, Laabs FC (2000) J Mater Res 15:2296
14. Ott RT, Ralph RW (2021) Recovering rare earth metals using bismuth extractant. Ames lab. http://isurftech.technologypublisher.com/technology/21024. Accessed 17 May 2021
15. Archambault S (2020) Economic analysis of rare earth elements extraction from clay waste. University of Tennessee Honors Thesis Projects. https://trace.tennessee.edu/cgi/viewcontent.cgi?article=3086&context=utk_chanhonoproj Accessed 23 Sept 2020
16. Das S, Gaustad G, Sekar A, Williams E (2018) J Clean Prod 189:539
17. Thompson VS, Gupta M, Jin H, Vahidi E, Yim M, Jindra MA, Nguyen V, Fujita Y, Sutherland JW, Jiao Y, Reed WD, Sustain ACS (2018) Chem Eng 6:1602
18. Nguyen RT, Diaz LA, Imholte DD, Lister TE (2017) JOM 69:1546
19. Matuszak D (2013) ARPA-E metals tool. U.S. Department of Energy
20. Rutherford M, Telgerafchi AE, Espinosa G, Powell AC, Dussault D (2021) Magnesium 2021:139

Extraction and Recovery of Rare-Earth Elements and Critical Materials from Coal Waste Using Low Cost Processing Methods: Acid Generation Using Acidithiobacillus Ferrooxidans Mediated Bio-Oxidation of Pyrite

Prasenjit Podder, Michael L. Free, and Prashant K. Sarswat

Abstract Acid is an essential component for effective leaching of REEs (rare earth elements) bearing minerals and other sources. Acid can be generated using pyrite found with REE bearing materials. In the presence of *A. Ferroxidans*, pyrite (FeS_2) is used as a source to generate ferric sulfate and sulfuric acid. Such microorganism-based bio-oxidation needs proper conditions for continuous acid generation. In this study, which involved pyrite to generate acid for leaching, a step-by-step analysis and associated details are presented. An automated bioreactor was used for bio-oxidation of pyrite of different levels of purity. Morphology and elemental composition of precipitates were analyzed using scanning electron microscopy equipped with energy-dispersive X-ray spectroscopy, and the phase purity was identified using X-Ray diffraction analysis. Eh and pH measurements have been performed for different pyrite powders. FeS_2 and Fe^{2+} reaction rate measurements were also done and pyrite usage efficiency was also estimated.

Introduction

This manuscript covers only parts of the broader topic of Extraction and recovery of Rare-earth elements and critical materials from coal waste using low cost processing methods. Pyrite oxidation is an essential technique for the geochemical circulations of iron and sulfur, and it is equally helpful to industrial mineral beneficiation processes [1]. When pyrite is exposed to the environment due to natural processes or anthropogenic activities, it is oxidized and forms sulfuric acid in the presence of humidity [2]. This process consequently contributes to the acidification of natural waters, such as rivers, streams, and lakes. Pyrite (FeS_2) is a very common sulfide mineral and can

P. Podder · M. L. Free (✉) · P. K. Sarswat
Department of Materials Science and Engineering, University of Utah, Salt Lake City, UT 84112, USA
e-mail: michael.free@utah.edu

T. Ouchi et al. (eds), *Rare Metal Technology 2022*, The Minerals, Metals & Materials Series, https://doi.org/10.1007/978-3-030-92662-5_5

be found at the Earth's surface. It contains a variety of trace metals and is one of the most extensively found minerals among all metal sulfide minerals [2, 3].

Pyrite oxidation is very complex and also a primary process in the leaching of metal sulfides. In most bioleaching operations, there are several strains of bacteria that perform different oxidation functions [4]. The most common bacteria associated with mineral leaching are *Leptospirillum ferrooxidans, Acidithiobacillus ferrooxidans, Acidithiobacillus trioxidanes, Acidithiobacillus acidophilus*, and Sulfurous-like bacteria. In the bio-oxidation of pyrite, *Acidithiobacillus ferroxidans* play an important role [5, 6]. *Acidithiobacillus ferrooxidans* is a gram-negative, autotrophic, and rod-shaped bacterium. Energy gain is possible when the oxidation of Fe takes place and proper oxygen supply is available for bacterial growth when sulfur availability is limited. This gain of the energy can be used in 'bio-oxidation' or 'bioleaching process' which can help to control the acid mine drainage environment by utilizing the material which provides nourishment for the growth and maintenance of life. To get a very favorable bacterial enhancement and create an appropriate environment, it is important to know the iron and sulfur oxidation pathway [7, 8]. In the presence of *A. Ferroxidans*, pyrite (FeS_2) is used as the source to generate ferric sulfate and sulfuric acid [9]. In the first step of the pyrite digestion, the ferric ions react with the FeS_2 and generate ferrous ions, which is essential for the autotrophic metabolic process of the bacteria to provide enough energy. This ferric oxidation of the pyrite step is entirely independent of microbes. The role of the bacteria is the regeneration of the ferric ions that were utilized in the pyrite oxidation process. The rate of oxidation of pyrite is related to the concentrations of reactants in the solution, which in this case involves the concentrations of ferrous ferric and ferric ions. Ferrous ion simply works as a reactant for bacteria. The bacteria can also interact directly with minerals. This interaction is believed to take place through the extracellular secretion of an enzyme. Another method of interaction includes oxidation with an enzyme that is distinct from sulfide minerals present in the cell. Pyrite oxidation by ferric ions can be termed as an electrochemical process consisting of two steps; the steps include cathodic reaction, electron transport, and anodic reaction. This set of reactions is as follows:

$$14Fe^{3+} + FeS_2 + 8H_2O \leftrightarrow 2SO_4^{2-} + 15\,Fe^{2+} + 16H^+ (\text{abiotic}) \quad (1)$$

$$14Fe^{2+} + 3.5O_2 + 14H^+ \leftrightarrow 14Fe^{3+} + 7H_2O\,(\text{biotic}) \quad (2)$$

The overall reaction is as follows:

$$FeS_2 + H_2O + 3.5O_2 \leftrightarrow Fe^{2+} + 2SO_4^{2-} + 2H^+ \quad (3)$$

An increase in the pyrite oxidation rate can be observed with increasing ferric iron concentration; however, an increase in ferrous iron concentration inhibits the chemical oxidation rate. Pyrite dissolution depends on the presence of bacteria, and hence, the rate of dissolution increases with an increase in the concentration of ferric ions. Also, the reaction depends on pH. These bacteria are relatively slow-growing and need an environment without toxicity. *A. ferroxidans* is not only capable of oxidizing ferrous ions to ferric ions but also is capable of sulfur oxidation. Therefore, it plays a very important for the acid generation process from pyrite oxidation. In this piece of research, the authors have shown an overall pyrite bio-oxidation performance by *Acidithiobacillus ferrooxidans* [4, 10–14].

Materials and Methods

Pyrite Concentrate and Nutritions

Three different kinds of pyrite have been used and were acquired from the University of Kentucky in a slurry form and then it was dried at room temperature. Before doing characterization or analysis, pyrite powder was splitted and mixed very well to make the powder highly homogenous. While adding this pyrite concentrate in the reactor system, it was mixed with a solution made with ATCC 9 k medium and DI water in a 50–50 ratio because 9 k medium helps bacteria to grow [15].

Bioreactor Experimental Setup

A 3 L automatic bioreactor was purchased from Chemglass Life Sciences. Then necessary equipment of reactor for instance agitator heat exchanger, air flow tube automatic feeding system are assembled in the laboratory and setup. During the setup, necessary precautions were taken and the scientific procedure has been followed. The picture of the setup of the 3L bioreactor after feeding 9 K medium and pyrite powder is shown in Fig. 1.

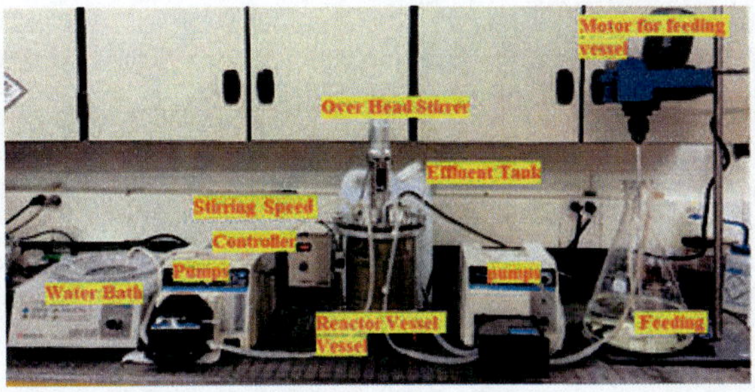

Fig. 1 A 3L bioreactor setup pyrite bio-oxidation

Analytical Method

Eh and pH were measured by Sper Scientific-860031 pH-mV meter-0 top 14pH range. Acid concentrations in the solution were measured by titration test. Because titration helps to know the obscure concentration of base or acid. To know about micro morphological structure and concentration of pyrite and solid, SEM imaging and EDS mapping were performed. Also, XRD analysis was also conducted to see the phase purity and other salts in pyrite powder and effluent solids. ICP-MS analysis also has been conducted on pyrite and effluent solids to acquire more information about their concentration in the pyrite sample.

Bio-Oxidation Experiment

The bacterial activity was examined by performing ferrous oxidation tests [16]. These tests were conducted on different days and utilized different amounts of $FeSO_4{\cdot}7H_2O$. Ferrous sulfate heptahydrate was added to the 3L bacterial solution in the reactor while oxidation–reduction potential Eh and pH were continuously monitored to see the changes. The Eh vs. time was measured and recorded for every test. By following the Eh data, the ferrous concentration vs. time, and the ferrous ion consumption rate vs. time, the rate of bacterial activity can be determined. When a high level of bio-oxidation rate was observed at a consistent level for approximately two residence times, appropriate samples were collected and saved for further analysis.

ICP-MS Analysis

ICP-MS analysis can help to track elemental concentrations in a sample. So, to analyze the pyrite and effluent solids, ICP-MS analyses were done. ICP-MS samples were prepared from 0.1 g of pyrite concentrate feeding material and 0.1 effluent solid from the 3L bioreactor. These materials were dissolved in 10 ml of aqua regia, then diluted to 1000 ml. ICP samples were prepared for the two digestion solutions and the bioleaching solutions.

Experimental Results and Discussion

Pyrite Powder and Effluent Solid Characterization

Particle Size Distribution Analysis

A particle size distribution analysis is essential for understanding each set of particles. Results from this analysis can help to determine particle properties and how they can react in different environments, which is needed to achieve goals. Experiments were conducted using a set of six sieves. The size analysis was done by analyzing 100 g of mixed powder from three different pyrite samples. Most of the powders from 100% pure pyrite and 60% pure pyrite were between 100 and 200 Mesh sizes. In contrast, the material that consisted of 20% pure pyrite had more fine silica or clay minerals. Table 1 shows the results of particle size distribution analysis.

SEM and Digital Imaging of Pyrite Samples and EDS Mapping Analysis of Pyrite Sources

Digital and SEM imaging as well as EDS mapping were conducted for different types of pyrite powder and effluent sold particles (see Fig. 2). It can be noticed that

Table 1 Particle size distribution of the pyrite feed sources

Mesh size	100% pure pyrite (g)	60% pure pyrite (g)	20% pure pyrite (g)
≤30	0	0.14	3.45
≤65	6.35	5.30	44.67
≤100	14.2	70.98	29.61
≤200	69.8	14.61	17.57
≤250	9.26	6.67	3.24
≤325	0.11	1.74	1.3
Total	99.7	99.4	99.8

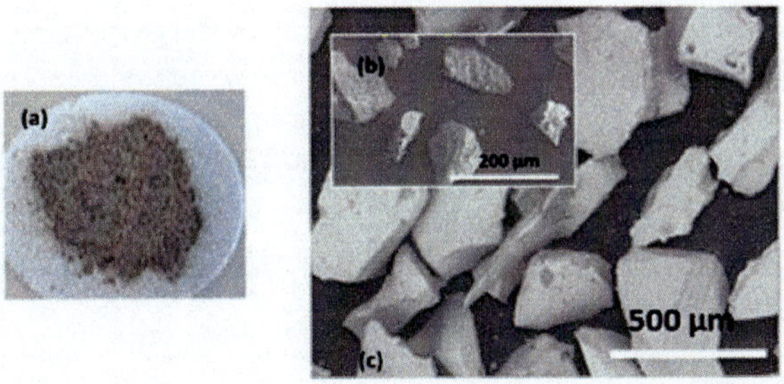

Fig. 2 **a** Digital image and **b** SEM image of 20% pyrite; **c** SEM image of concentrated pyrite

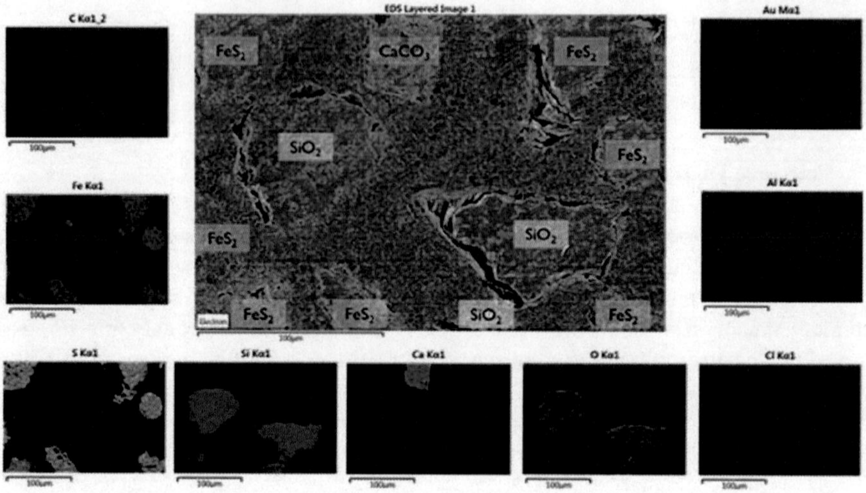

Fig. 3 EDS mapping on the cross section of the pyrite concentrate particles spot 1

20% pure pyrite particles clearly show distinct scattered particles of irregular shape. The size of these particles was ~100–200 μm (see Fig. 2b). Figure 2c shows the SEM images of concentrated pyrite with different magnifications. EDS elemental mapping was also conducted. In order to increase the conductivity of the sample, gold sputtering was conducted on the samples to coat an 8 nm gold conducting layer. Then the SEM and EDS were conducted on these samples. The cross sections of the pyrite concentrate particles were exposed by polishing. EDS mapping analysis on the elemental composition of these particles is shown in Fig. 3. Based on the elemental composition of each particle, the possible chemical compositions of the particles were labeled in the SEM image. A significant number of FeS_2 particles were observed with some SiO_2 and $CaCO_3$ particles. These observations confirmed the existence of metal sulfide other than iron.

X-Ray Diffraction Analysis

Figure 4a, b show the XRD patterns of the pyrite concentrate and the effluent solids from the 3L bioreactor. As can be seen, the main composite in the pyrite concentrate was pyrite, calcite, and silica. While in the effluent solid, the main composites are calcium sulfate, jarosite, and silica. There are no peaks associated with pyrite, which indicates that most of the jarosite has been consumed in the bioreactor. However, pyrite residues might be covered with the jarosite or calcium sulfate precipitate formed in the bioreactor. To confirm this, the solid should be ground to remove possible precipitate shells on the residue pyrite.

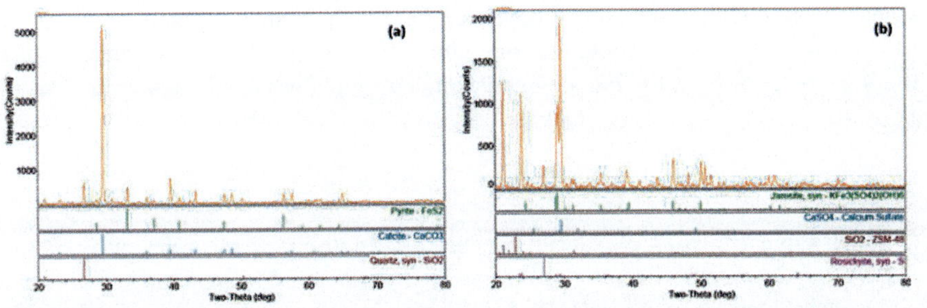

Fig. 4 **a** XRD results of the minus 100 Mesh pyrite concentrate particles; **b** XRD results of the effluent solids from the 3L bioreactor

Fig. 5 ICP-MS results of the pyrite concentrate digestion solution converted to solid composition (%)

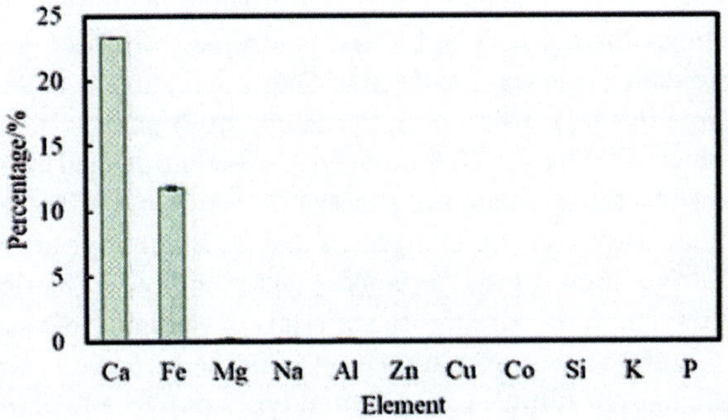

ICP-MS Analysis

The analytical ICP-MS results of the pyrite concentrate digestion solution are shown in Fig. 5. As shown, the main elements in the pyrite concentrates are Fe and Ca. Other elements with minor concentrations are Mg, Zn, Al, Na, Cu, and Co. No Si, K, and P were detected in the digestion solution.

Analysis of Bio-Oxidation Performance and Oxidation–Reduction Potential (ORP) Values

The rate of ferrous ion oxidation by bacteria can be measured by adding ferrous sulfate to the solution, followed by measuring the change in oxidation–reduction potential (ORP) as a function of time. The Nernst Equation for the $Fe^{3+} + e^- \leftrightarrow Fe^{2+}$ half-cell reaction can be written as follows:

$$E = E_o - \frac{RT}{nF} ln \frac{a_{Fe^{2+}}}{a_{Fe^{3+}}} \tag{4}$$

where 'E' is the electrochemical potential, 'E_o' is the standard potential, 'R' is the gas constant, 'T' is the absolute temperature, 'n' is the number of electrons, 'a' is the activity, and 'F' is the Faraday constant. This equation can be rearranged to

$$E = E_o + \frac{RT}{nF} ln \frac{\gamma_{Fe^{3+}} + m_{Fe^{3+}}}{\gamma_{Fe^{2+}} + m_{Fe^{2+}}} \quad (5)$$

For a typical ferrous oxidation test, bacterial activity was evaluated by adding $FeSO_4 \cdot 7H_2O$ into the bacterial solution in the reactor. The resulting ORP value change was measured with time. By knowing an initial potential from an ORP probe and measuring the new ORP value after adding a small amount of ferrous or ferric sulfate, the concentration of ferrous ions initially present can be calculated and the concentration of free ferric ions initially present can be calculated [16]. Bio-oxidation rate data has been analyzed by following the procedure mentioned above. Figure 6a shows the changes in bio-oxidation rate for different feed materials when the bio-oxidation rate was most stable. Experimental data collection was done with a one-day time interval. 100% pure pyrite and 60% pure pyrite have the higher bio-oxidation rates, whereas the 20% pure pyrite concentrate had a comparatively lower rate. These results can evaluate the success of pyrite and effluent characterization. From there it already has been interpreted that 20% pure pyrite has the lowest concentration of ferrous ions, which severely affected the bio-oxidation rate. In comparison, 100% pure and 60% pure pyrite materials have higher bio-oxidation rates.

ORP values were tracked during the operation, every day. Figure 6b shows the changes of Eh(mV) for different types of pyrite feed materials. During the operation period, three different types of pyrite concentrate and an identical operation system

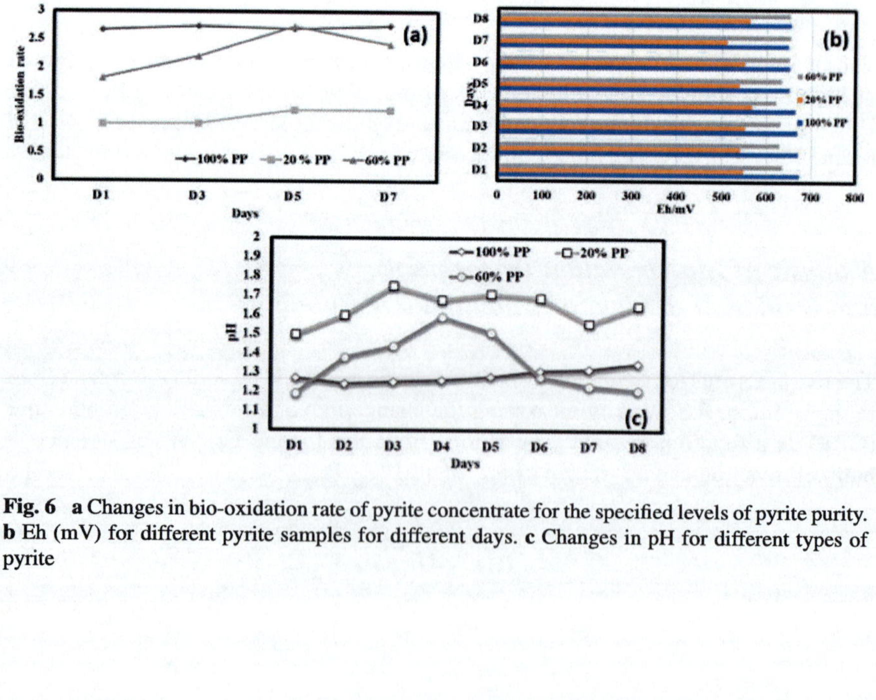

Fig. 6 **a** Changes in bio-oxidation rate of pyrite concentrate for the specified levels of pyrite purity. **b** Eh (mV) for different pyrite samples for different days. **c** Changes in pH for different types of pyrite

were maintained with a residence time of 15 days; other parameters were maintained with 6% solids feeding and the same stirring speed and air flow rate. Eh measurements for the sample with 20% pure pyrite were relatively lower than that for concentrated pyrite and 60% pure pyrite. An interesting observation was the steadiness of the Eh value for each test series involving 20 or 60% pyrite.

Analysis of pH Data Measurement and Acid Generation

pH values were monitored carefully in order to monitor the environment for *Acidithiobacillus ferrooxidans*. This microorganism can thrive [17] in a pH range of 1–2. Figure 6c shows the changes in pH during different types of pyrite feeding.

Acid generation from bio-oxidation of coal waste pyrite was analyzed by titration. Titration tests were used to determine the available acid in the leaching solution. The test procedures are listed below.

(1) Preparation of 0.1 M NaOH solution by dissolving NaOH pellets into DI water.
(2) 20 ml of the leaching solution was taken from the bioreactor and filtered to remove solid particles.
(3) Preparation of a burette with the 0.1 M NaOH solution, and recording of the initial reading on the burette.
(4) Putting 10 mL of the filtered leaching solution into an Erlenmeyer flask and putting it under the burette.
(5) Opening of the valve of the burette and starting titration while tracking the pH of the solution with a pH meter until the pH reading reached 2.5, at which point the valve was closed.
(6) Recording of the final reading on the burette.

The acid concentration can be determined with the calculations shown below:

$C_{NaOH} = 0.1$ M
$V_{NaOH} = 25$ ml
$V_{Acid} = 10$ ml

The mass balance is as follows:

$C_{NaOH} \cdot V_{NaOH} = C_{Acid} \cdot V_{Acid}$.
As an example, for a test in which 15 ml of NaOH was consumed, this leads to:
$0.1\ \text{M} \times 25\ \text{mL} = C_{Acid} \times 10\ \text{mL}$
$C_{Acid} = 0.25$ M (available relative to pH 2.5 in solution)

During this operation period of the bioreactor with automatic feeding and effluent, 5 g pyrite (FeS_2) was fed to the bioreactor with 9 K medium to the bioreactor 3 times a day. Simultaneously, there is 100 ml of the leaching solution in the bioreactor that was pumped to the effluent vessel 3 times per day. The acid generation rate and efficiency can be estimated based on the mass balance of the main input and output materials. So, the acid generation rate (amount of H^+ generated per day) is as follows:

$$0.25 \text{ mol/L} \times 0.3\text{L} = 0.075 \text{ mol/day}$$

24 grams of pyrite, which are fed each day, contain

$$m_{FeS_2} = \frac{24\text{g}}{120(g/\text{mol})} = 0.2 \text{ mol}$$

In this example, the effective efficiency for production of acid from per mole of FeS_2 is as follows:

$$\eta = \frac{0.075(mol/day)}{0.2(mol/day)} \times 100\% = 37.5\%$$

FeS_2 and Fe^{3+} Reaction Rate Measurement

During the bioreactor operation (when pyrite is added to the reactor at the very beginning of the pyrite oxidation), the ferric ions react with the FeS_2 and Fe^{3+} in the solution is reduced to Fe^{2+}, which is the reactant for the bacterial oxidation. The reaction rate of the FeS_2 and Fe^{3+} reaction could be measured to show how fast Fe^{2+} can be provided in the bioreactor [4]. pH measurements help to determine the acid consumption by feed material and production through bio-oxidation and subsequent pyrite leaching.

Pyrite source material characterization confirms that pyrite concentrate has some calcite and magnetite, which are acid consuming materials. The back-titration method was used to measure the calcite concentration in the pyrite concentrate and final acid generation data was evaluated. This should be related to the reaction between $CaCO_3$ and HCl then the formation of CO_2. About ~22.5% calcite equivalent was found in the pyrite concentrate, based on feed material consumption of acid.

Pyrite and ferric ion reaction rates were measured using two tests at two different levels of ferric concentration (10 g/L Fe^{3+} test and 20 g/l Fe^{3+}) using regular and small size pyrite particles. Smaller pyrite particles were attained by grinding the regular pyrite particles, followed by sieving using a 200 Mesh sieve.

10 G/l Test

The Eh versus time plots of the two tests and the concentration of the Fe^{3+} in the solution were determined and plotted as a function of time as shown in Fig. 7. As can be seen, the Eh value drops very fast for the first 100 min. After that, the change in Eh drops slowly for the regular pyrite particles after 474 min, and for the smallest pyrite after 621 min, no changes were observed. Red dots in the figure show the end points. In this test, the amount of Fe^{3+} consumed by the reaction can be calculated by the concentration change and the volume of the solution, and the amount of FeS_2

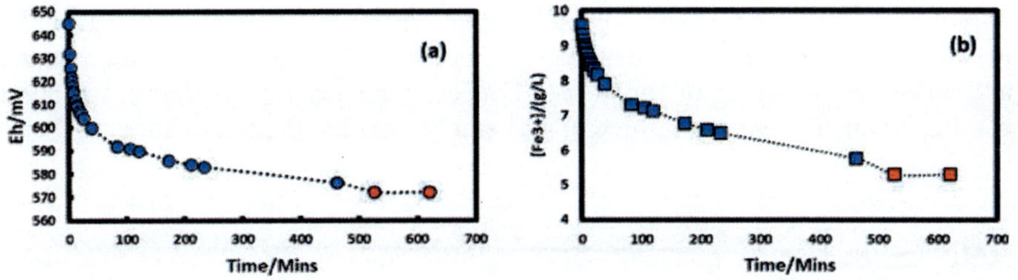

Fig. 7 Eh versus Time curve and Fe^{3+} concentration as a function of time in the FeS_2 and 10 g/l Fe^{3+} reaction tests

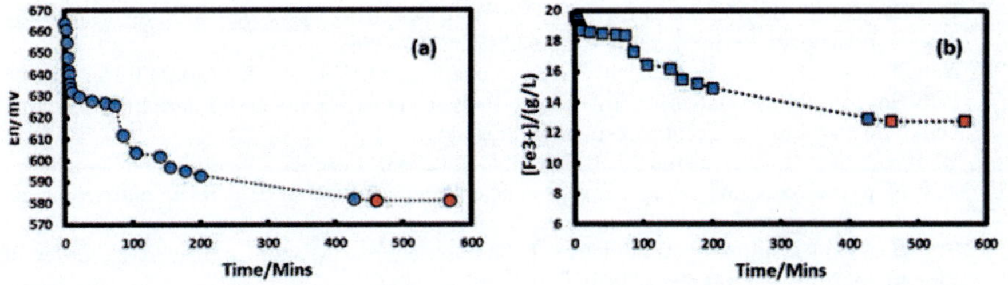

Fig. 8 Eh versus Time curve and Fe^{3+} concentration as a function of time in the FeS_2 and 20 g/l Fe^{3+} reaction tests

is also calculated and compared with the Fe^{3+} consumption. The ratio of Fe^{3+}/FeS_2 was found to be11.34.

20 G/l Test

During the 20 g/l Fe^{3+} test, the Eh value drops very fast during the first 80–95 min. After that, the change in Eh was not significant, but it kept dropping. No change was observed after 569 min. Red dots in the figure shows the end points. For regular size pyrite particle, the Fe^{3+} concentration dropped from 20 to 12.749 g/L after 461 min. The concentration of the Fe^{3+} in the solution is calculated and plotted as a function of time, as shown in Fig. 8. The ratio of Fe^{3+}/FeS_2 was calculated to be 14.40.

Conclusions

Overall, the authors have demonstrated pyrite bio-oxidation for acid generation. In this research, systematic information that includes bioreactor setup, pyrite characterization, morphology of samples, phase purity of samples and effluent solid, analysis of bio-oxidation as well as acid generation rate have been evaluated. The pH and Eh data for pure pyrite were found to be relatively constant for individual sets of parameters after a steady-state was achieved. However, tests with feed material containing

20% pyrite were less constant, although variations were within pH 1.3 to 1.8. The net acid production of 0.37 mol of acid per mole of FeS_2 was deduced based on the available data for one set of conditions. The reaction rates of the FeS_2 consumption and Fe^{3+} reduction were measured to evaluate bio-oxidation performance.

References

1. Zhou Y, Short MD, Li J, Schumann RC, Smart RSC, Gerson AR, Qian G (2017) Control of acid generation from pyrite oxidation in a highly reactive natural waste: a laboratory case study. Minerals 7:89
2. Dos Santos EC, de Mendonça Silva JC, Duarte HlA (2016) Pyrite oxidation mechanism by oxygen in aqueous medium. J Phys Chem C 120:2760–2768
3. Yadollahi A, Abdollahi H, Ardejani FD, Mirmohammadi M, Magdouli S (2021) Bio-oxidation behavior of pyrite, marcasite, pyrrhotite, and arsenopyrite by sulfur-and iron-oxidizing acidophiles. Bioresour Technol Rep 15:100699
4. Bouffard SC, Rivera-Vasquez BF, Dixon DG (2006) Leaching kinetics and stoichiometry of pyrite oxidation from a pyrite–marcasite concentrate in acid ferric sulfate media. Hydrometallurgy 84:225–238
5. Tupikina O, Rassulov V, Kondrat'eva T (2009) Patterns of pyrite oxidation by different microorganisms. Microbiology 78:165–169
6. Sarswat PK, Leake M, Allen L, Free ML, Hu X, Kim D, Noble A, Luttrell GH (2020) Efficient recovery of rare earth elements from coal based resources: a bioleaching approach. Mater Today Chem 16:100246
7. Zhan Y, Yang M, Zhang S, Zhao D, Duan J, Wang W, Yan L (2019) Iron and sulfur oxidation pathways of *Acidithiobacillus ferrooxidans*. World J Microbiol Biotechnol 35:1–12
8. Tu Z, Guo C, Zhang T, Lu G, Wan J, Liao C, Dang Z (2017) Investigation of intermediate sulfur species during pyrite oxidation in the presence and absence of *Acidithiobacillus ferrooxidans*. Hydrometallurgy 167:58–65
9. Zhang Z, Allen L, Podder P, Free ML, Sarswat PK (2021) Recovery and enhanced upgrading of rare earth elements from coal-based resources: bioleaching and precipitation. Minerals 11:484
10. Ye J, Wang S, Zhang P, Nabi M, Tao X, Zhang H, Liu Y (2020) L-cysteine addition enhances microbial surface oxidation of coal inorganic sulfur: Complexation of cysteine and pyrite, inhibition of jarosite formation, environmental effects. Environ Res 187:109705
11. Wang H, Dowd PA, Xu C (2019) A reaction rate model for pyrite oxidation considering the influence of water content and temperature. Miner Eng 134:345–355
12. Wang G, Liu X, Wu Y, Zeng T, Li S, Liu J, Liu Y, Xie S (2020) Bio-oxidation of a high-sulfur refractory gold concentrate with a two-stage chemical-biological approach. Hydrometallurgy 197:105421
13. Fowler T, Holmes P, Crundwell F (1999) Mechanism of pyrite dissolution in the presence of *Thiobacillus ferrooxidans*. Appl Environ Microbiol 65:2987–2993
14. Sarswat PK, Zhang Z, Free ML (2021) Rare earth elements extraction from coal waste using a biooxidation approach. In: Proceedings of the rare metal technology 2021. Springer International Publishing, Berlin/Heidelberg, Germany, pp 211–216
15. Hoffman ND (2020) Cultivation and use of *Acidithiobacillus ferrooxidans* in Tellurium biorecovery. Idaho National Lab. (INL), Idaho Falls, ID (United States)
16. Free M (2013) Hydrometallurgy: fundamentals and applications. John Wiley & Sons, New York
17. Valdés J, Pedroso I, Quatrini R, Dodson RJ, Tettelin H, Blake R, Eisen JA, Holmes DS (2008) *Acidithiobacillus ferrooxidans* metabolism: from genome sequence to industrial applications. BMC Genomics 9:1–24

Extraction of Rare Earth Metals from Coal Ash Using Mild Lixiviants in a Single Step Process

Riya Banerjee, Saswati Chakladar, and Sanchita Chakravarty

Abstract The redemption of Rare Earth Elements from coal ash has multifarious advantages in the utilization of waste and also as a way to minimize unnecessary mining. There is a significant escalation in the quantity of coal ash production, as well as the need for REE in modern-day advanced technological products. The utilization of coal ash, knowing its REE enrichment, has been a persistent area of research and development. Our approach is to extract REEs from coal ash in a non-destructive way utilizing the least quantity of relatively eco-friendly lixiviant. The novelty lies in efficient and selective leaching of REE using milder and less corrosive reagents compared to mineral acids. Our research group is currently engaged in exploring the leaching potential and kinetics of novel organic lixiviants which can leach REEs (~60–70%) using lower quantities (2–10%), preferably in a single step. Interesting findings from our experimental work will be discussed.

Keywords Coal ash · Rare earth elements · Leaching

Introduction

Coal ash has emerged as an attractive resource for REEs, owing to their enrichment in critical REEs and low-to-no-cost availability as waste products. Typically, if coal contains 60 ppm of REE, a power plant ash contains on an average 470 ppm of REEs [1]. The large differences between the global demand and their fewer resources have made these rare earth elements indispensable raw materials [2]. A number of high-tech products depend on these rare earth elements for example cell phones, flat-screen TVs, electric cars, wind turbines, satellites, defense aircraft, and magnets. According to the recent industrial classification, REEs are broadly divided into three categories (1) Critical—Nd, Eu, Tb, Dy, Y, and Er; (2) Uncritical—La, Pr, Sm, and Gd; and (3) Excessive—Ce, Ho, Tm, Yb, and Lu.

R. Banerjee · S. Chakladar · S. Chakravarty (✉)
Analytical and Applied Chemistry Division, CSIR-National Metallurgical Laboratory, Jamshedpur 831007, India
e-mail: sanchita@nmlindia.org

T. Ouchi et al. (eds), *Rare Metal Technology 2022*, The Minerals, Metals & Materials Series, https://doi.org/10.1007/978-3-030-92662-5_6

An understanding of the natural occurrence of REEs with coal minerals is fundamentally important to develop a suitable chemical and physical extraction process. Based on literature reports, a common consensus seems to be that most of the REEs are primarily distributed in the glassy amorphous aluminosilicate matrix of coal ash [3]. Hence, wet leaching is the preferred technique to extract them out from coal ash. Leaching using strong mineral acids such as HCl or HNO_3 at elevated temperatures is the most prevailing method till date [4 and references therein].

In nutshell, the extraction of REE from coal ash heavily depends on the usage of strong mineral acids which are harmful to the environment. Employment of strong acids also impairs the selectivity of REEs over transition metals during the leaching process. Co-leaching of both REEs and transition metals necessitates the introduction of additional solvent extraction step for their separation.

Organic acids, on the other hand, are non-corrosive, readily available, and environment friendly compared to mineral acids. However, they have been primarily implicated as a co-lixiviant for REE extraction from various sources. In combination with mineral acid treatment and organic extraction, organic acids have been reported to be used for REE extraction from coal ash [5]. In another report, formic acid was shown to be utilized for the extraction of REE from NdFeB magnets [6]. However, their leaching potential when used as the sole lixiviant has not been reported.

In the present study, we performed a detailed mechanistic investigation of the leaching process for REEs from coal ash using organic acids in a single step process. A systematic study on their leaching ability varying the crucial process parameters like acid concentration, temperature, and leaching duration was performed. Furthermore, a detailed investigation of the leaching potential and reaction kinetics of organic acids was compared with mineral acids, which is crucial to understand the commercialization aspect of it.

Experimental

Materials and Methods

In the present study, the coal ash which was examined was derived from coal of Indian origin. Carboxylic acids such as tartaric acid and lactic acid (purity >95%) were purchased from Loba Chemie, Mumbai, India, and used without any modification. For fusion experiments, the lithium metaborate and lithium tetraborate which were needed were also purchased from Loba Chemie. Nitric acid (Analytical Grade, 65%) was procured from Rankem, India. Deionized (DI) water was produced inhouse using a Milli Q water purification system. Preparation of coal ash sample was done using a muffle furnace in presence of air at 850 °C for 1 h. The elemental composition of the coal ash was determined using an X-ray fluorescence (XRF) spectrophotometer (S8 Tiger, Bruker) using the press pellet method. The concentration of metal ions in solution was determined using an inductively coupled plasma-mass spectrometer (ICP-MS) (Perkin Elmer, Model ELAN DRC-e with Axial Field Technology), inductively coupled plasma-optical emission spectrophotometer (ICP-OES) (Thermo Scientific, i CAP 7000 series), and atomic absorption spectrophotometer (AAS) (Thermo Scientific iCE3000) instruments.

Procedure to Estimate REE Content in Coal Ash

The total REE content in coal ash was determined using the borate fusion method wherein 0.1 g of coal ash was mixed with 0.15 g of lithium tetraborate and 0.2 g of lithium metaborate followed by fusion at 950 °C for 1 h in a muffle furnace. The fused sample obtained was then dissolved in 100 ml of 10% nitric acid solution to ensure complete dissolution of the sample. Finally, the REE content in the solution was determined using ICP-MS instrument.

Leaching Procedure Using Organic Acids and Mineral Acids

For each leaching experiment, 0.1 g of coal ash was taken and 15 mL of aqueous organic acid solution/ mineral acid was added to it. The organic structure of the two acids is shown in Fig. 1. The mixture was stirred at 600 rpm at the desired temperature for a specific duration using a magnetic hot plate stirrer. The resultant leachate was vacuum filtered using Whatman 42 filter paper to separate out the residual coal ash from the leached solution. The concentration of REE in leached solution was measured using ICP-MS after dilution (to 250 ml) and acidification (using 5% HNO_3).

Results and Discussion

Leaching of REEs from coal ash is conventionally done using strong mineral acids solely or by using mineral acids in combination with solvent extraction. Despite their corrosive nature, mineral acids are predominantly used owing to their high efficiency and lower cost. Organic acids are emerging to be potential alternatives to mineral

Fig. 1 Molecular structure of the two organic acids

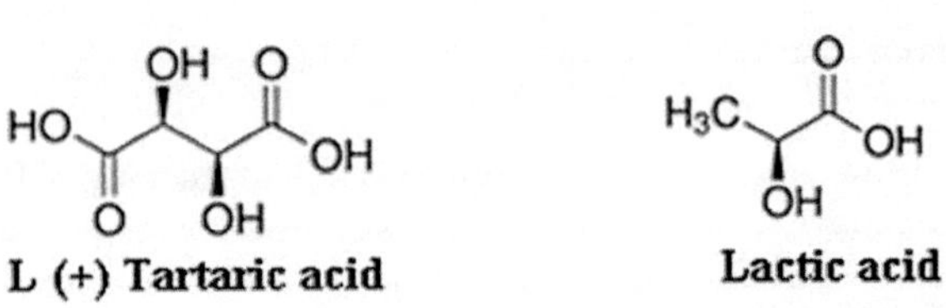

Table 1 Mineralogical composition of coal ash being examined

XRF analysis		ICP-MS analysis			
Composition	Concentration (Wt %)	LREE	Concentration (ppm)	HREE	Concentration (ppm)
SiO_2	56.61	La	106.04	Tb	1.69
Al_2O_3	28.29	Ce	195.03	Dy	10.98
Fe_2O_3	4.95	Pr	16.57	Ho	1.91
K_2O	3.26	Nd	75.22	Er	6.08
TiO_2	1.68	Sm	13.99	Tm	0.85
CaO	1.61	Eu	3.33	Yb	4.17
MgO	1.38	Gd	14.43	Lu	0.69
SO_3	1.25				
P_2O_5	0.44				
BaO	0.21				
Na_2O	0.15				

acids for REE extraction due to their mild nature and easy availability. In the present study, we have chosen two low molecular weight organic acids (Fig. 1) and examined their leaching behavior from coal ash on the basis of crucial process parameters. Last but not the least, the leaching potency and reaction kinetics of organic acids were compared with mineral acids to gain an in-depth understanding of the two processes.

Mineralogical Characterization of Coal Ash

The coal ash sample which was examined was thoroughly characterized first to know the composition of major oxides as well as rare earth elements using XRF and ICP-MS analysis. The results are shown in Table 1. On the basis of XRF analysis, Si, Al, Fe, Ti, Ca, Mg, and S were present as the major elements, and P, Ba, Sr, Mn, Zr, Cr, Cu, Ni were present as the minor elements in the coal ash sample. ICP-MS of the fused coal ash sample indicated the presence of a total of 450.98 ppm of REE in the coal ash. Among LREEs, concentration of Ce and La was reasonably higher than the rest of the REEs.

Acid Concentration

The quantity of acid required for efficient leaching of REEs from coal ash is a crucial parameter and quite important towards a process's economic evaluation. The variations in leaching potency of carboxylic acids are governed by the following factors:

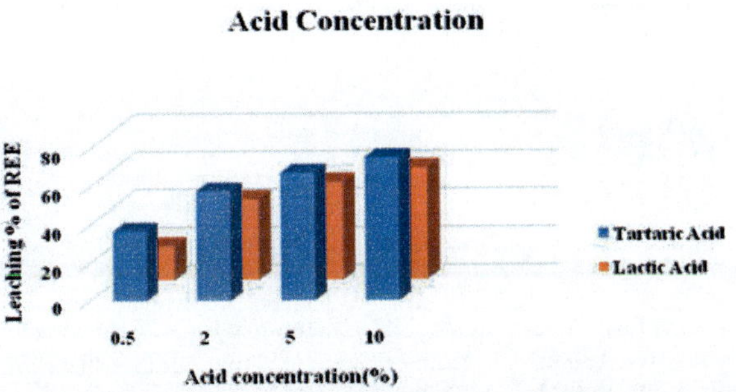

Fig. 2 Effect on acid concentration on leaching efficiency of REEs (Pulp density: 150 mL: 1 g; Time: 1 h; Temperature: 90 °C)

(1) pKa values of ionisable protons; (2) Number of functional groups in chemical structure which can aid in complexation; (3) steric factors; and (4) stability constants of the complexes formed between the REE metals as substrates and carboxylic acid ligands. Lower the pKa of the ionisable proton, stronger is the acid. In this case, tartaric acid (pKa 2.98, 4.34) was found to be superior to lactic acid, both with respect to acid strength and complexing ability (Fig. 1). The results we obtained for both under different acid concentrations (0.5, 2, 5, 10, and 15%) are demonstrated in Fig. 2. It clearly shows that tartaric acid was more efficient in complexation and subsequent leaching of REEs from coal ash matrix as compared to lactic acid. The optimum acid concentration based on leaching efficiencies for both the organic acids was determined to be 5% (w/v).

Effect of Temperature and Duration of Leaching

Temperature and duration of leaching are very crucial parameters that need to be examined while optimizing a leaching process. During optimization of our leaching experiments, we investigated a wide range of durations (5, 10, 20, 30, and 60 min) and temperatures (25, 60, 70, 80, and 90 °C). The results we obtained are shown in Fig. 3. The optimum leaching duration was deduced to be 60 min while 90 °C was selected to be the optimum temperature for the best efficiency. Higher temperatures are typically known to ameliorate the dissolution of rare earth elements, thereby improving the leaching potency. This observed trend was comparable to previous literature reports on REE leaching using a high concentration of mineral acids [7]. Under optimized reaction conditions of temperature (90 °C) and duration of leaching (60 min), the total REE which was leached by tartaric acid was 62%, whereas that for lactic acid was 56%.

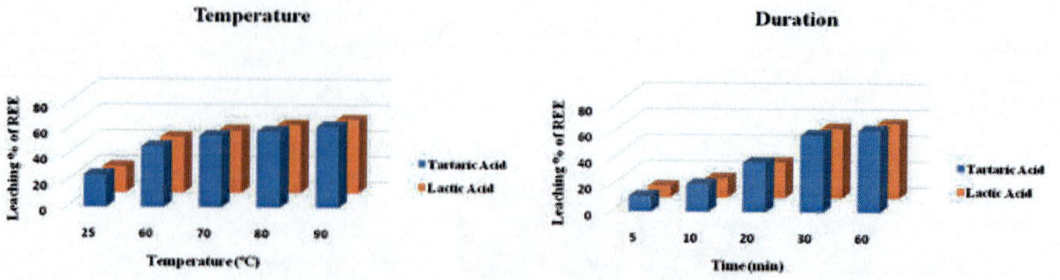

Fig. 3 Variation in leaching efficiency under different temperatures (Acid concentration: 5% (w/v); Duration: 1 h; Pulp density: 150 mL: 1 g) and different leaching durations (Acid concentration: 5% (w/v); Temperature: 90 °C; Pulp density: 150 mL: 1 g)

Reaction Kinetics

The leaching kinetics of REEs from coal and coal ash using minerals acids has been extensively studied in literature [8]. However, the reaction kinetics of organic acids is still unknown. We performed some preliminary studies on tartaric acid only till date. Based on the results obtained, the leaching kinetics of organic acids seems to be in fair agreement with mineral acids where the process is controlled by both diffusion and chemical reaction.

Comparison With Mineral Acids

A comparative analysis of the leaching efficiency of REEs from coal ash using organic acids with that of mineral acids was essential. We decided to perform the leaching experiments from coal ash using 0.25 M of acid concentration, which is a much lower concentration than what is typically employed for mineral acids in the conventional leaching process (1 M). The results we obtained for the leaching of LREEs (which is comparatively more abundant than HREEs) are shown in Fig. 4. It is clearly evident that the leaching performance of tartaric acid (~60% yield) is quite comparable to that of mineral acids (~72% yield).

Selectivity of REEs Over Transition/Main Group Elements

Selective leaching of elements is one of the most crucial aspects of any leaching process; in particular, for rare earth extraction, it is the selectivity over co-leaching of transition metals. Mineral acids are strongly ionic and hence impart less selectivity during REE leaching resulting in co-leaching of both REEs and transition elements. The selectivity results obtained in this study (Fig. 5) were encouraging with respect to the selectivity obtained for Al, Si, and Fe. Tartaric acid leached out the total REEs

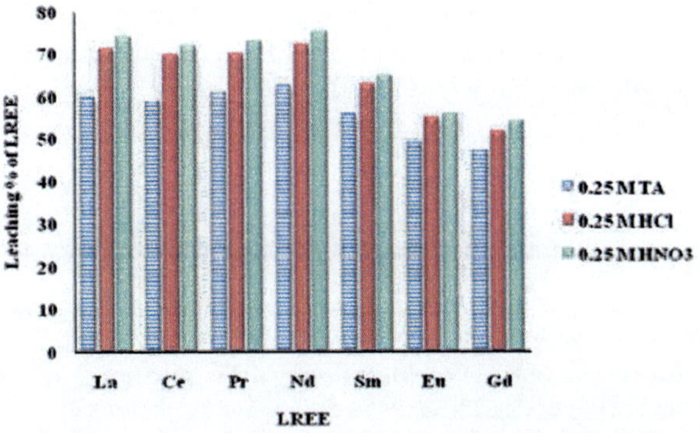

Fig. 4 Comparison of leaching efficiency of tartaric acid with that of mineral acids at 0.25 M concentration (Pulp density: 150 mL: 1 g; Time: 1 h; Temperature: 90 °C)

Fig. 5 Selectivity of REE leaching over Al, Si, and Fe

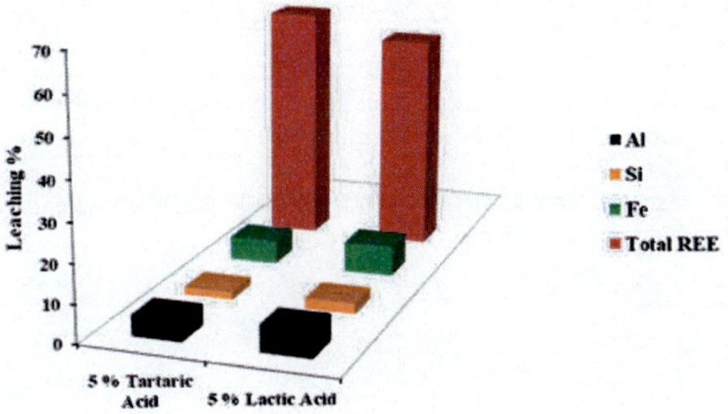

in much higher preference (60–70%) over the predominant elements present in coal ash such as Al (5%), Si (2%), and Fe (6%), which is noteworthy. Lactic acid also performed with commendable selectivity.

Conclusion

Organic carboxylic acids, like tartaric acid and lactic acid, were found to possess noteworthy leaching potential for rare earth elements from coal ash at considerably lower concentrations (5% w/v). The optimum conditions for REE leaching using organic acids were found to be at 90 °C and for 60 min duration. When compared, organic acids demonstrated comparable leaching potential and leaching kinetics with that of mineral acids. Owing to their mild nature, organic carboxylic acids were further advantageous in leaching out REEs selectively over transition elements. To our knowledge, this is the first detailed study where organic carboxylic acids have been demonstrated to have comparable efficiency to that of mineral acids to leach out REEs from coal ash, when used on its own. Experimental results obtained along with future work will be discussed in detail.

References

1. Seredin VV, Dai S (2012) Coal deposits as potential alternative sources for lanthanides and yttrium. Int J Coal Geol 94:67–93
2. Franus W, Wiatros-Motyka MM, Wdowin M (2015) Coal fly ashs a resource for rare earth elements. Environ Sci Pollut Res 22(12):9464–9474
3. Zhang W, Aaron N, Yang X, Honaker R (2020) A comprehensive review of rare earth elements recovery from coal-related materials. Minerals 10(5):451
4. Peiravi M, Ackah L, Guru R, Mohanty M, Liu J (2017) Chemical extraction of rare earth elements from coal ash. Miner Metall Process 34:170–177
5. Joshi PB, Preda DV, Skyler DA, Tsinberg A, Green BD, Marinelli WJ (2015) Recovery of rare earth elements other publications and compounds from coal ash. US Patent No.: 8,968,688 B2
6. Gergoric M, Ravaux C, Steenar BT, Espegren F, Retegan T (2018) Leaching and recovery of rare-earth elements from neodymium magnet waste using organic acids. Metals 8:721
7. Kumari A, Parween R, Chakravarty S, Parmar K, Pathak DD, Leed J, Jha MK (2019) Novel approach to recover rare earth metals (REMs) from Indian coal bottom ash. Hydrometallurgy 187:1–7
8. Yang X, Honaker RQ (2020) leaching kinetics of rare earth elements from fire clay seam coal. Minerals 10:491 and references therein

Separation of Scandium by Phosphinic Acid Extractant from Sulfuric Acid Solution

Amilton Barbosa Botelho Junior, Denise Crocce Romano Espinosa, and Jorge Alberto Soares Tenório

Abstract The main challenge for scandium extraction from bauxite residue by hydrometallurgy is the separation step. After sulfuric acid leaching, the solvent extraction process is required for a high separation degree. The presence of contaminants such as iron and aluminum in a concentration 1000 times higher than scandium is a problem to be overcome. The present study aims at the separation of scandium from the sulfuric acid solution using Cyanex 923 in kerosene. First, zirconium was removed using Aliquat 336. Then, Cyanex 923 was used to remove scandium over impurities, such as iron, aluminum, and titanium. Results have demonstrated that the pH does not affect scandium separation, while temperature, organic extractant concentration, and TBP affect the separation. Experiments were carried out for 15 min under magnetic stirring, and samples were analyzed in EDXRF and ICP-OES.

Keywords Sc(III) · Hydrometallurgy · Cyanex 923 · D2EHPA · Bauxite residue

Introduction

The increased demand to supply technologies for renewable energy and e-mobility resulted in a risk of the supply chain in the medium and middle-term. The list of necessary materials varies for each country or region, where the rare earth elements are included in all lists [1, 2], including scandium, yttrium, and lanthanides. Among them, scandium sources are rarely found, which impacts its price [3]. The element is almost two thousand more valuable than lanthanum, cerium, and neodymium [4–6].

The extraction of scandium from primary sources occurs in a few thortveitite and lolbeckite ores ($(Sc, Y)_2Si_2O_7$) sites in Madagascar, Norway, and the USA. There are several sources unexplored, where bauxite residues are the most important due to the amount of waste stored in dams (4 billion tons in the world) and the scandium content (from 20 to 180 g/ton) [3, 7–9].

A. B. Botelho Junior (✉) · D. C. R. Espinosa · J. A. S. Tenório
Department of Chemical Engineering, Polytechnic School of University of São Paulo, São Paulo, Brazil
e-mail: amilton.junior@usp.br

T. Ouchi et al. (eds), *Rare Metal Technology 2022*, The Minerals, Metals & Materials Series, https://doi.org/10.1007/978-3-030-92662-5_7

The hydrometallurgical process has several advances in comparison to pyrometallurgical, as high-pure products may be obtained. On the other hand, the acid leaching to extract scandium from the solid phase into a solution result in a solution concentrated in contaminants. All iron and aluminum are extracted in the direct leaching to obtain scandium—the concentration of impurities would be over 3,000 times higher than scandium [10–14].

However, there is a lack in the literature aimed at selective separation of scandium. Solvent extraction is the primary technique, and phosphinic acids are the main extractants for rare earth elements [15–17]. For this reason, it is necessary to explore the technique for scandium obtained from bauxite residue leach solution.

The goal of the present study was the extraction of scandium by solvent extraction. The synthetic solution was obtained after the acid leaching of bauxite residue, as previously reported [10]. Then, zirconium is removed using amine-based extractant Alamine 336. Experiments were carried out using Cyanex 923 and D2EHPA. The effect of TBP was explored to improve scandium separation [18].

Materials and Methods

Materials

The synthetic solution was prepared using iron and aluminum sulfate salts; lanthanum, cerium, yttrium, and neodymium oxides; and a standard solution of scandium, zirconium, and titanium. The concentration of each element is presented in Table 1. Sulfuric acid concentrated and sodium hydroxide (8 and 1 mol/L) were prepared with ultrapure water for pH adjustment. The pH was measured using Ag/AgCl electrode 3 mol/L (Sensoglass). All reagents were of analytical grade.

Organic extractants Cyanex 923 (mixture of four trialkylphosphine acids) and D2EHPA (di-2-Ethylhexyl phosphoric acid), and TBP (tributylphosphate) were used without any treatment. Commercial kerosene was used as a diluent. Samples were diluted in nitric acid 4% prepared with ultrapure water when required. Chemical analyzes were carried out in EDXRF (PANalytical Epsilon 3 XL) to quantify major elements (Fe, Al, and Ti) and ICP-OES (Agilent Technologies 7000 series) for minor elements.

Table 1 Concentration of the elements present in the synthetic solution for scandium extraction experiments (mg/kg) [10]

	Al	Fe	Sc	Ti	Zr	La	Ce	Nd	Y
mg/L	5059.8	11,804.3	2.7	200.4	43.4	5.9	27.6	2.2	1.5

Table 2 Parameters studied in solvent extraction experiments

Organic extractant concentration	5–25%
TBP	1–10%
A/O ratio	5/1–1/5

Methodology

First, the organic extractant was diluted in kerosene in the desired concentration. Next, the aqueous phase (Table 1) at pH 0.5, as the real leach solution, was mixed with the organic phase under magnetic stirring and pH control for 15 min. Then, the phases were separated and filtered to ensure the organic phase was removed before chemical analysis. Experiments were performed with an A/O ratio (aqueous phase/organic phase) equals 1/1 (15 mL of each solution).

The parameters studied are presented in Table 2. The effect of the A/O ratio was studied to improve the separation factor of scandium. The effect of organic extractant concentration was studied as 5, 10, 15, 20, and 25%, and TBP concentration was studied for the values 1, 2.5, 5 and 10%.

The extraction efficiency (E), distribution ratio (D), and separation factor (β) were calculated as depicted in Eqs. (1)–(3), where M_i and M_a are initial and final concentrations of the metallic ions in the aqueous phase; and D_1 and D_2 are the distribution ratio of metallic elements.

$$E = \frac{M_i - M_a}{M_i} \tag{1}$$

$$D = \frac{M_i - M_a}{M_a} \tag{2}$$

$$\beta_{1,2} = \frac{D_1}{D_2} \tag{3}$$

Results and Discussion

As reported by the literature, scandium extraction is highly efficient using organophosphorus extractants [3]. However, zirconium is also extracted by the same compounds. According to Banda et al. [19], the extraction of zirconium in sulfuric acid media by organophosphorus extractants, including D2EHPA, is over 80% [19]. For this reason, its removal before scandium extraction is essential.

Zirconium was removed from the solution using amine-based extractant Alamine 336. The same parameters presented in Table 2 were studied, and results demonstrated that all zirconium is removed using 10% of organic extractant in kerosene, A/O ratio 1/1, at pH 1.0, and 25 °C for 15 min with only one theoretical contact.

For scandium extraction, preliminary experiments showed that the pH or temperature did not improve its extraction nor increased the separation factor among the other elements (mainly iron, aluminum, and titanium). For this reason, experiments were carried out at pH 0.5 and 25 °C.

The resulting solution was studied for scandium removal by Cyanex 923 and D2EHPA. First, the effect of organic extractant concentration was studied. Figure 1 shows the distribution ratio of the elements varying the concentration of D2EHPA. The extraction of scandium achieved 100% in all conditions evaluated. Extraction of titanium and iron increased from 57 and 21% (D2EHPA 5%) to 76 and 47% (D2EHPA 25%), respectively.

In the case of Cyanex 923 (Fig. 2), similar behavior was observed. Scandium extraction achieved 100% in concentration from 10 to 25%. In titanium and iron, the extraction efficiency was 56% and 28%, respectively, in an experiment carried out with Cyanex 923 25%.

As observed by the distribution ratio, both organic extractants are more selective for scandium than other elements, where Cyanex 923 was more selective for the target element. The distribution ratio of scandium was over 80 in both cases, while titanium, iron, aluminum, and the rare earth elements were lower than 3.5 for D2EHPA and 1.5 for Cyanex 923.

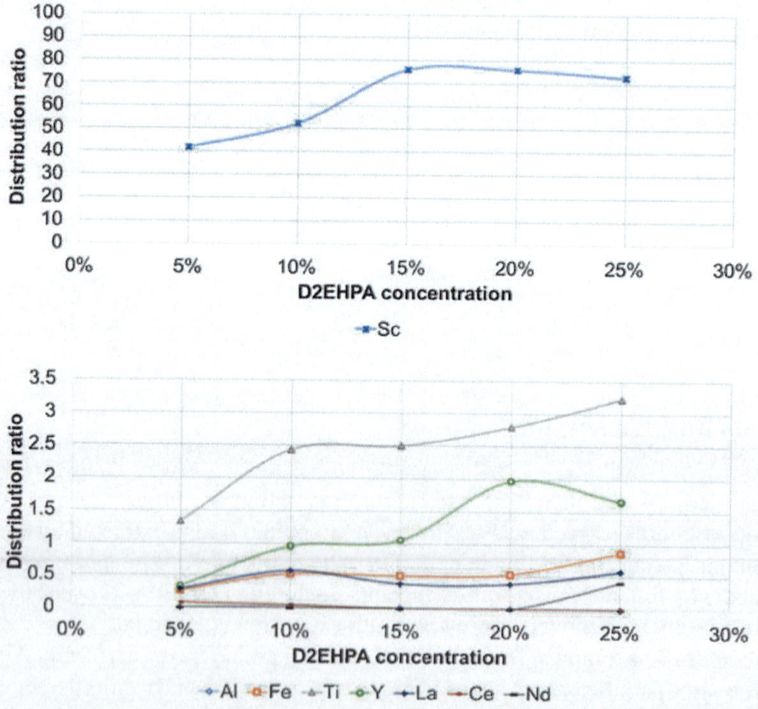

Fig. 1 Distribution ratio varying the concentration of D2EHPA. Experimental conditions: A/O ratio = 1:1; pH 0.5; 15 min; 25 °C

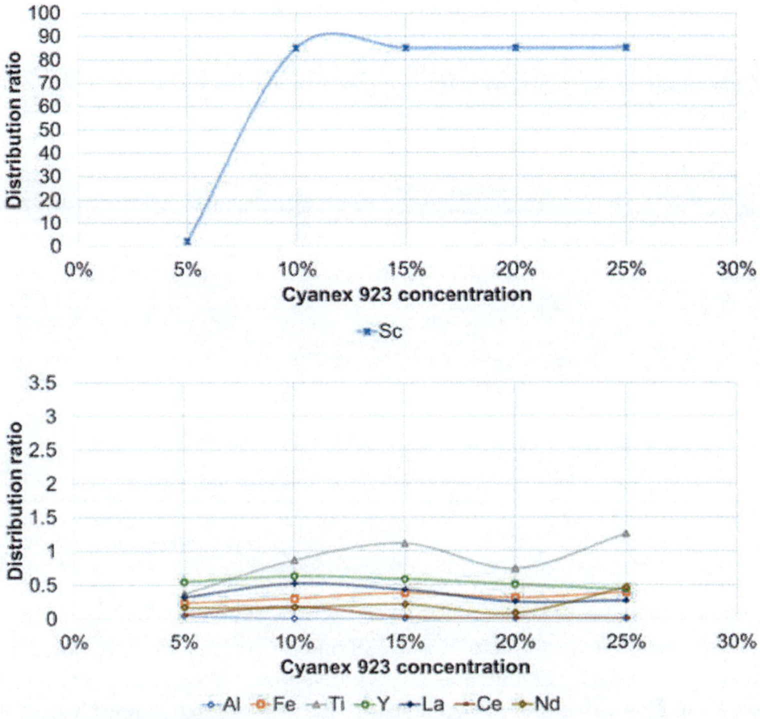

Fig. 2 Distribution ratio varying the concentration of Cyanex 923. Experimental conditions: A/O ratio = 1:1; pH 0.5; 15 min; 25 °C

Such results are important to obtain a high purity scandium product. As observed, the separation of scandium from other elements was higher using Cyanex 923. The effect of TBP was also explored. In both cases, the extraction of scandium remained constant—100%. However, the extraction of all other elements declined using D2EHPA as the concentration of TBP increased. The TBP did not affect extraction efficiency in experiments performed with Cyanex 923. The distribution ratio, extraction efficiency, and separation factor between scandium and other elements achieved similar results with the mixture of D2EHPA 10% + TBP 10% as Cyanex 10% (without TBP) (Fig. 3).

The effect of the A/O ratio was also evaluated. Experiments were carried out at 25 °C, pH 0.5, and organic extractant concentration 10% for 15 min. As the amount of organic phase increased, the extraction of the elements also increased. In the case of D2EHPA, the extraction of scandium, titanium, and iron increased from 80%, 37%, and 31% (A/O ratio = 5/1) to 100%, 71%, and 35%, respectively. For Cyanex 923, the extraction increased from 85, 29, and 30% to 100, 46, and 26% for scandium, titanium, and iron, respectively.

The extraction of scandium was more selective using Cyanex 923 than D2EHPA. The separation factor of Sc/Fe and Sc/Ti for D2EHPA was 99 and 21, while for

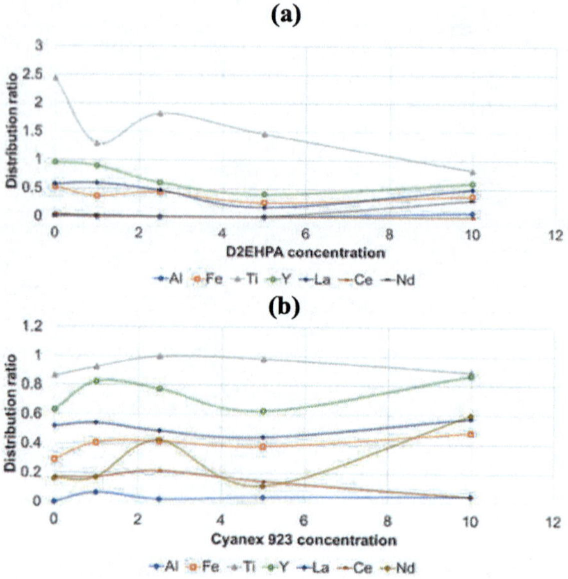

Fig. 3 Distribution ratio varying the concentration of TBP using **a** D2EHPA and **b** Cyanex 923. Experimental conditions: organic extractant concentration = 10%; A/O ratio = 1:1; pH 0.5; 15 min; 25 °C

Cyanex 923, the separation factor were 288 and 98, respectively, for experiments performed with an A/O ratio equal to 1/1. As the amount of organic phase increased, the separation factor declined. In conclusion, the organic extractant Cyanex 923 is better for scandium separation than D2EHPA. Similar results were obtained by Souza et al. [17].

Conclusion

The goal of the present study was the scandium separation using phosphinic acid extractant from bauxite residue leach solution. D2EHPA and Cyanex 923 were studied, varying the effect of concentration, TBP, and A/O ratio. Results demonstrated that Cyanex 923 was better for scandium separation than D2EHPA. The mixture D2EHPA 10% + TBP 10% achieved the same separation factor as Cyanex 10% without TBP. The distribution ratio of scandium was over 80 by both organic extractants, while titanium, iron, aluminum, and the rare earth elements were lower than 3.5 for D2EHPA and 1.5 for Cyanex 923. The best conditions for scandium separation were: A/O ratio equals 1/1, Cyanex 923 10%, pH 0.5, and 25 °C for 15 min.

Acknowledgements The authors acknowledge the University of Sao Paulo for supporting this project and Sao Paulo Research Foundation and Capes (FAPESP—process number: 2012/51871-9, 2018/03483-6, 2018/11417-3, and 2019/11866-5) for the financial support.

References

1. Bobba S, Carrara S, Huisman J, Mathieux F, Pavel C (2020). Critical raw materials for strategic technologies and sectors in the EU—a foresight study. https://doi.org/10.2873/58081
2. Martins LS, Guimarães LF, Botelho Junior AB, Tenório JAS, Espinosa DCR (2021) Electric car battery: an overview on global demand, recycling and future approaches towards sustainability. J Environ Manage 295:113091. https://doi.org/10.1016/j.jenvman.2021.113091
3. Botelho Junior AB, Espinosa DCR, Vaughan J, Tenório JAS (2021) Recovery of scandium from various sources: a critical review of the state of the art and future prospects. Miner Eng 172:107148. https://doi.org/10.1016/j.mineng.2021.107148
4. Gambogi J (2021) Yttrium, USGS, p 2. https://pubs.usgs.gov/periodicals/mcs2021/mcs2021-yttrium.pdf. Accessed 21 Mar 2021
5. Gambogi J (2021) Scandium, USGS, p 2. https://pubs.usgs.gov/periodicals/mcs2021/mcs2021-scandium.pdf. Accessed 21 Mar 2021
6. Gambogi J (2021) Rare earths, USGS, p 2. https://pubs.usgs.gov/periodicals/mcs2021/mcs2021-rare-earths.pdf. Accessed 21 Mar 2021
7. Junior ABB, Espinosa DCR, Tenório JAS (2020) Characterization of bauxite residue from a press filter system: comparative study and challenges for scandium extraction. Mining Metall Explor. https://doi.org/10.1007/s42461-020-00333-3
8. Wang W, Pranolo Y, Cheng CY (2011) Metallurgical processes for scandium recovery from various resources: a review. Hydrometallurgy 108:100–108. https://doi.org/10.1016/j.hydromet.2011.03.001
9. Borra CR, Blanpain B, Pontikes Y, Binnemans K, Van Gerven T (2016) Recovery of rare earths and other valuable metals from bauxite residue (red mud): a review. J Sustain Metall 2:365–386. https://doi.org/10.1007/s40831-016-0068-2
10. Junior ABB, Espinosa DCR, Tenório JAS (2021) Extraction of scandium from critical elements-bearing mining waste: silica gel avoiding in leaching reaction of bauxite residue. J Sustain Metall. https://doi.org/10.1007/s40831-021-00434-3
11. Borra CR, Pontikes Y, Binnemans K, Van Gerven T (2015) Leaching of rare earths from bauxite residue (red mud). Miner Eng 76:20–27. https://doi.org/10.1016/j.mineng.2015.01.005
12. Davris P, Balomenos E, Panias D, Paspaliaris I (2015) Leaching rare earth elements from bauxite residue using brønsted acidic ionic liquids. In: Rare earths industry technological, economic, and environmental implications. Elsevier Inc., pp 183–197. https://doi.org/10.1016/B978-0-12-802328-0.00012-7
13. Economou-Eliopoulos M, Frei R, Megremi I (2016) Potential leaching of Cr(VI) from laterite mines and residues of metallurgical products (red mud and slag): an integrated approach. J Geochem Explor 162:40–49. https://doi.org/10.1016/j.gexplo.2015.12.007
14. Fulford GD, Lever G, Sato T (1991) Recovery of rare earth elements from Bayer process red mud. US Patent US5030424
15. Chen Y, Ma S, Ning S, Zhong Y, Wang X, Fujita T, Wei Y (2021) Highly efficient recovery and purification of scandium from the waste sulfuric acid solution from titanium dioxide production by solvent extraction. J Environ Chem Eng 9:106226. https://doi.org/10.1016/j.jece.2021.106226
16. Ye Q, Li G, Deng B, Luo J, Rao M, Peng Z, Zhang Y, Jiang T (2019) Solvent extraction behavior of metal ions and selective separation Sc3+ in phosphoric acid medium using P204. Sep Purif Technol 209:175–181. https://doi.org/10.1016/j.seppur.2018.07.033
17. Souza AGO, Aliprandini P, Espinosa DCR, Tenório JAS (2019) Scandium extraction from nickel processing waste using Cyanex 923 in sulfuric medium. JOM 71:2003–2009. https://doi.org/10.1007/s11837-019-03427-6

18. Botelho Junior AB, Espinosa DCR, Tenório JAS (2021) Selective separation of Sc(III) and Zr(IV) from the leaching of bauxite residue using trialkylphosphine acids, tertiary amine, tributyl phosphate and their mixtures. Sep Purif Technol 279:119798. https://doi.org/10.1016/j.seppur.2021.119798
19. Banda R, Min SH, Lee MS (2014) Selective extraction of Hf(IV) over Zr(IV) from aqueous H_2SO_4 solutions by solvent extraction with acidic organophosphorous based extractants. J Chem Technol Biotechnol 89:1712–1719. https://doi.org/10.1002/jctb.4249

Part II
Solvent Extraction for Rare Metals

Physicochemistry in Service of Process Design: Case of Uranium Recovery by Liquid–Liquid Extraction

Alexandre Chagnes

Abstract Although solvent extraction seems to be easy to operate in hydrometallurgical processes, its implementation can be difficult and costly. In order to reduce capital and operation expenditures, process optimization is mandatory. It consists in the use of appropriate equipment, flowsheet, and parameters optimization in order to reduce energy and reactive consumption, effluent generation, etc. For this goal, two approaches can be implemented: (i) an empiric approach that relies on feedbacks and experiment design for optimizing operational parameters, and (ii) a more rigorous approach that relies on the comprehension of the physicochemistry taking place during process operation. The first approach is often used in industry, but it is never sure that the process could always operate under optimized conditions. The second approach is interesting as the comprehension of the physicochemistry can get the process more flexible towards external constraints, and it can lead to the development of breakthrough solutions. This paper will show how the comprehension of the physicochemistry involved in the solvent extraction process can contribute to improve process performances. Two examples are presented in this paper for the sake of illustration: (i) the comprehension of the physicochemical phenomena responsible for the chemical degradation of the extraction solvent during liquid–liquid extraction of uranium from acidic sulfate media in order to keep high uranium extraction efficiency, and (ii) the development of a new extractant for the recovery of uranium from wet phosphoric acid by means of a rational approach combining modelling of uranium (IV) extraction and investigation of uranium speciation in the extraction solvent.

Keywords Solvent extraction · Physicochemistry · Modelling · Uranium · Phosphoric acid · Chemical degradation

A. Chagnes (✉)
Université de Lorraine, CNRS, GeoRessources, 54000 Nancy, France
e-mail: alexandre.chagnes@univ-lorraine.fr

T. Ouchi et al. (eds), *Rare Metal Technology 2022*, The Minerals, Metals & Materials Series, https://doi.org/10.1007/978-3-030-92662-5_8

Introduction

Solvent extraction is a common technology for the recovery and separation of base and strategic metals [1]. For instance, the separation of cobalt from nickel is an example for which solvent extraction has proven to be technically and commercially successful. The production of primary uranium is also widely based on solvent extraction with tertiary amines, while solvent extraction with tributyl phosphate is a key step of the PUREX process for the recovery of uranium and plutonium from used nuclear fuels. Solvent extraction is a simple technology but the design of the extraction solvents, i.e., the formulation of the organic phases, must take into account technological, economic, and environmental constraints. The feed solutions containing the metal values to be recovered are characterized by many parameters including the nature of the aqueous phase (sulfate, chloride, nitrate, phosphate, etc.), their elemental composition, their oxidation potential, the level of acidity, their ionic strength, etc. All these parameters constitute a series of constraints that induces a physicochemistry of the extraction systems much more complex than the simplified view of the transfer of a given solute from one phase to the other. The formulation of an extraction solvent should obviously be focused primarily on the selection/design of an extractant or a synergistic mixture of extractants exhibiting strong, reversible, and selective interactions with the target element(s). This is a necessary but not sufficient task, as many other phenomena such as salt-out effects, supramolecular structure of the solvent, etc., take place and can affect favorably or unfavorably the extraction processes [2–4].

Likewise, the performance of solvent extraction plants can be drastically affected by external constraints such as aging of the extraction solvent, losses by solubility, entrainment, evaporation, and modification of the ore composition strongly, which can influence the performances of the extraction processes in terms of its efficiency, selectivity, formation of cruds, etc. [5]. It is of great interest to control these phenomena to avert a decrease in production to design extractants or the formulation of new extraction solvents that can reduce the impact of these external constraints on the process performances. New extractants can also be designed to improve the selectivity and the extraction ability of the process towards targeted metals over impurities.

This paper illustrates the importance of the physicochemistry to improve the performances of solvent extraction processes by addressing two topics: (i) the comprehension of the physicochemical phenomena responsible for the chemical degradation of the extraction solvent during liquid–liquid extraction of uranium from acidic sulfate media in order to keep high uranium extraction efficiency, and (ii) the development of a new extractant for the recovery of uranium from wet phosphoric acid by means of a rational approach combining modelling of uranium (IV) extraction and investigation of uranium speciation in the extraction solvent.

Experimental and Computation

Products

Alamine® 336 was provided by Cognis; 1-tridecanol (purity = 97%), *n*-dodecane, dioctylamine, and sodium carbonate were from Aldrich and used without further purification. The feed solution was prepared from analytical grade concentrated sulfuric acid (95–98%, Fluka), uranyl sulfate (500 g · L^{-1}, Fluka), and water (resistivity > 18 MW · cm) purified with a milli-Q Gradient system from Millipore Corporation. In all experiments, organic phases correspond to 0.14 M Alamine® 336 or 0.14 M dioctylamine modified with 5% (wt) of 1-tridecanol diluted in *n*-dodecane.

Commercial bis-(2-ethylhexyl)phosphoric acid (Aldrich, purity >97%), tri-*n*-octylphosphine oxide (Fluka, purity >90%) as the extractants, and Isane IP 185 (Total Fluids, a 100% aliphatic diluent with flash point and boiling point equal to 66 °C and 187 °C, respectively) as the diluent were used as received. Concentrated phosphoric acid (VWR, AnalanR Normapur, purity >85%) and uranyl nitrate (Fluka) were also used as delivered. Synthetic aqueous H_3PO_4 solutions were prepared from commercial concentrated phosphoric acid and water purified with a milli-Q Gradient system from Millipore Corporation (resistivity > 18 MW · cm). They initially contained 0.34 g. L^{-1} uranium (VI) [1.4×10^{-3} mol L^{-1}] or 1 g. L^{-1} iron (III) and variable concentrations of phosphoric acid. WPA obtained from AREVA contained 0.144 g · L^{-1} uranium and 2.2 g · L^{-1} iron.

Prior liquid–liquid extraction experiments with WPA, organic matter was removed by a series of contacts with pentane. After each extraction, pentane present in WPA was removed under vacuum and the presence of organic matter was checked by ICP-MS. Di-n-HMOPO was synthesized by Rescoll. The BiDiBOPP analogs Fig. 5a–g were synthesized as reported in previous papers [6].

Methods

Solvent Extraction of Uranium(VI) from Acidic Sulfate Media

Extraction isotherm was obtained by mixing aqueous solutions containing 737 mg/L uranium (VI) in 5 g/L sulfuric acid with organic phases for a series of phase volume ratios (O/A) ranging from O/A = 0.05 to O/A = 20. In these experiments, organic phases were preconditioned with 100 g/L sulfuric acid.

Stripping isotherm was obtained by equilibrating organic phases initially containing 7950 mg/L uranium (VI) with 199 g/L sodium carbonate aqueous solutions for a series of phase volume ratios (O/A) ranging from 0.1 to 10.

Uranium concentration in aqueous solutions was determined with a Varian Vista Pro ICP-AES and uranium (VI) concentration in organic phases was deduced from mass-balance.

Experimental extraction distribution isotherms of uranium (VI) between aqueous and organic phases fit well the following equation:

$$[U]_{org} = A \times \left(1 - e^{-B \cdot [U]_{aq}}\right) \quad (1)$$

where $[U]_{org}$ and $[U]_{aq}$ ($mg \cdot L^{-1}$) denote uranium concentrations in organic and aqueous phases, respectively. $A = 6.92.10^3$ g/L, $B = 0.03\ g \cdot L^{-1}$ in the presence of Alamine® 336 in organic phase and sulfuric acid solution as aqueous phase. $A = 2.3.10^3$ g/L, $B = 0.02\ g \cdot L^{-1}$ in the presence of dioctylamine in organic phase and sulfuric acid solution as aqueous phase.

The experimental stripping isotherm of uranium (VI) between aqueous and organic phases is described by Eq. (2):

$$[U]_{aq} = A \times \left(1 - e^{-B \cdot [U]_{org}}\right) \quad (2)$$

with $A = 6.0 \times 10^4$ g/L and $B = 3.5 \times 10^{-4}\ g \cdot L^{-1}$ in the presence of Alamine® 336 or dioctylamine in organic phase and 199 g/L sodium carbonate aqueous solution as aqueous phase.

Solvent Extraction of Uranium(VI) from Phosphoric Acid

Liquid–liquid extraction experiments were carried out by mixing the feed aqueous solutions (i.e., synthetic H_3PO_4 solutions or organic matter free WPA) with mixtures of $0.5\ mol \cdot L^{-1}$ acidic cationic extractant and $0.125\ mol \cdot L^{-1}$ synergistic co-extractant diluted in Isane IP 185 at a phase volume ratio equal to 1. Uranium concentration in aqueous solutions was determined by ICP-AES using a Varian Vista Pro spectrometer. Before analyses, the aqueous solutions were filtered with a Minisart AC STER filter (28 mm, 0.45 μm, hydrophilic cellulose acetate membrane). The concentration of uranium (VI) in organic phases was deduced from mass-balance calculation, i.e., from the difference between the uranium concentration in the aqueous phase before and after extraction.

Computing

Computer simulations were carried out on a PC with the free mathematical software Scilab. The program was written under Scilab language, and the set of non-linear equations derived from mass-balance was solved by using the Newton–Raphson method [7]. The critical parameter for the identification of the best flowsheets is the residual fraction (f) defined as the uranium concentration in the raffinate (x_4) over its concentration in the inlet feed solution (x_0), $f = x_4/x_0$ (see definition of x_4 and x_0 in Fig. 1).

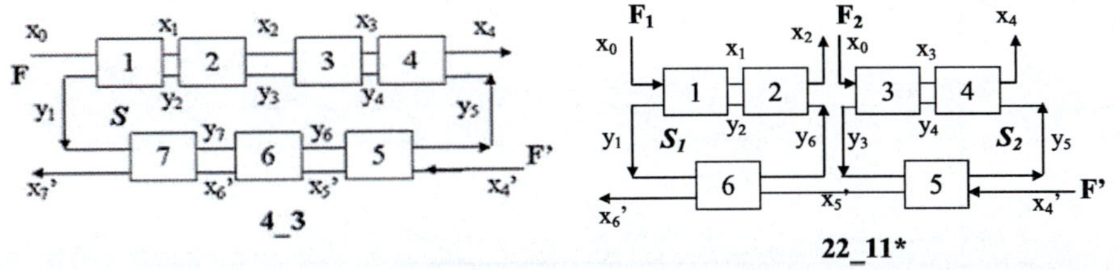

Fig. 1 Classical counter-current and unconventional two feed inlets flowsheets, 4_3 and 22_11*, respectively. x_i and y_i represent the uranium concentration in the aqueous and organics phases of the mixer-settler *i*, respectively

The residual fraction represents the uranium losses due to chemical degradation of the extraction solvent, bad operating conditions (feed flow rate, solvent flow rate, or stripping flow rate), or bad optimization of the flowsheets.

Results and Discussion

Physicochemistry of the Chemical Degradation of an Extraction Solvent Used for the Recovery of Uranium (VI) from Acidic Sulfate Media

Solvent extraction processes may undergo stresses such as radiolysis or chemical degradation. For example, chemical degradation of trioctylamine and tridecanol in *n*-dodecane as a diluent may occur due to the presence of strong oxidant metal ions in the feed solution such as vanadium(V) [8] or chromium(VI) [9]. In the case of vanadium(V), the degradation occurs as follows: (i) Vanadium(V) is extracted by Alamine® 336 from aqueous phase to organic phase; (ii) the strong oxidation power of Vanadium(V) is responsible for the oxidation of the modifier (1-tridecanol) to carboxylic acid according to a radical mechanism; and (iii) the preceding radicals react quickly with Alamine® 336 (composed mainly of tri-n-octylamine) to form di-n-octylamine (DOA) as a degradation product [10]. Furthermore, the degradation of Alamine® 336 to di-n-octylamine is responsible for a drop in extraction efficiency as di-n-octylamine is a less efficient extractant of uranium (VI) than tri-n-octylamine.

Figure 2 displays the sturdiness of the 4_3 and 22_11* flowsheets against the degradation progress, i.e., DOA molar fraction in the Alamine® 336-tridecanol-*n*-dodecane system.

There is no influence of the chemical degradation on the extraction efficiency for the classical counter-current flowsheet 4_3 reported in Fig. 1 until the molar fraction of DOA is lower than 0.1. At a higher DOA molar fraction, the residual fraction increases linearly. The use of DOA instead of Alamine® 336 is responsible for an extraction loss equal to 8.8%.

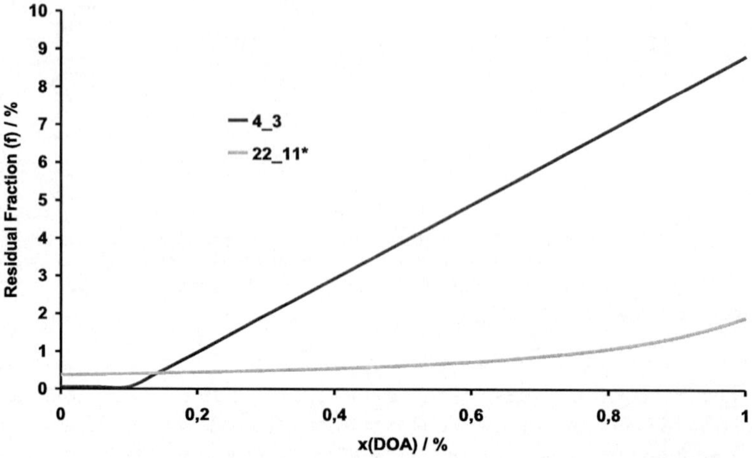

Fig. 2 Influence of the di-n-octylamine molar fraction (x(DOA)) on the residual fraction (f) for the flowsheets 4_3 and 22_11*. The operating conditions for 4_3 and 22_11* are S = 29 $m^3 \cdot h^{-1}$, F = 115 $m^3 \cdot h^{-1}$, initial uranium concentration x_0 = 1677 $mg \cdot L^{-1}$, and F′ = 4.80 $m^3 \cdot h^{-1}$, feed flow rate F = 90 $m^3 \cdot h^{-1}$, initial uranium concentration x_0 = 400 $mg \cdot L^{-1}$, respectively

The implementation of the 22_11* flowsheet (Fig. 1) in a solvent extraction process for the recovery of uranium (VI) from sulphuric acid media permits to enhance the resistance of the solvent extraction process towards the chemical degradation. In fact, a slow increase of the residual fraction is observed when the molar fraction of DOA in the organic phase increases (Fig. 2). The use of a second loop in the extraction process permits to strongly increase the resistance of the process against degradation. It may be supposed that the effect of the solvent extraction degradation may be counterbalanced by the increase of the extraction capacity of unconventional flowsheets.

Physicochemistry of the Liquid–Liquid Extraction of Uranium(VI) from Phosphoric Acid

Depending on the concentration of phosphoric acid, uranium (VI) can exist as the uranyl cation UO_2^{2+} and a series of phosphate complexes [11]. As a result, the total concentration of uranium (VI) in aqueous phosphoric acid solutions is given by Eq. (3):

$$[U(VI)] = \left[UO_2^{2+}\right](1 + f([X_i], \beta_i)) \tag{3}$$

where f is a speciation function of uranium (VI), depending on the nature and the concentration of the complexes which can be formed in the presence of the ligands X_i (e.g., $H_2PO_4^{2-}$) and b_i denotes the associated formation constants expressed in terms of concentrations. In the present case, the value of the function f increases with the concentration of phosphoric acid (i.e., the complexation of the uranyl cations is favored by an increase of the concentration of phosphoric acid).

The extraction of uranium (VI) by Di-(2-ethylhexyl) phosphoric acid (D2EHPA)/tri-n-octylphosphine oxide (TOPO) and bis(1,3-dibutyloxypropan-2-yl) phosphoric acid (BiDiBOPP)/di-68 n-hexyl octyl methoxy phosphine oxide (di-n-HMOPO) systems can be represented by the following equation [12]:

$$UO_2^{2+} + 2\overline{(HL)_2} + \overline{T} = \overline{UO_2(HL)_2L_2T} + 2H^+ \tag{4}$$

where UO_2^{2+}, $\overline{T}$, $\overline{HL}$, and $\overline{UO_2(HL)_2L_2T}$ denote the free uranyl cation in the aqueous phase, the phosphine oxide (TOPO or di-n-HMOPO), the acidic extractant (D2EHPA or BiDiBOPP or new phosphorus extractants), and the metal-extractant complex in the organic phase, respectively.

The following equation can be deduced from Eq. (4):

$$\begin{aligned}\log(D_{U(VI)}) = \; & \log(K^o_{ex}) + 2\log([\overline{(HL)_2}]) + \log([\overline{T}]) + 2pH \\ & - \log(1 + f([X_i], \beta_i)) + \log(\gamma_{UO_2^{2+}})\end{aligned} \tag{5}$$

In the presence of a large excess of T and $(HL)_2$ compared to the amount of uranium (VI) to be extracted, as it is the case in the present study, the term $2\log([\overline{(HL)_2}]) + \log([\overline{T}])$ can be considered as constant. As a result, the variation of $\log(D_{U(VI)})$ with the concentration of phosphoric acid arises mainly from three effects as discussed below.

(1) Equation (5) indicates that the distribution coefficient of uranium (VI) decreases when the pH decreases, i.e., when the concentration of phosphoric acid increases.
(2) As the function f increases with the concentration of phosphoric acid, Eq. (5) also indicates that the distribution coefficient of uranium (VI) should decrease with the concentration of phosphoric acid.
(3) Finally, the term $\log(g_{UO2+})$ corresponds to the salting-out effect due to the variation of activity coefficients of the species present in the aqueous phase.

The plot of $\log(D_{U(VI)})$ versus H_3PO_4 concentration shows that the distribution coefficient of uranium (VI) decreases sharply when the phosphoric acid concentration increases (Fig. 3) for the two considered systems.

To investigate the influence of pH as expressed in Eq. (5), the curve of $\log(D_{U(VI)})$ versus pH was plotted for the D2EHPA/TOPO system (Fig. 4). The value of pH of each aqueous phosphoric acid solution was taken in Elmore et al.'s work [13]. The plot of $\log(D_{U(VI)}) = f(pH)$ gives a straight line with a slope of 2.1 as expected from Eq. (5) if the term $-\log(1 + f([X_i], \beta_i))$ compensates the term $+\log(g_{UO2+})$, i.e., if the

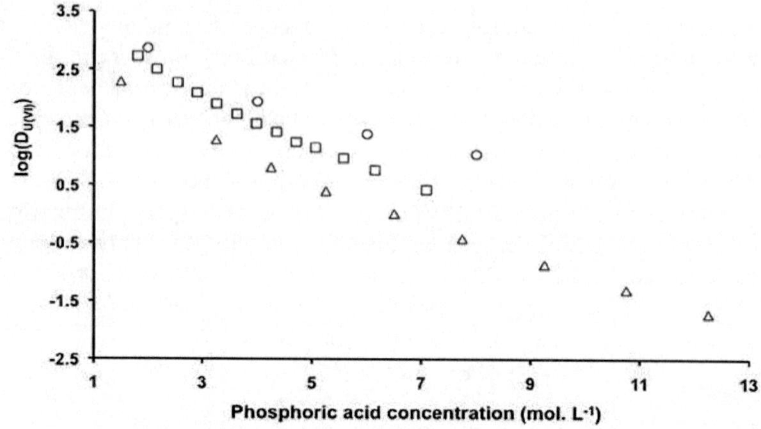

Fig. 3 Logarithm of the distribution coefficient of uranium (VI) as a function of the concentration of phosphoric acid, at 25 °C. Organic phase: 0.5 M extractant + 0.125 M phosphine oxide diluted in Isane IP 185. □: BiDiBOPP/di-n-HMOPO; O: D2EHPA/TOPO. Δ: Hurst et al. values (1972) determined for 0.18 M D2EHPA + 0.05 M TOPO diluted in refined Amsco Odorless 450 at 25 °C

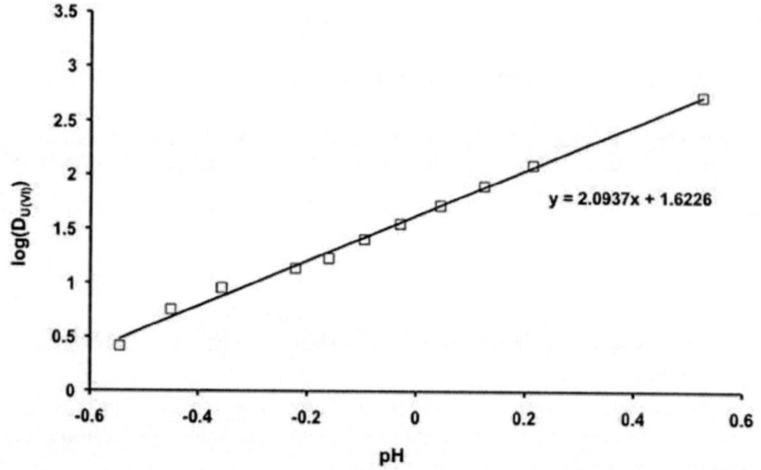

Fig. 4 Logarithm of the distribution coefficient of uranium (VI) as a function of pH, at 25 °C. Organic phase: 0.5 M D2EHPA + 0.125 M TOPO diluted in Isane IP 185. Aqueous phase: synthetic H_3PO_4 solutions

positive influence of the salting-out effect roughly compensates the negative effect arising from the increasing complexation of the uranyl cations by the phosphoric acid medium. The same tendency appears for the system BiDiBOPP/di-n-HMOPO.

The use of 0.5 mol.L^{-1} D2EHPA and 0.125 mol.L^{-1} TOPO diluted in Isane gave a distribution coefficient of uranium (VI) of 11.8 and a selectivity coefficient against

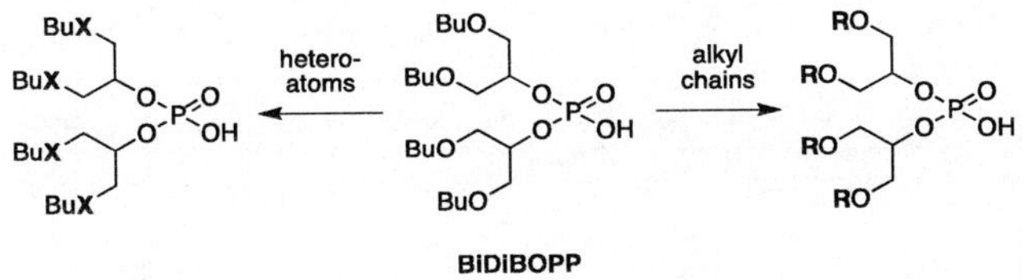

Scheme 1 Molecules synthesized to investigate theinfluence of the chemical structure on the extraction performances

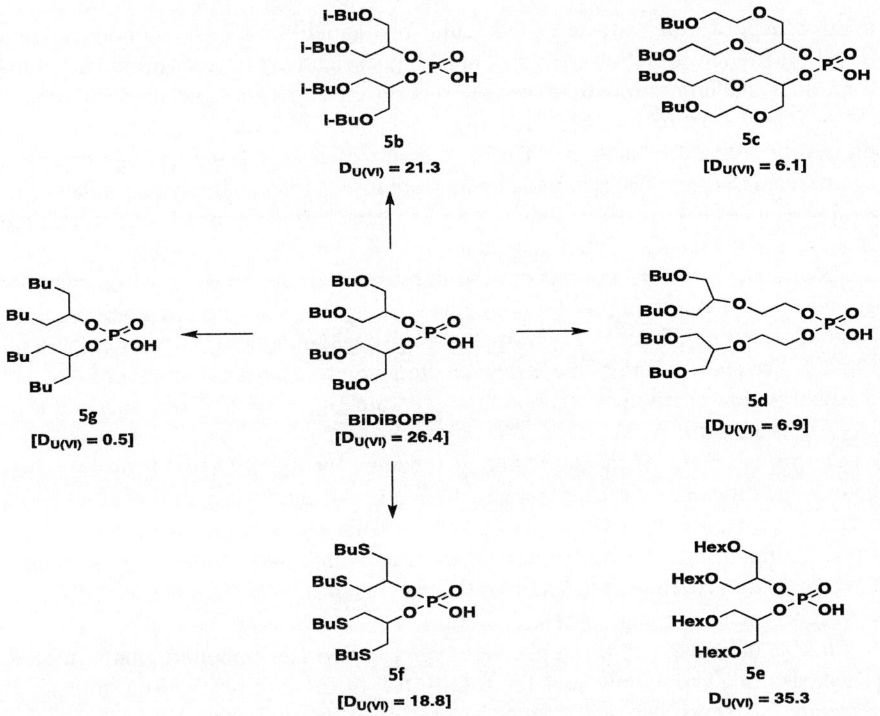

Fig. 5 Distribution coefficient of uranium (VI) between a mixture of 0.5 mol · L^{-1} extractant (BiDiBOPP or its analogues) and 0.125 mol · L^{-1} TOPO in Isane IP 185 and an aqueous phase containing initially 0.34 g · L^{-1} uranium (VI) (1.4 × 10^{-3} mol · L^{-1}) in 5.3 mol · L^{-1} H_3PO_4. Temperature = (25.0 ± 0.2) °C

iron(III) of 150 (Table 1), i.e., the ratio between distribution coefficient of uranium (VI) on the distribution coefficient of iron(III).

Under the same conditions, the mixture of 0.5 mol · L^{-1} BiDiBOPP and 0.125 mol · L^{-1} di-n-HMOPO permitted to reach a distribution coefficient and a selectivity coefficient against iron(III) of 40.0 and 70, respectively. The decrease of the selectivity coefficient against iron(III) corresponds to a reinforced extraction of iron(III) in organic phase and this causes troubles during the stripping phase,

Table 1 Distribution coefficient of uranium (VI) ($D_{U(VI)}$) and iron (III) ($D_{Fe(III)}$) between 0.5 mol · L^{-1} extractant and 0.125 mol · L^{-1} phosphine oxide diluted in Isane IP 185 and a 30%wt wet phosphoric acid containing 0.144 g · L^{-1} uranium and 2.2 g · L^{-1} iron. Volume ratio: $V_o/V_a = 1$, T = 25 °C. s = selectivity coefficient

Entry	Extractant	Co-extractant	$D_{U(VI)}$	$D_{Fe(III)}$	s
1	D2EHPA	TOPO	11.8	0.08	150
2	BiDiBOPP	di-n-HMOPO	40.0	0.57	70

so that iron should be removed before the stripping step. As a consequence, the development of new extractants molecules that should be robust, economical, and have both higher extraction efficiency and higher selectivity towards uranium against impurities, including iron, than the previous systems appears to be very important.

In order to try to improve the extraction ability of synergistic systems based on phosphorus extractants and TOPO, a series of BiDiBOPP's analogs has been synthesized (Scheme 1). Two parameters were investigated, namely the influence of the presence of hetero-atoms (noted X in Scheme 1) in the aliphatic chains and the influence of the length of the latter (noted R in Scheme 1).

Examination of Fig. 5 points out the importance of the presence of four butoxy groups in BiDiBOPP. In the absence of these oxygen atoms, the distribution coefficient of uranium (VI) decreases from 26.4 for BiDiBOPP down to 0.5 for compound Fig. 5g. The substitution of these oxygen atoms by four sulfur atoms Fig. 5f lowers the distribution coefficient of uranium (VI) from $D_{U(VI)} = 26.4$ for BiDiBOPP to $D_{U(VI)} = 18.8$ for Fig. 5f.

The replacement of the four butoxy functions of the BiDiBOPP by four *iso*-butoxy functions Fig. 5b causes a slight decrease of the distribution coefficient of uranium (VI), i.e., from $D_{U(VI)} = 26.4$–21.3. This phenomenon may be due to a steric hindrance effect. The increase of the steric hindrance significantly decreases the distribution coefficient of uranium (VI) as it is observed when D2EHPA ($D_{U(VI)} = 9.2$) is replaced by Fig. 5g ($D_{U(VI)} = 0.5$).

Finally, an increase of the number of oxygen atoms in the lipophilic chains implies a decrease of $D_{U(VI)}$ from 26.4 for BiDiBOPP to 6.1 and 6.9 for Fig. 5c and d, respectively. Nevertheless, it cannot be precluded that the decrease of $D_{U(VI)}$ can also arise from the bulkiness effect and acid aggregation effects on the extraction performance. An increase of carbon atoms number in the aliphatic chain led to an increase of $D_{U(VI)}$, i.e., the distribution coefficient of uranium (VI) varies from $D_{U(VI)} = 26.4$–35.3 when BiDiBOPP is replaced by Fig. 5e.

Conclusions

The investigation of new flowsheets for solvent extraction processes may be a way to decrease the number of mixers-settlers in the process with the attention of lowering the processing cost. In this work, the classical flowsheet, corresponding to a counter-current configuration with 4 mixers-settlers in the extraction stage and 3 mixers-settlers in the stripping stage, is more efficient than the unconventional flowsheet with two feed inlets, two mixers-settlers in the extraction stage, and one mixer-settler in the stripping stage. This latter one permits to work at high feed flow without saturating the organic phase. Implementation of new flowsheets in solvent extraction processes is a new route to delay the chemical or radiolysis degradation of extraction solvent. In fact, this work highlights the importance of the flowsheets on the sturdiness of a solvent extraction process against the degradation of solvents. It appears that the unconventional flowsheet (22_11*) is a relevant configuration to decrease the impact of the degradation on the extraction efficiency.

Regarding the physicochemistry involved in the extraction of uranium (IV) from concentrated phosphoric acid, the influence of phosphoric acid concentration was studied for the U-H_3PO_4-D2EHPA/TOPO and U-H_3PO_4-BiDiBOPP/di-n-HMOPO systems. Investigation on the variation of $\log(D_{U(VI)})$ versus pH suggests that the positive influence of the salting-out effect compensates for the negative effect arising from increasing complexation of the uranyl cations by the phosphoric acid medium as the concentration of H_3PO_4 is increased.

Furthermore, six BiDiBOPP analogs were synthesized to investigate the influence of the structure of the alkyl chains on the distribution coefficient of uranium. The first tendency is that the presence of two hetero-atoms (S or O) in each of the two aliphatic chains of phosphoric acid diesters, as in BiDiBOPP, enhances the affinity of the resulting molecules for uranium (VI), compared to D2EHPA. However, the increase of the number of oxygen atoms beyond 2 in each of the two aliphatic chains does not further improve the affinity of the resulting extractants for uranium (VI). The second tendency is that the replacement of hydrophobic chains in BiDiBOPP by longer aliphatic chains is responsible for an increase of the distribution coefficient of uranium (VI).

References

1. Cote G (2000) Hydrometallurgy of strategic metals. Solvent Extr Ion Exch 18(4):703–727
2. Cote G, Jakubiak A, Bauer D, Szymanowski J, Mokili B, Poitrenaud C (1994) Modelling of extraction equilibrium for copper (II) extraction by pyridine carboxylic acid esters from concentrated chloride solutions at constant water activity and constant total concentration of ionic or molecular species dissolved in the aqueous solution. Solvent Extr Ion Exch 12(1):99–120
3. Cote G, Jakubiak A (1996) Modelling of extraction equilibrium for zinc(II) extraction by a bibenzimidazole type reagent (ACORGA ZNX 50) from chloride solutions. Hydrometallurgy 43(1–3):277–286
4. Cote G (2003) The supramolecular speciation: a key for improved understanding and modeling of chemical reactivity in complex systems. Radiochim Acta 91(11):639–643
5. Rydberg J (2004) Solvent extraction: principles and practices, 2nd edn. Marcel Dekker Inc., New York, NY, USA

6. Beltrami D, Chagnes A, Haddad M, Laureano H, Mokhtari H, Courtaud B, Jugé S, Cote G (2013) Development of new cationic exchangers for the recovery of Uranium (VI) from concentrated phosphoric acid. Sep Sci Technol 48(3):480–486
7. Ypmat TJ (1995) Historical development of the Newton-Raphson method. SIAM Rev 37(4):531–551
8. Chagnes A, Rager MN, Courtaud B, Thiry J, Cote G (2010) Speciation of vanadium (V) extracted from acidic sulfate media by trioctylamine in n-dodecane modified with 1-tridecanol. Hydrometallurgy 104:20–24
9. Chagnes A, Cote G (2018) Chemical degradation of a mixture of tri-n-Octylamine and 1-Tridecanol in the presence of Chromium(VI) in Acidic Sulfate Media. Metals 8:57
10. Chagnes A, Fossé C, Courtaud B, Thiry J, Cote G (2011) Chemical degradation of the mixture of trioctylamine (extractant) and 1-tridecanol (phase modifier) in acidic sulfate media in the presence of vanadium (V). Hydrometallurgy 105(3–4):328–333
11. Thamer BJ (1957) Spectrophotometric and solvent-extraction studies of uranyl phosphate complexes. J Am Chem Soc 79:4298–4305
12. Beltrami D, Chagnes A, Mokhtari H, Courtaud B, Cote G (2012) Modelling of the extraction of uranium(VI) from concentrated phosphoric acid by synergistic mixtures of bis-(2-ethylhexyl)phosphoric acid and tri-n-octylphosphine oxide. Hydrometallurgy 129–130:118–125
13. Elmore KL, Hatfield JD, Dunn RL, Jones AD (1965) Dissociation of phosphoric acid solutions at 25°. J Phys Chem 69:3520–3525

Solvent Extraction Separation of Tungsten and Vanadium from Simulated Leaching Solution of Spent SCR Catalyst

Hanyu Wang, Guihong Han, Yanfang Huang, and Shengpeng Su

Abstract It is of great significance for the resource utilization of spent SCR catalysts to separate tungsten and vanadium from leaching solutions. In this study, selective separation of vanadium from a mixed solution of tungsten and vanadium was investigated by solvent extraction using ammonium salts as extractant. The effect of different parameters including initial pH, phase ratio, temperature, and time was studied, respectively. The results demonstrated tungsten and vanadium mainly exist in the form of WO_4^{2-} and $V_4O_{12}^{4-}$ in solutions with pH ranging from 8.5 to 9.5 by simulation calculation. In addition, it was found that vanadium can be easily extracted over tungsten by methyl trioctyl ammonium chloride (Aliquat 336). More than 93.5% of vanadium was extracted, while most of the tungsten remained in raffinates under the optimal conditions (initial pH 8.4, 10 vol.% N263, contact time 30 min, temperature 25 °C and phase ratio (O/A) 1/6).

Keywords Tungsten · Vanadium · Selective separation · Solvent extraction

Introduction

Selective catalytic reduction (SCR) technology is the main technique for reducing nitrogen oxide (NOx) emissions from coal-fired power plants around the world [1–3]. Nowadays, the $V_2O_5-WO_3/TiO_2$ catalyst is the most popular commercial SCR catalyst due to its excellent catalytic properties, such as durability from poisons, sulfur resistance, and wide operating temperature window [4]. In general, the composition of spent V_2O_5-WO_3/TiO_2 catalyst is approximately 80–90 wt% TiO_2, 0.5–1.5 wt% V_2O_5, and 7–10 wt% WO_3 [5–7]. The typical chemical lifetime of SCR catalyst is approximately 2–3 years [8, 9]. In China, it is estimated that 250,000–300,000 m^3/year of the catalyst after 2020 becomes solid wastes [10], In addition, both vanadium and tungsten are toxic to human and environmental health [11]. Therefore, it is a great challenge to treat spent SCR catalyst [12].

H. Wang · G. Han (✉) · Y. Huang · S. Su
School of Chemical Engineering, Zhengzhou University, Zhengzhou 450001, P. R. China
e-mail: hanguihong@zzu.edu.cn

T. Ouchi et al. (eds), *Rare Metal Technology 2022*, The Minerals, Metals & Materials Series, https://doi.org/10.1007/978-3-030-92662-5_9

Various approaches have been developed for recovering spent SCR catalysts, such as sodium roasting, acid leaching, alkali leaching, and bioleaching [13, 14]. However, to date, only a few investigations have been performed on the deep separation of tungsten and vanadium from the leaching solution of spent SCR catalysts. It is of great significance for the resource utilization of spent SCR catalysts to separate tungsten and vanadium from leaching solutions [15]. During the past decade, some processes and technologies for the separation of vanadium and tungsten have been developed and reported. The main methods include precipitation, ion exchange, and solvent extraction [16]· Among these methods, solvent extraction, as a clean and effective technique, has been become the most applicable method in the separation and concentration of vanadium [17].

In this study, selective separation of vanadium from a mixed solution of tungsten and vanadium was investigated by solvent extraction using ammonium salts as extractant. The effect of different extractants, initial pH, phase ratio, temperature, and contact time on the separation of tungsten and vanadium was studied and the species distribution of tungsten and vanadium ions was simulated under different pH values. This work provides a reference for the selective separation of tungsten and vanadium by solvent extraction.

Experiment

Materials and Methods

The organic phase that was used in this experiment consisted of the extractant of primary amine salt (N1923), trioctylamine (TOA, N235), methyl trioctyl ammonium chloride (Aliquat 336), and the modifier of sec-octanol in sulfonated kerosene. All organic reagents were purchased from Shanghai Aladdin Company and used as received. The simulated solution containing $10.6\ g \cdot L^{-1}$ W(VI) and, $1.1\ g \cdot L^{-1}$ V(V) was synthesized by dissolving sodium tungstate and sodium vanadate (shanghai Macklin Company) in deionized water.

The tungsten and vanadium concentrations were determined by ICP-OES (Thermo Scientific) and the pH of the solution was measured by pH meter (Shanghai Mettler Instruments Factory, China).

Separation Theory Basis

During the solvent extraction process, the extraction and separation behavior of metals strongly depended on the speciation of the metals in an aqueous solution. The existence of the various vanadate and tungstate species mainly depends on the solution pH and concentration [18]. The main ion forms of vanadium and tungsten

Table 1 Main existing forms of tungsten and vanadium ions under different pH conditions (derived from Visual MINTEQ software simulation calculation)

pH	Species	pH	Species
0–4	$HW_7O_{24}{}^{5-}$	0–2	$VO_2{}^{+}$
4–10	$H_2W_{12}O_{42}{}^{10-}$ $W_7O_{24}{}^{6-}$ $H_2W_6O_{22}{}^{6-}$	2–6	$H_2V_{10}O_{28}{}^{4-}$ $HV_{10}O_{18}{}^{5-}$ $V_4O_{12}{}^{4-}$
10–12	$H_2W_6O_{22}{}^{6-}$ $WO_4{}^{2-}$	6–12	$V_4O_{12}{}^{4-}$ $V_5O_{15}{}^{5-}$
12–14	$WO_4{}^{2-}$	12–14	$VO_4{}^{3-}$ $HVO_4{}^{2-}$ $V_2O_7{}^{4-}$

in solution were summarized in Table 1 under concentrations of 10 $g \cdot L^{-1}$ tungsten and 1 $g \cdot L^{-1}$ vanadium. As can be seen, tungsten and vanadium mainly exist in the form of $WO_4{}^{2-}$ and $V_4O_{12}{}^{4-}$ at the solution initial pH range of 8–9.

Solvent Extraction Experiment

The influence of extractant concentration on separation efficiency was explored by fixing the extractant concentration at 10% and changing the phase ratio (O/A). The pH value of the aqueous phase was adjusted to the desired value by adding HCl/NaOH solution, the aqueous phase and organic phase with different volume ratios were transferred into a conical bottle, and the extraction experiments were conducted in a shaker with a speed of 200 rpm/min. In addition, aqueous and organic solutions were equilibrated using a separating funnel for 10 min. The concentration of metal ions is calculated by subtraction and all experiments were carried out in duplicate to ensure accuracy.

Results and Discussion

Effect of Extractants and Initial pH

The extractants used for the recovery of tungsten and vanadium mainly include Aliquat 336, TOA, N1923, and N235 the effect of the four extractants and initial pH on extraction of vanadium and tungsten were depicted in Fig. 1. Among the four extractants, only Aliquat 336 showed high selectivity in the separation of tungsten and vanadium. The extraction reaction using Aliquat 336 as extractant can be expressed in Eqs. (1)–(2).

$$4\overline{(R_4N)Cl} + V_4O_{12}^{4-} \leftrightarrow \overline{(R_4N)V_4O_{12}} + 4Cl^- \quad (1)$$

$$2\overline{(R_4N)Cl} + WO_4^{2-} \leftrightarrow \overline{(R_4N_2)WO_4} + 2Cl^- \quad (2)$$

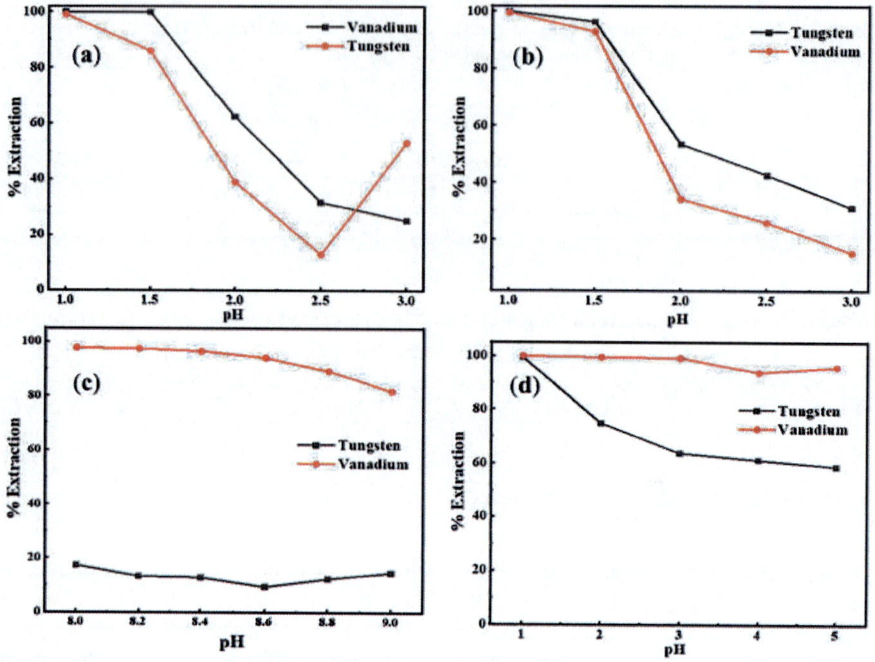

Fig. 1 Effect of pH and extractants on extraction of N235 (**a**). TOA (**b**) and Aliquat 336 (**c**) and N1923 (**d**)

It can be observed from Fig. 1c that vanadium has a high extraction ratio as the solution initial pH of range of 8–9. The extraction rate of vanadium gradually decreased with the increase of pH, and the extraction rate of tungsten showed a trend of first decreasing and then increasing. The main reason is that the existence of tungsten and vanadium was not unique at the solution initial pH of range 8–9 [18]. As the pH value is 8.6, the single-stage extraction ratio of tungsten and vanadium was 9.42% and 94.16%, respectively, reaching the best separation effect. Thus, the optimum pH value of the initial aqueous phase is determined to be 8.6.

Effect of Modifier

The effects of different modifiers (sec-octanol, TBP, Isodecyl alcohol) on the separation of tungsten and vanadium were investigated by using the organic solution consisting of 10% (m/m) N263 in sulfonated kerosene with the O/A ratio of 1/6. As can be seen from Fig. 2, the three modifiers had little effect on the extraction rate of vanadium, but the loss rate of tungsten reached 30% when isodecyl alcohol was used as the modifier, it may be due to the separation factor reached a maximum of

Fig. 2 Effects of different modifiers on metal extraction ratios

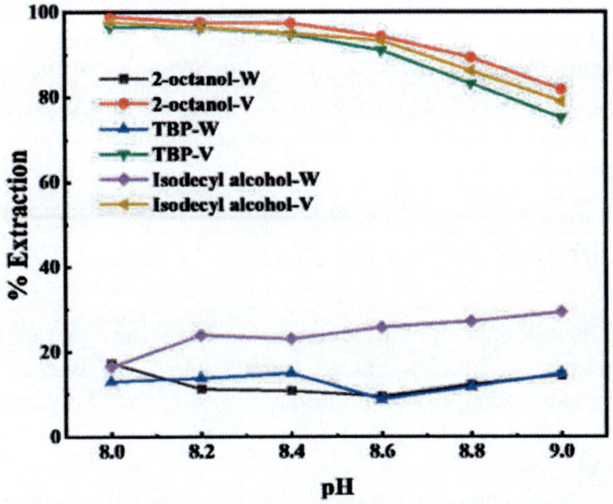

140 when secondary octanol was used modifier. Therefore, sec-octanol was selected as the modifier in subsequent experiments.

Effect of the Phase Ratio

To determine the effect of different phase ratios for stripping, Stripping experiments with phase ratios (O/A) ranging from 1/4 to 1/10 were carried out with the solution initial pH of 8.6 and 30 min contact time, and the results were displayed in Fig. 3. As the phase ratio (O/A) increased from 1/10 to 1/4, the striping percentage of tungsten and vanadium increased by 15% (from 4.72 to 19.72%) and 4.92% (from 88.96 to

Fig. 3 Effects of O/A ratio on metal extraction ratios

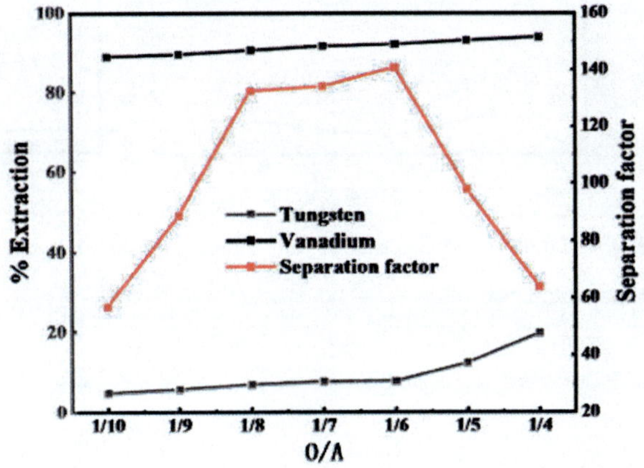

93.88%), respectively. Further, the tungsten-to-vanadium separation factor reached a maximum of 140.6 when the phase ratio (O/A) was 1/6. With the increase of phase ratio (O/A), the extraction rate of tungsten and vanadium was increased, but the loss rate of tungsten increased sharply when the phase ratio (O/A) exceeds 1/6. Therefore, the optimal phase ratio (O/A) was determined to be 1/6.

Effect of the Temperature and Contact Time

The influence of contact time and temperature on tungsten and vanadium extraction was carried out by shaking aqueous phases of 10.6 $g \cdot L^{-1}$ tungsten and 1.1 $g \cdot L^{-1}$ vanadium and organic phases of 10% (m/m) Aliquat 336 in sulfonated kerosene at the solution initial pH of 8.6 and phase ratio O/A = 1/6, and the results were shown in Fig. 4.

As can be seen, contact time and temperature had little effect on the separation of tungsten and vanadium. The extraction rate for both metals was faster with Aliquat 336, and within 10 min of contact time, 9.81% of tungsten and 89.82% of vanadium were extracted. The reaction gradually tends to equilibrium during the extraction process of tungsten and vanadium, and the loss of tungsten decreased first and then increased and reached the peak at 30 min. The separation efficiency of tungsten and vanadium showed a slight increasing trend with the increase of temperature. Considering comprehensively, the best separation temperature and time are 35 °C and 30 min.

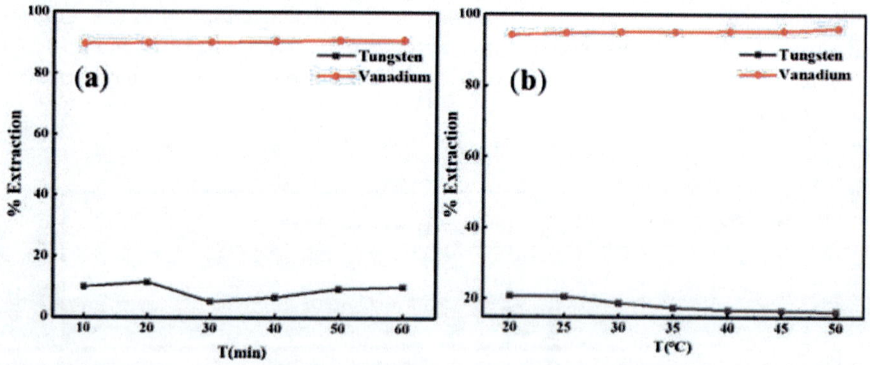

Fig. 4 Effects of contact time (**a**) and temperature (**b**) on metal extraction ratios

Conclusions

The comprehensive liquid–liquid extraction of tungsten and vanadium by Aliquat 336 in sulfonated kerosene from the spent SCR catalyst simulated leachate containing approximately 10 $g \cdot L^{-1}$ of tungsten and 1 $g \cdot L^{-1}$of vanadium was investigated. The organic solution consisting of 10% (m/m) Aliquat 336 and 10% (v/v) sec-octanol in sulfonated kerosene with a feed solution at pH 8.74 and an O/A ratio of 1/6 can be used for selective separation of tungsten and vanadium solutions based on the differences in the existing state. After one extraction operation, the removal rate of vanadium reached 92.02%, the loss rate of tungsten was <8%, and the separation factor of tungsten-vanadium reached 140.6, which demonstrated the feasibility of selective separation of tungsten and vanadium.

Acknowledgements This work was financially supported by the Natural Science Foundation of China (U2004215, Nos. 51974280 and 51774252) and, the Educational Commission Fund of Henan Province of China (Nos. 20HASTIT012 18A450001 and 17A450001).

References

1. Choi IH, Moon G, Lee JY, Jyothi RK (2018) Extraction of tungsten and vanadium from spent selective catalytic reduction catalyst for stationary application by pressure leaching process. J Clean Prod 197:163–1699
2. Choi IH, Kim HR, Moon G, Jyothi RK, Lee JY (2018) Spent V_2O_5-WO_3/TiO_2 catalyst processing for valuable metals by soda roasting-water leaching. J Hydrometallurgy 175:292–299
3. Cao YJ, Yuan JF, Du H, Dreisinger D, Li M (2021) A clean and efficient approach for recovery of vanadium and tungsten from spent SCR catalyst. J Minerals Eng 165(2):106857
4. Marberger A, Elsener M, Ferri D, Kröcher D (2015) VOx surface coverage optimization of V_2O_5/WO_3-TiO_2 SCR catalysts by variation of the V loading and by aging. J Catalysts 5(4):1704–1720
5. Zhang QJ, Wu YF, Yuan HR (2020) Recycling strategies of spent V_2O_5-WO_3/TiO_2 catalyst: a review. J Resourc Conservat Recycling 161:104983
6. Chen JP, Yang RT (1992) Role of WO_3 in Mixed V_2O_5–WO_3/TiO_2 catalysts for selective catalytic reduction of nitric oxide with ammonia. Appl Catalysis A: General 80(1):135–148
7. Dai ZJ, Wang LL, Tang H, Sun ZJ, Liu W, Sun Y, Su S, Xiang J (2018) Speciation analysis and leaching behaviors of selected trace elements in spent SCR catalyst. J Chemosphere 207:440–448
8. Wang SH, Xie YL, Yan WF, Wu XE, Wang CT, Zhao F (2018) Leaching of vanadium from waste V_2O_5-WO_3/TiO_2 catalyst catalyzed by functional microorganisms. J Sci Total Environ 639:497–503
9. Li JH, Peng Y, Chang HZ, Li X, Rittenden JC, Hao JM (2016) Chemical poison and regeneration of SCR catalysts for NOx removal from stationary sources. J Front Environ Sci Eng 10(3):413–427
10. Cao LM, Wang Q, Zhang W, Qiu ZF, Yang J (2018) Spent SCR catalysts and environmental management in typical coal-fired power plant. J Equip Environ Eng (in Chinese)
11. Liu ZG, Ottinger NA, Cremeens CM (2015) Vanadium and tungsten release from V-based selective catalytic reduction diesel aftertreatment. J Atmos Environ 104:154–161

12. Wu WF, Wang CY, Bao WJ, Li HQ (2018) Selective reduction leaching of vanadium and iron by oxalic acid from spent V_2O_5-WO_3/TiO_2 catalyst. J Hydrometallurgy 179:52–59
13. Liu NN, Xu XY, Liu Y (2020) Recovery of vanadium and tungsten from spent selective catalytic reduction catalyst by alkaline pressure leaching. J Physicochem Prob Mineral Process 56(3):405–418
14. Vyas S, Ting YP (2020) Microbial leaching of heavy metals using Escherichia coli and evaluation of bioleaching mechanism. J Bioresourc Technol Rep 9:100368
15. Sola ABC, Parhi PK, Lee JY, Kang HN, Jyothi RK (2020) Environmentally friendly approach to recover vanadium and tungsten from spent SCR catalyst leach liquors using Aliquat 336. J RSC Adv 10(34):19736–19746
16. Wang LP, Zhang GQ, Guan WJ, Zeng L, Zhou Q, XIA Y, CAO ZY (2018) Complete removal of trace vanadium from ammonium tungstate solutions by solvent extraction. J Hydrometallurgy 179:268–273
17. Xiong P, Zhang YM, Huang J, Bao SX, Yang X, Shen C (2017) High-efficient and selective extraction of vanadium (V) with N235–P507 synergistic extraction system. J Chem Eng Res Des 120:284–290
18. Nguyen TH, Lee MS (2016) A review on the separation of molybdenum, tungsten, and vanadium from leach liquors of diverse resources by solvent extraction. J Geosyst Eng 19(5):247–259

Extraction Behaviors of Vanadium(V) with Unacidified and Acidified N1923 from a Real Leachate of Vanadium-Titanomagnetite

Lei Liu, Zhaobo Liu, Nianwen Pu, Yunfeng Fu, Zhongyu Zhang, Shangchao Du, Guoshan Du, Ninglei Sun, Dehua Wang, and Xiaoyan Li

Abstract Recently, vanadium extraction has received special attention owing to the application of vanadium in the energy storage. In this study, a vanadium-containing leachate obtained from vanadium-titanomagnetite was employed for the solvent extraction. The primary amine N1923 is selected for vanadium recovery from the feed solution with a high vanadium concentration of 29.21 g/L. The effects of initial pH of feed solution on the vanadium extraction with unacidified N1923 and HCl acidified N1923 are first investigated, respectively. Subsequently, the vanadium extraction performance by HNO_3 acidified N1923 is presented. Furthermore, the comparative analysis of vanadium stripping by NaOH from both loaded HCl and HNO_3 acidified N1923 is also presented. These experimental results provide a significant reference for the vanadium extraction by N1923 in the industrial production.

Keywords Vanadium · Chromium · N1923 · Acidification · Solvent extraction

Introduction

Vanadium is widely applied in the field of specialty steel alloys owing to its physical properties of tensile strength, hardness, and fatigue resistance [1]. In the sulfuric acid industry, vanadium oxides can also be used as an effective catalyst for the oxidation of sulfur dioxide to sulfur trioxide [2]. Recently, of particular interest is the application of vanadium in the energy storage. The vanadium redox flow batteries can avoid the problem of cross-contamination by osmotic diffusion and can easily be scaled up by simply building much bigger tanks [3].

Vanadium resources generally could be extracted from vanadium-titanomagnetite, phosphate rock, bitumen, carbonaceous shale, and secondary materials such as fly

L. Liu · N. Pu · Z. Zhang
Sichuan Xingming Energy and Environmental Protection Technology Co., Ltd., Chengdu 610106, Sichuan, China

Z. Liu (✉) · Y. Fu · S. Du · G. Du · N. Sun · D. Wang · X. Li
ENFI R&D Institute, China ENFI Engineering Corporation, No. 12 Fuxing Avenue, Haidian District, Beijing 100038, China
e-mail: liuzhaobo@enfi.com.cn; liu_zhaobo@sina.com

T. Ouchi et al. (eds), *Rare Metal Technology 2022*, The Minerals, Metals & Materials Series, https://doi.org/10.1007/978-3-030-92662-5_10

ash produced from oil industry and petroleum coke [1, 4]. The typically process for vanadium producing is salt roasting-water leaching coupled with ammonium metavanadate precipitation. Solvent extraction is another effective way to recover vanadium and also a common method for the further purification of vanadium solutions. Moreover, this method is particularly effective for the high concentration of vanadium solutions owing to its large capacity [5]. So far, various extractants have been reported and employed for the vanadium extraction. The organic phases consist of organophosphorus extractants like D2EHPA, HEHEHP, TBP, Cyanex 272, amine extractants like Alamine 336, trialkylamine, primary amine N1923, LK-N21, N1519, N235, and LIX 63 of hydroxyoximes and Aliquat 336 of quaternary ammonium salt extractant [6–8].

Among these extractants mentioned before, the most commonly used organic phase for the vanadium recovery in the industrial process is primary amine extractants. This organic phase shows a higher selectivity and affinity for V(V) than Cr(VI), and the corresponding extraction behavior follows a hydrogen bond association mechanism [9]. Zhang et al. [10] pointed out that the 20% N1923 dissolved in sulfonated kerosene shows a good extraction ability for V(V) at initial aqueous pH 1.4 and the addition of 5% octanol as diluent is beneficial for the phase separation process. Jiang et al. [1] revealed that the N1923 coupled with organophosphorus extractants like D2EHPA, HEHEHP, and Cyanex272 has synergistic effects on vanadium extraction from the chloride medium. Besides, attention should be the occurrence of interfacial crud when using N1923 for the vanadium extraction, especially during the industrial process [11]. The N1923-isooctyl alcohol system acidified by HNO_3 can be employed for the separation of vanadium from a strongly alkaline aqueous solution with the pH even above 12 [12]. Moreover, the N1923 acidified by H_2SO_4 also shows an excellent extraction performance for vanadium [13].

The separation performance of V(V) from Cr(VI) containing solution could be further improved by the addition of trisodium citrate or sodium tartrate into the alkaline aqueous solution in advance [14, 15]. On the other hand, the use of diluents like TBP and *n*-hexane can also be effective to improve the separation performance [16]. After mixing the acidified N1923 with ionic liquid of [A336] [NO_3], the organic compounds exhibit an obvious synergistic effect for the separation of V from Cr-containing solution, and the maximum separation coefficient ($\beta^{V}{}_{Cr}$) can be achieved at pH 9.0 [17]. The coextraction of Cr can be effectively avoided by adding sodium sulfate into the alkaline aqueous solutions. Wen et al. [18] suggested that the structure of vanadium extracted by N1923 was possibly present as (RNH_2)·(H_3VO_4), (RNH_2)·($H_3V_3O_9$), or (RNH_2) · ($H_6V_{10}O_{28}$) complexes. While chromium was only extracted as $HCrO_4^-$ complexes [19]. As compared with N1923, the N1519 not only exhibits excellent extraction performance for vanadium but is also cheaper and more accessible [20]. The only drawback is the fact that these vanadium-organic compounds purification requires multiple solvent-out crystallization.

So far, much work has been carried out on the extraction of vanadium with N1923 from low concentration vanadium solutions. However, little work has been focused on the vanadium extraction behaviors from a high concentration vanadium feed solution.

Experimental

The commercially available organic phases of N1923 and sulfonated kerosene were directly used in this study without any further purification. All the extraction and stripping experiments were carried out in a beaker under certain conditions. The stirring and temperature required for these experiments were provided by a magnetic heating mixer (85-2A KEXI Instrument, China). The subsequent phase separation process was completed in a separation funnel. The chemical agents including HCl, HNO_3, NaOH, and $NaNO_3$ were purchased from Sinopharm Chemical Reagent Co., Ltd.

The pH values of aqueous solutions were tested by a Mettler Toledo meter (Pro2Go, Switzerland). The designed pH of aqueous solution was adjusted by the addition of NaOH or HCl solutions. The concentrations of investigated elements in the solution were determined by an inductively coupled plasma optical emission spectrometer (ICP-OES, Optima 7000 DV, Perkin Elmer, USA), the result is shown in Table 1. Prior to the ICP test, the aqueous samples need to be filtered through a membrane with pore size of 0.45 μm.

The extraction efficiency (*E*) and stripping efficiency (*S*) were calculated based on the following equations:

$$E = \frac{C_0 - C_1}{C_0} \times 100\% \tag{1}$$

$$S = \frac{C_2 V_2}{(C_0 - C_1) V_0} \times 100\% \tag{2}$$

wherein C_0 is the concentration of investigated ion in the aqueous solution before the extraction process, C_1 is the concentration of investigated ion in the aqueous solution after the extraction process, C_2 is the concentration of investigated ion in the striping solution after the stripping process, V_2 is the volume of the striping solution, and V_0 is the volume of the original aqueous solution before the extraction process.

The separation and purification of investigated element are characterized by the distribution ratio (D_M) and separation factor (β_N^M), which are derived from

$$D_M = \frac{C_{M,O}}{C_{M,A}} \tag{3}$$

Table 1 Major chemical position of the real feed solution originated from the V–Cr slag tested by ICP-OES

Elements	V	Si	Na	K	Cr	Ca
Concentrations (g/L)	29.21	0.68	27.37	0.34	3.70	0.79

$$\beta_N^M = \frac{D_M}{D_N} \tag{4}$$

where $C_{M,O}$ is the concentration of M ion in the organic phase when the extraction is balanced and $C_{M,A}$ is the concentration of M ion in the aqueous phase when the extraction is balanced.

Results and Discussion

Effects of Initial pH on Vanadium Extraction with Unacidified N1923

The results of vanadium extraction with unacidified N1923 are listed in Table 2. The extraction efficiency of vanadium first increases from 53.76% at initial pH value of 1.91 and to 59.57% at pH 4.01 and then drops rapidly to 28.79% at pH 7.03. In respect of chromium extraction efficiency, it primarily shows a gradual decline from 14.50 to 1.21% with the increase of pH from 1.91 to 6.03. With continuing to increase the pH to 7.03, the extraction efficiency of chromium sharply increases to 6.55%. The separation factors of β_{Cr}^V over the pH range of 1.91–7.03 are also calculated, which are shown on the right side of Table 2. The smaller and larger initial pH values are both adverse to the separation of V(V) from Cr(VI) with unacidified N1923.

Effects of Initial pH on Vanadium Extraction with N1923 Acidified by HCl

In order to improve the extraction performance of vanadium, the extraction behaviors of V(V) and Cr(VI) with acidified N1923 were explored over the initial pH range

Table 2 Effects of initial pH values on the separation of V(V) from Cr(VI) with unacidified N1923

pH		Extraction efficiency (%)		β_{Cr}^V
Initial	Equilibrium	Cr(VI)	V(V)	
1.91	7.96	14.50	53.76	3.71
2.80	8.19	12.84	54.44	4.24
4.01	6.62	8.96	59.57	6.65
5.03	6.77	6.17	56.69	9.19
6.03	8.63	1.21	29.66	24.51
7.03	8.81	6.55	28.79	4.40

Table 3 Effects of initial pH values on the separation of V(V) from Cr(VI) with N1923 acidified by HCl

pH		Extraction efficiency (%)		β_{Cr}^{V}
Initial	Equilibrium	Cr(VI)	V(V)	
6.00	6.03	68.94	94.34	1.37
6.94	6.40	40.64	89.37	2.20
7.96	6.82	25.91	91.56	3.53
9.03	7.57	6.68	79.19	11.85
9.72	8.83	10.94	46.78	4.28

of 6.00–9.72. Prior to the extraction procedure, the organic phase is first acidified by hydrochloride acid under the condition of HCl concentration 1 M, O/A 1:1 and shaking time 30 min. As shown in Table 3, the vanadium extraction efficiencies are greatly improved by the acidification process and as high as 94.34% at initial pH 6.00. Unfortunately, the extraction efficiencies of chromium are also improved to some extent, especially in the pH range of 6.00–7.96. The decrease of corresponding separation factor β_{C}^{V} will definitely lead to the increase of final stages of crossflow extraction. The extraction selectivity for vanadium with acidified N1923 needs to be further optimized.

Effects of HCl Acidified N1923 Concentration and O/A Ratio on Vanadium Extraction

The results of HCl acidified N1923 concentration effects on separation of vanadium from chromium are summarized in Table 4. It can be observed that the N1923 concentration has an apparent effect on the separation factor β_{Cr}^{V}. When the N1923 concentration is adjusted to 20%, the β_{Cr}^{V} is changed to 26.95 and seems much better than other cases. The extraction efficiency of vanadium and chromium are closely related to the acidified N1923 concentration. Table 5 presents the results of vanadium and chromium separation under various O/A ratios. Generally, the extraction efficiencies of Cr(VI) and V(V) are both increased with the increase of O/A ratio

Table 4 Effects of HCl acidified N1923 concentration on vanadium extraction

N1923 concentration (%)	pH		Extraction efficiency (%)		Distribution ratio		β_{Cr}^{V}
	Initial	Equilibrium	Cr(VI)	V(V)	Cr(VI)	V(V)	
20	6.99	6.56	54.06	96.95	1.18	31.80	26.95
15	7.06	6.56	30.26	78.41	0.43	3.63	8.44
10	6.96	6.55	18.65	58.21	0.23	1.39	6.04

Table 5 Effects of O/A ratio on vanadium extraction with HCl acidified N1923

O/A ratio	pH		Extraction efficiency (%)		Distribution ratioβ^{V}_{Cr}		β^{V}_{Cr}
	Initial	Equilibrium	Cr(VI)	V(V)	Cr(VI)	V(V)	
1:2	6.99	6.66	45.53	72.98	0.84	2.70	3.21
1:1	7.02	6.63	48.99	94.87	0.96	18.50	19.27
2:1	7.03	6.58	85.91	96.06	6.10	24.37	3.60

from 1:2 to 2:1. Moreover, the V(V) extraction efficiency increases rapidly before the O/A ratio increases up to 1:1 and then slowly with continuing increase the O/A ratio. While for Cr(VI) extraction efficiency, the opposite is true.

Effects of NaOH Concentration and O/A Ratio on the Stripping of Vanadium from N1923

In addition to considering extraction behavior, it is also necessary to test the vanadium stripping characteristics of this extraction system. The loaded organic phase is obtained under the conditions of initial pH 7, O/A 1:1, reaction time 5 min, 20% N1923 acidified by 1 M HCl, and washed by deionized water. As listed in Table 6, the NaOH concentration has an apparent effect on the stripping results of V(V) and Cr(VI). The stripping efficiencies of V(V) and Cr(VI), respectively, fluctuate around 20 and 25% over the NaOH concentration range of 0.5–2.0 M. As the O/A ratio is set at 2, there is a rapid increase in the stripping efficiencies of Cr(VI) and V(V). Moreover, the stripping efficiency of impurity chromium can go as high as 86.66% and the stripping efficiency of objective element vanadium is <55%. In order to maximize the vanadium stripping efficiency, the higher concentration of NaOH and $NaNO_3$ solutions were thus applied to stripping process and the results indicate that neither the higher concentration of NaOH solutions nor the $NaNO_3$

Table 6 Effects of NaOH concentration and O/A ratio on the stripping of vanadium from HCl acidified N1923

O/A ratio	NaOH concentration (M)	Stripping efficiency (%)	
		Cr(VI)	V(V)
1:1	0.5	25.07	18.21
1:1	1.0	28.73	23.27
1:1	2.0	22.66	20.01
2:1	2.0	86.66	53.25
1:1	4.0	33.78	27.71
1:1	1.0 $NaNO_3$	0.30	0.71
1:1	1.0 NaOH + 1 $NaNO_3$	21.04	15.16

solutions can effectively improve the stripping performance of vanadium from HCl acidified N1923. In fact, it can be concluded that the NaOH concentration of 0.5 M is enough for the stripping of V(V) based on the testing results.

Vanadium Extraction Behaviors with N1923 Acidified by HNO_3

Although the acidification process is beneficial for the V(V) extraction, it is difficult to carry out the back striping of V(V) from the HCl acidified N1923 completely. The low stripping efficiency of vanadium could be assigned to the strong complexing ability of chloride ions after N1923 acidified by HCl. The ionic complex originated is difficult to be destroyed by NaOH solution, even concentrated NaOH solution. On the other hand, a higher extraction efficiency of V(V) can also be achieved by HNO_3 acidified N1923. The corresponding extraction results are given in Table 7. It is obvious that the HNO_3 acidification process plays an important role in improving the vanadium extraction performance. As compared with former extraction results, the chromium extraction efficiency is higher than untreated N1923 but lower than HCl acidified N1923. Besides, it is worthwhile to note that the V(V) stripping process from HNO_3 acidified N1923 is much easier than from HCl acidified organic phase.

The results of vanadium stripping from HNO_3 acidified N1923 with various conditions are shown in Table 8. Level of 92.19% V(V) stripping efficiency could be obtained under the condition of 1 M NaOH, O/A ratio 1, and reaction time 10 min. As for this extraction system, the augmenting NaOH concentration doesn't bring a positive effect on the V(V) stripping efficiency. The $NaNO_3$ solution can also be applied for the vanadium stripping from HNO_3 acidified N1923, but the stripping performed by 1 M $NaNO_3$ is poor. Besides, the improvement of stripping efficiency is very low even if the reaction time is extended to 30 or 60 min. This fact indicates that the vanadium stripping process could be balanced within 10 min. A slight increase in reaction temperature like 45 °C is of little benefit to the vanadium stripping process.

Table 7 V(V) extraction with N1923 acidified by HNO_3

pH		Extraction efficiency (%)		Distribution ratio		β_{Cr}^{V}
Initial	Equilibrium	Cr(VI)	V(V)	Cr(VI)	V(V)	
7.01	6.90	43.09	74.33	0.76	2.90	3.82
7.02	6.94	41.39	74.98	0.71	3.00	4.23
6.99	6.87	41.04	74.40	0.70	2.91	4.16

Table 8 Results of vanadium stripping from HNO_3 acidified N1923

O/A ratio	Stripping liquor	Reaction time (min)	Stripping efficiency (%)	
			Cr(VI)	V(V)
1:1	1 M NaOH	10	89.02	92.19
1:1	2 M NaOH	10	90.88	91.55
1:1	1 M $NaNO_3$	10	39.96	12.25
1:1	1 M NaOH	30	76.81	95.46
1:1	1 M NaOH	60	81.56	93.74
1:1	1 M NaOH	30 + 45 °C	82.35	90.40

Conclusions

The primary amine N1923 is effective for the recovery of vanadium from a real leachate of vanadium-titanomagnetite in a pH range of 1.91–6.77. The acidification of N1913 is beneficial for the vanadium extraction in a neutral environment, but not for the improvement of separation of vanadium from chromium. Moreover, the vanadium extraction efficiency by HCl acidified N1923 is generally higher than the HNO_3 acidified organic phase. However, the stripping of vanadium by NaOH from the loaded HNO_3 acidified N1923 is much easier than the HCl acidified phase. A high stripping efficiency of >90% of V(V) could be easily achieved in a single stage by using 1 M NaOH solution. The HNO_3 is thus recommend for the acidification process of N1923 prior to the vanadium solvent extraction. The $NaNO_3$ solution is not suitable for the vanadium stripping from neither the HNO_3 nor HCl acidified N1923. The vanadium stripping process is not sensitive to the reaction temperature.

Acknowledgements The authors are grateful for financial support from Enterprise Project (Z2006) and National Key R&D Program of China (Grant No. 2019YFC1907402).

References

1. Jiang D et al (2015) Study on the synergistic extraction of vanadium by mixtures of acidic organophosphorus extractants and primary amine N1923. Sep Purif Technol 156:835–840
2. García-Labiano F et al (2016) Sulphuric acid production via chemical looping combustion of elemental sulphur. Appl Energy 178:736–745
3. Gilligan R, Nikoloski AN (2020) The extraction of vanadium from titanomagnetites and other sources. Minerals Eng 146:106106
4. Li W et al (2013) Comparison of ion exchange and solvent extraction in recovering vanadium from sulfuric acid leach solutions of stone coal. Hydrometallurgy 131–132:1–7
5. Ning P et al (2016) High-efficient extraction of vanadium and its application in the utilization of the chromium-bearing vanadium slag. Chem Eng J 301:132–138
6. Liu Z et al (2020) Separation and recovery of vanadium and iron from oxalic-acid-based shale leachate by coextraction and stepwise stripping. Separat Purif Technol 244:116532

7. Xiong P et al (2017) High-efficient and selective extraction of vanadium (V) with N235-P507 synergistic extraction system. Chem Eng Res Des: Trans Inst Chem Eng Part A 120:284–290
8. Ning P et al (2014) Selective extraction and deep separation of V(V) and Cr(VI) in the leaching solution of chromium-bearing vanadium slag with primary amine LK-N21. Sep Purif Technol 137:109–115
9. Jing X et al (2017) Separation of V(V) and Cr(VI) in leaching solution using annular centrifugal contactors. Chem Eng J 315:373–381
10. Zhang Y, Zhang TA (2015) Extraction of vanadium from the leaching solution from converter vanadium slag without roasting using N1923. Nonferrous Metals Sci Eng 6(6):14–19
11. Ning P et al (2009) Characterization and prevention of interfacial crud produced during the extraction of vanadium and chromium by primary amine. Hydrometallurgy 97(1):131–136
12. Sun P, Huang K, Liu H (2016) Separation of V and Cr in alkaline aqueous solution using acidified primary amine N1923. Hydrometallurgy 165:370–380
13. Liu Y, Yang C (2010) Research on direct precipitate of vanadium in vanadiferous leaching solution by solvent extraction. Hydrometall China 29(4):263–266
14. Sun P et al (2017) Separation of V from alkaline solution containing Cr using acidified primary amine N1923 with the addition of trisodium citrate. Sep Purif Technol 179:504–512
15. Sun P, Huang K, Liu H (2016) Separation of V and Cr from alkaline aqueous solutions using acidified primary amine A-N1923-sodium tartrate system. Chin J Nonferrous Metals 26(7):1559–1567
16. Ning P et al (2009) Behavior of interfacial crud produced in extraction separation of vanadium and chromium. Chinese J Nonferrous Metals 19(4):773–778
17. Zhao J et al (2015) Efficient separation of vanadium from chromium by a novel ionic liquid-based synergistic extraction strategy. Chem Eng J 264:487–496
18. Wen J et al (2017) Insights into the extraction of various vanadium species by primary amine. Hydrometallurgy 173:57–62
19. Sun Y (2018) Applied fundamental research on the resourceful recovery of chromium-bearing vanadium slag based on extraction methods. Tianjin University, Tianjin
20. Jing X et al (2017) High-performance recovery of Vanadium(V) in leaching/aqueous solution by a reusable reagent-primary amine N1519. ACS Sust Chem Eng 5(4):3096–3102

Novel Environmentally Friendly Leaching Process for Vanadium and Tungsten Recovery from Spent SCR Catalyst

Ana Belen Cueva Sola, Jong Hyuk Jeon, Rajesh Kumar Jyothi, and Jin-Young Lee

Abstract Selective catalytic reduction (SCR) is the best technology to reduce the nitrogen oxides (NOx) emissions in industries. Due to the increasing demand and the limited lifespan of them, after deactivation, they usually end up in landfills posing a major environmental issue. Most catalysts for stationary applications contain around 0.5–1.5% wt V_2O_5 and 7–10% wt of WO_3 in a TiO_2 glass fiber matrix and considering the necessity of replacing primary with secondary sources makes the recovery of vanadium and tungsten a major motivation for this research. During this investigation, spent SCR catalyst soda roasting process with dissolved NaOH compared with the traditional NaOH dry roasting and its influence in the subsequent water leaching was studied. With the ideal parameters obtained after optimization, there is a reduction of roasting temperature, leaching time, and the amount of silicon impurities leached alongside the title metals.

Keywords SCR catalyst · Vanadium · Tungsten · Leaching · De-silication

Introduction

Lately, environmental policies have become stricter towards emissions of solid, liquid, and gaseous effluents from industrial and urban sites. In the case of air pollutants, one of the most important noxious waste is nitrogen oxides (NOx) [1–3] known for the health and environmental threat they pose. For industrial applications, the most advanced, effective, and widely used SCR catalyst is the one contain around 0.5–1.5% wt V2O5 and 7–10% wt of WO3 in a TiO_2 glass fiber matrix (mainly composed of silicon, aluminium, and calcium oxides). The main benefit of these type of catalysts is their durability, ability for regeneration, and resistance to poison. However,

A. B. Cueva Sola · J. H. Jeon · R. K. Jyothi (✉) · J.-Y. Lee
Convergence Research Center for Development of Mineral Resources (DMR), Korea Institute of Geoscience and Mineral Resources (KIGAM), Daejeon 34132, Korea
e-mail: rkumarphd@kigam.re.kr

A. B. Cueva Sola · R. K. Jyothi · J.-Y. Lee
Department of Resource Recycling, Korea University of Science and Technology (UST), Daejeon 34113, Korea

T. Ouchi et al. (eds), *Rare Metal Technology 2022*, The Minerals, Metals & Materials Series, https://doi.org/10.1007/978-3-030-92662-5_11

after they cannot be regenerated anymore, they must be disposed. Currently, the most common disposal method for spent SCR catalysts is in landfills, which are specially designated depending on the regulations of each country to consider if the waste is hazardous or not [4, 5].

The toxicity of vanadium compounds leached when landfilled coupled with the environmental necessity to replace primary metal sources with secondary has led to the necessity and the increasing importance of the recycle of spent SCR catalyst as an important and valuable sources for vanadium, tungsten and titanium. The recovered metals can be used as primary raw materials in catalysts or in any other industrial application [6–8].

At this time, the studies developed to separate and recover metals from the spent SCR catalyst are based in the usage of alkali reagents, sometimes in a pyrometallurgical way (roasting) or sometimes for direct leaching (hydrometallurgy). The main advantage of this approach is the great solubility of vanadium and tungsten compounds while titanium compounds remain insoluble [3, 9, 10].

A variety of research efforts have been done in this topic and Choi et al. [10] have developed a method to treat and recovery the valuable metals in the spent catalyst with the advantage of using Na_2CO_3, a rather inexpensive reagent in comparison to NaOH, for roasting. Under the optimum conditions, the extraction efficiency of vanadium and tungsten reached more than 99%, while titanium can be recovered as sodium titanate to be further processed to TiO_2 [10]. In addition, the same research group studied the optimum conditions for the pressure leaching of spent SCR catalyst to obtain a leach liquor with a 91.5% recovery of vanadium and 87% recovery of tungsten. However, the main issue with pressure leaching is the high energetic burden [9]. In addition, most of the research efforts have been dedicated to obtain a high leaching rate of vanadium and tungsten without considering the impurities that can be leached alongside the tittle metals, especially silicon compounds. If a large amount of silicon is leached together with vanadium and tungsten for separation of the title metals by ion-exchange or solvent extraction, silicon must be removed adding an extra step to the recovery and purification process [10–13] (Fig. 1).

The study presented in the paper deals with a water leaching for vanadium and tungsten recovery from spent SCR catalyst after roasted using a NaOH solution instead of solid NaOH as the roasting agent. The leaching efficiency of the title metals and amount of silicon leached are compared and the agent amount, temperature, and roasting time alongside the pulp density, temperature, and leaching time are optimized to minimize the amount of silicon leached while maximizing leaching rate of vanadium and tungsten.

Experimental

The raw spent SCR catalyst was obtained from the thermoelectric power plant in Samcheonpo, South Korea, cleaned, and grinded to a size of less than 100 μm. The spent SCR catalyst was mixed in the first experimental condition with solid sodium

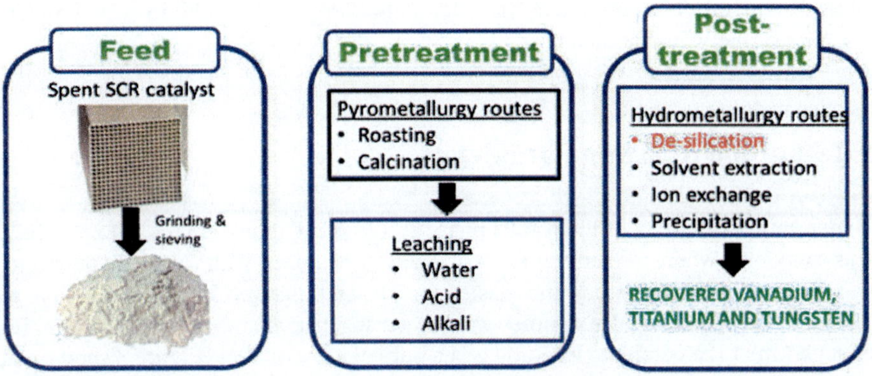

Fig. 1 Common recovery scheme for vanadium and tungsten from spent SCR catalyst

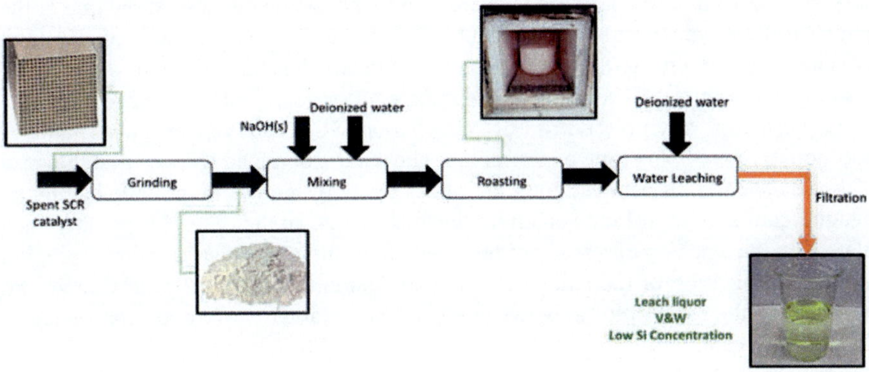

Fig. 2 Leaching process proposed in this study

hydroxide (Junsei Chemical, Japan) in a ratio NaOH/spent catalyst = 0.5 and in the second experimental condition NaOH was dissolved in sufficient water and mixed with spent SCR catalyst in the same ratio as mentioned above. Both roasting agents were optimized for roasting time, temperature, and roasting agent. After optimizing the roasting parameters, the solids were leached at different pulp densities, times, and temperatures. The overall process done in this study is shown in Fig. 2.

Results and Discussion

In the roasting process, vanadium and tungsten in the spent catalyst react with NaOH to form $NaVO_3$ and Na_2WO_4 (water-soluble). However, in the spent SCR catalysts, there is also a glass fiber matrix to improve the mechanical properties of the catalyst and it is mainly formed of SiO_2. Therefore, Na_2SiO_3 is generated alongside and subsequently leached due to its water solubility possibly inhibiting a complete leaching of the valuable metals.

Effect of the Roasting Parameters

To determine the effect of the NaOH phase and roasting temperature, an experiment was conducted where the same amount of NaOH, one mixed with water and one solid, was mixed with the feedstock and roasted for 2 h at temperatures varying from 773 to 1173 K. After roasting, the samples were water leached at ambient temperature, for enough time for a complete leaching and a pulp density of 30%. Figure 3 shows that when roasting temperature increases the leaching efficiency of vanadium, tungsten, and silicon increase too. However, tungsten reaches a plateau after 973 K, therefore, due to the low vanadium concentration and the necessity to reduce silicon leached that is the optimum temperature chosen. Moreover, the results also show that at the same roasting temperature when using NaOH solution as roasting agent the amount of vanadium and tungsten leached increase, whilst the leaching efficiency decreases, leading us to use NaOH solution as the roasting agent for further experiments.

Experiments varying the ratio of NaOH/spent SCR catalyst were also made in addition to the roasting time experiments. The results obtained showed that the ratio of NaOH/spent SCR catalyst that maximizes the tungsten leaching while having an insignificant increase of the vanadium leaching is 0.4. In the case of time, there is a plateau in the leaching efficiency of tungsten after 120 min. The goal of this research is the minimization of the silicon leached alongside the valuable metals, therefore minimizing the reaction time between silicon and NaOH is of great importance to achieve this goal.

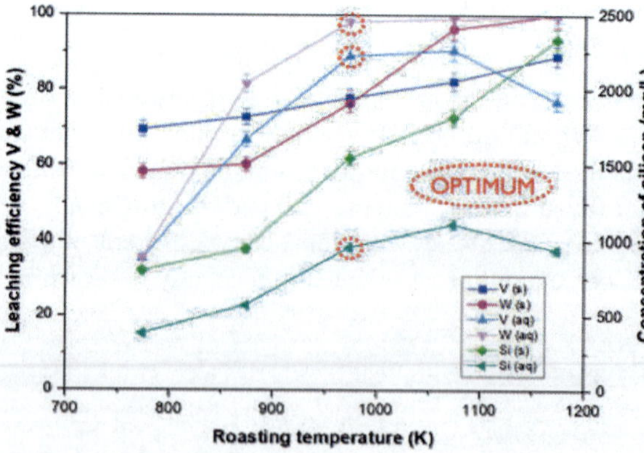

Fig. 3 Leaching efficiency of V, W, and Si depending on the roasting temperature and NaOH phase. (s) Represents NaOH solid and (aq.) represents solution as roasting agent

Effect of the Leaching Parameters

The effect of leaching time in the efficiency was also studied. Time was varied from 10 to 180 min, while the other conditions: soda roasting with NaOH solution 0.4 NaOH/feedstock ratio, 973 K for 2 h, and pulp density 30% at 298 K. As observed in Fig. 4, increasing the time does not have a great effect in the leaching of the valuable metals (vanadium and tungsten). However, when leaching time increases, there is a constant increase in the amount of silicon leaching, which leads us to choose 30 min as the ideal leaching time. Various studies done to leach vanadium and tungsten from spent SCR catalyst show a long leaching time to reach efficient conditions, usually more than 1 h [11, 14–16], yet the method presented in this paper reaches maximum leaching efficiency in 30 min with a minimization of the impurities leaching (Si).

In addition to the leaching time, pulp density and leaching temperature were also studied. In the case of the pulp density, it does not influence greatly in the leaching efficiency of vanadium and tungsten, though the highest efficiency is obtained for a 30% pulp density, which will be considered the optimum. Leaching temperature has a great influence in the mass transfer of the water-soluble compounds formed during the roasting process, therefore minimizing temperature is one of the main goals of this study in order to minimize the amount of silicon leached alongside the valuable metals [1, 15, 17, 18]. Results showed that vanadium and tungsten reach a maximum leaching efficiency from 298 K when using the optimum conditions described above, therefore for an economic standpoint as well as the minimization of silicon leached 298 K was set as the optimum leaching temperature.

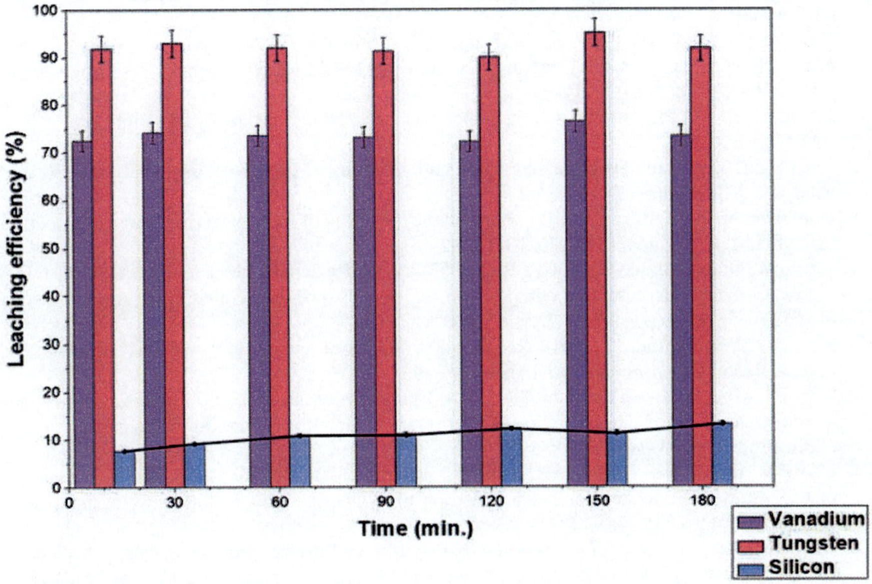

Fig. 4 Leaching time versus efficiency

Conclusion

- Spent SCR catalyst are a serious risk for environmental and human health when being landfilled. In this investigation, spent SCR catalyst is used as a secondary source for vanadium and tungsten leaching while minimizing the amount of silicon impurities leached.
- The common roasting process done using solid NaOH as a roasting agent is compared with an innovative process using dissolved NaOH with the purpose to maximize leaching of valuable metals while reducing the amount the impurities leached.
- After optimization, roasting done by NaOH solution proved a more effective method to reduce the amount of silicon leached to almost half while maximizing the vanadium ($80.2 \pm 3\%$) and tungsten leaching ($95.4 \pm 3\%$).

Acknowledgements This study was supported by the R&D Center for Valuable Recycling (Global-Top R&D Program) of the Ministry of Environment (Project Number: 2019002230001), Korea.

References

1. Jyothi RK, Moon G, Kim H-R, Lee J-Y, Choi IH (2017) Spent V_2O_5-WO_3/TiO_2 catalyst processing for valuable metals by soda roasting-water leaching. Hydrometallurgy 175(December):292–299
2. Radojevic M (1998) Reduction of nitrogen oxides in flue gases. Environ Pollut 102(1, Suppl 1):685–689
3. Cueva Sola AB, Parhi PK, Lee JY, Kang HN, Jyothi RK (2020) Environmentally friendly approach to recover vanadium and tungsten from spent SCR catalyst leach liquors using Aliquat 336. RSC Adv 10(34):19736–19746
4. Forzatti P (2001) Present status and perspectives in de-NOx SCR catalysis. Appl Catal A Gen 222(1–2):221–236
5. Ferella F (2020) A review on management and recycling of spent selective catalytic reduction catalysts. J Clean Prod 246:118990
6. Choi IH et al (2020) Recent developments in the recycling of spent selective catalytic reduction catalyst in South Korea. Catalysts 10(2):182
7. Zhang W, Qi S, Pantaleo G, Liotta LF (2019) WO_3–V_2O_5 active oxides for NOx SCR by NH_3: preparation methods, catalysts' composition, and deactivation mechanism—a review. Catalysts 9(6):527
8. Zhang Q, Wu Y, Yuan H (2020) Recycling strategies of spent V_2O_5-WO_3/TiO_2 catalyst: a review. Resourc Conserv Recycl 161(May):04983
9. Choi IH, Moon G, Lee JY, Jyothi RK (2018) Extraction of tungsten and vanadium from spent selective catalytic reduction catalyst for stationary application by pressure leaching process. J Clean Prod 197:163–169
10. Choi IH, Moon G, Lee JY, Jyothi RK (2019) Alkali fusion using sodium carbonate for extraction of vanadium and tungsten for the preparation of synthetic sodium titanate from spent SCR catalyst. Sci Rep 9(1):1–8
11. Choi IH, Moon G, Lee JY, Jyothi RK (2018) Hydrometallurgical processing of spent selective catalytic reduction (SCR) catalyst for recovery of tungsten. Hydrometallurgy 178(April):137–145

12. Cueva Sola AB, Parhi PK, Thenepalli T, Jyothi RK (2020) Rare-earth metal recovery for green technologies. Springer International Publishing, Cham
13. Ferella F, Innocenzi V, Maggiore F (2016) Oil refining spent catalysts: a review of possible recycling technologies. Resour Conserv Recycl 108:10–20
14. Yao J et al (2021) Successive calcination-oxalate acid leaching treatment of spent SCR catalyst: a highly efficient and selective method for recycling tungsten element. Hydrometallurgy 201:105576
15. Wu W, Wang C, Bao W, Li H (2018) Selective reduction leaching of vanadium and iron by oxalic acid from spent V_2O_5-WO_3/TiO_2 catalyst. Hydrometallurgy 179(May):52–59
16. Cao Y, Yuan J, Du H, Dreisinger D, Li M (2021) A clean and efficient approach for recovery of vanadium and tungsten from spent SCR catalyst. Miner Eng 165:106857
17. Su Q, Miao J, Li H, Chen Y, Chen J, Wang J (2018) Optimizing vanadium and tungsten leaching with lowered silicon from spent SCR catalyst by pre-mixing treatment. Hydrometallurgy 181:230–239
18. Luo L, Miyazaki T, Shibayama A, Yen W, Fujita T (2003) A novel process for recovery of tungsten and vanadium from a leach solution of tungsten alloy scrap. Miner Eng 16(7):665–670

Effect of Antisolvent Type and Concentration on Morphology and Crystal Size of $(NH_4)_3ScF_6$ Obtained by Antisolvent Crystallization

Edward Michael Peters, Michael Svärd, and Kerstin Forsberg

Abstract Scandium is a metal of value with increasing demand but limited supply. During the valorization of bauxite residue, scandium becomes concentrated in an NH_4F strip liquor, from which it can be recovered as $(NH_4)_3ScF_6$ by antisolvent crystallization. This study investigates the use of different antisolvents and their concentrations on the morphology and crystal size of the $(NH_4)_3ScF_6$ crystals produced. The antisolvents include alcohols, ketones, and a sulfoxide. These were added all at once, either pure or diluted, to the aqueous solution to attain final concentration of 2, 4, or 8 mol/L total solution in separate experiments. Changes in the functional group and concentration of the antisolvents were observed to induce morphological and crystal size modifications of $(NH_4)_3ScF_6$. The differences are likely due to the specific interactions of the antisolvent molecules with the atoms at the different faces of the crystals.

Keywords $(NH_4)_3ScF_6$ · Antisolvent crystallization · Morphology · Crystal size

Introduction

Scandium is a metal that has gained research interest in the past years due to its increasing applications and limited supply. It can be used to produce high strength, lightweight aluminum–scandium alloys containing <2 wt. % Sc as well as in solid oxide fuel cells, ceramics, electronics, lasers, lighting, and radioactive isotopes [1, 2]. However, the demand is exceeding supply due to the increased applications of the metal, yet there is a scarcity of scandium bearing ores since scandium does

E. M. Peters (✉) · M. Svärd · K. Forsberg (✉)
Department of Chemical Engineering, KTH Royal Institute of Technology, SE-100 44 Stockholm, Sweden
e-mail: edwpet@kth.se

K. Forsberg
e-mail: kerstino@kth.se

M. Svärd
e-mail: micsva@kth.se

T. Ouchi et al. (eds), *Rare Metal Technology 2022*, The Minerals, Metals & Materials Series, https://doi.org/10.1007/978-3-030-92662-5_12

not combine with common ore-forming anions [2]. This makes it challenging to recover scandium from the earth's crust since it is widely dispersed although its average concentration of 22 ppm is higher than the abundance of some common metals such as lead, tin, mercury, silver, and cobalt [1–3]. Therefore, the current research focuses on recovering scandium from waste streams of other mining and metallurgical operations such as the bauxite residue from the alumina industry and titanium dioxide acid waste.

It has been shown that scandium can be recovered from such waste streams by a combination of unit operations which include leaching, solvent extraction, crystallization, calcination, and metallothermic reduction [4–12]. The solvent extraction stage produces strip liquors containing >2 g/L Sc and some metal impurities such as Fe, Al, and Zr in 3 mol/L NH_4F from which scandium can be recovered by antisolvent crystallization as ammonium scandium hexafluoride ($(NH_4)_3ScF_6$) using alcoholic solvents [12, 13]. Methanol and ethanol have proven to be the most favorable antisolvents due to their low costs and the ease of recovery of these solvents by distillation making it possible to recycle the antisolvent, thereby promoting favorable process economics [14]. Other antisolvents such as n-propanol, isopropanol, and acetone have also shown competitive performance in the recovery of $(NH_4)_3ScF_6$ by antisolvent crystallization [12]. However, although some antisolvents show competitive performance, they are not preferable from economic and safety perspectives since their higher boiling temperatures makes it difficult to recover and reuse them, they are generally more expensive, or because of their toxicity.

The type and concentration of antisolvent can affect the characteristics of the final product, such as morphology and crystal size [15–19]. Different antisolvents have different polarity indices which show the extent to which the molecules of the antisolvent can form H-bonds [19]. The higher the polarity index of a solvent, the greater is its H bonding propensity. This implies that the type and extent of interactions between the antisolvent molecules and the solvent molecules or solute molecules differ from one antisolvent to another [18, 20] and knowledge of the antisolvent properties can be used to fine-tune the characteristics of the final crystal product. In addition, the functional group of the antisolvent has been reported to play an important role for the crystal morphology for solvents with almost similar polarity indices [19]. The mixing properties of the antisolvents also influence the initial bulk and local supersaturation and could have an impact on morphology and crystal size of the product. The polarity indices of the antisolvents used in this study are shown in Table 1.

The extent to which the antisolvent molecule adsorbs on crystal surfaces depends on the type and strength of interactions between the antisolvent molecules and the solute molecules on the crystal surface [20]. This is determined by the chemical environment of the crystal surface. This implies that antisolvent molecules can adsorb on crystal faces selectively, thereby blocking growth of some crystal faces and altering the morphology [25, 26]. In some cases, a mixture of two or more antisolvents can provide the necessary antisolvent adsorptive properties required to yield crystals of the desired quality [20, 27]. The antisolvent can also alter the interactions between the impurity molecules present in solution and the solute molecules on the crystal surface,

Table 1 Polarity indices of water and of the antisolvents used [21–24]

Antisolvent	Polarity index
Water	9.0
Dimethylsulfoxide (DMSO)	7.2
Methanol	6.6
Ethanol	5.2
Acetone	5.1
Methyl ethyl ketone (MEK)	4.5
N-propanol	4.0
Isopropanol	3.9
1,3-propane-diol	–

causing some impurities to desorb from specific crystal faces. This can occur either by displacement of impurity species by antisolvent molecules on the crystal surface or by the new interactions introduced between the antisolvent molecules and the impurity ions [26]. The antisolvent added could also change the pH and affect the solution speciation causing the concentration of ions and molecular complexes to change in solution, further affecting the interactions of impurities with crystal faces [26].

The agitation mechanism and degree of agitation also plays an important role in determining the bulk supersaturation homogeneity of the system [18]; therefore, it is important to employ similar agitation conditions to observe morphological and/or polymorphic modifications caused by the antisolvent. The role of initial supersaturation in batch crystallization systems in modifying crystal morphology has been reported in literature. In such systems, as the supersaturation decreases, the difference between the rate of growth of specific crystal faces diminishes which determines the morphology of the final crystal product [18].

This study investigates the effect of different antisolvents and their concentrations at the point of nucleation and subsequent growth in the system on the morphology and crystal size of $(NH_4)_3ScF_6$ during antisolvent crystallization. The antisolvents include alcohols, ketones, and an organosulfur compound. The choice of the antisolvents in this study is solely based on their different functional groups and not necessarily on their suitability for use from an industrial perspective. The purpose is to understand if different functional groups in addition to the antisolvent polarity index affect the morphology and crystal size differently and develop an understanding of the mechanisms by which different functional groups interact with solute molecules on crystal faces.

Table 2 Composition of strip liquor in 3 mol/L NH_4F (pH: 5.53)

Sc	Al	Zr	V	Ti	Nb	Y	U	Th	Nd	Ce	Fe	Dy
mg/kg												
2572.7	0.9	1.1	5.5	8.2	0.7	0.3	3.7	4.4	1.0	2.1	4.4	0.0

Methodology

Table 2 shows the composition of the strip liquor used. The strip liquor, supplied by MEAB Chemie Technik GmbH, was generated from a scandium hydroxide concentrate by acid leaching and solvent extraction of the pregnant leach solution followed by stripping the organic phase with a 3 mol/L NH_4F solution.

The antisolvents used are 99.9% v/v methanol, 99.85% v/v ethanol, 99.5% v/v n-propanol, 99.7% v/v isopropanol, >99.7% v/v 1,3-propane-diol, 99.8% v/v acetone (also known as dimethyl ketone), and >99.5% v/v methyl ethyl ketone (MEK) as well as an organosulfur compound, dimethyl sulfoxide (DMSO) of purity >99.9% v/v. The antisolvent was added all at once to strip liquor to attain final concentration of either 2, 4, or 8 mol/L on a total solution basis. In some cases, the antisolvent was diluted to 60% v/v and added to attain the same final concentration. Antisolvent addition was conducted under agitation at 500 rpm using a magnetic stirrer in all experiments. The suspension was filtered under vacuum and the solids dried under ambient conditions overnight. The dry solids were analyzed by powder XRD to determine the phase of the solid product and to ascertain that no polymorphic transformation had occurred. SEM–EDS was used to observe the morphology of the crystals and determine the surface composition of the solids, respectively.

Results and Discussion

Figure 1 shows the XRD diffractograms of the solid samples obtained from some of the experiments in comparison to the reference pattern of $(NH_4)_3ScF_6$.

In all cases, the XRD patterns correspond to that of monoclinic $(NH_4)_3ScF_6$ (PDF card 00-040-0595). This implies that any morphological modifications observed were caused by the changes in antisolvent type and/or concentration and are not a result of polymorphic transformation. For all alcohols, the final concentration of Sc in the solution when the antisolvent concentration is 8 mol/L will be very low and almost similar [28], so the initial supersaturation is almost the same. It is recommended to conduct more tests to check if this is true for the other antisolvents.

Figure 2 shows the micrographs for the solid product obtained from experiments in which the undiluted antisolvent was added to attain a final concentration of 8 mol/L.

The morphology of $(NH_4)_3ScF_6$ crystals obtained from undiluted alcohol antisolvents (Fig. 2a–e) is quite similar with isodimensional hexagonal shapes and a small

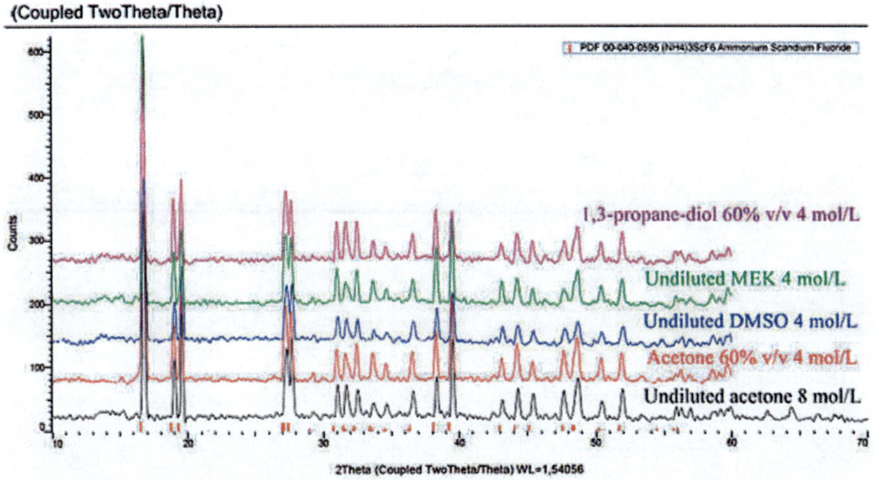

Fig. 1 XRD diffractograms of the solids

population of slightly elongated crystals. However, the size of crystals obtained when 1,3-propane-diol was used as the antisolvent is much smaller than the size of crystals obtained when monohydric alcohol antisolvents were used. This can be attributed to the presence of two hydroxyl groups in 1,3-propane-diol which could increase the number and strength of interactions between the antisolvent molecules and the crystal surface atoms, thereby hindering growth of the crystals. The residual supersaturation available is then consumed by nucleation. Since the morphology of the crystals is similar, it can be deduced that the action of the dihydric alcohol antisolvent molecules is even on all crystal faces.

On the contrary, rod-shaped $(NH_4)_3ScF_6$ crystals were obtained when acetone was used as the antisolvent (Fig. 2f). Interestingly, the crystals did not have discernible faces and edges and some crystals were T-shaped and L-shaped which could be evidence that some crystal faces were disappearing from the normal hexagonal shapes obtained when alcohol antisolvents were used. This implies that the functional group of the antisolvent determines the interactions between the antisolvent molecules and the atoms on specific crystal faces since acetone and ethanol have almost similar polarity indices. Since rod shapes were obtained, it is evident that acetone molecules preferentially adsorbed on some faces of the crystals, hindering growth of these faces, and promoting unidirectional growth of $(NH_4)_3ScF_6$ crystals. Bigger crystals were also obtained in the case of acetone, although they are elongated since adsorption of the antisolvent on some crystal faces means that solute units are added to the crystals on a smaller surface area.

Figure 3 shows the micrographs for the solid product obtained from experiments in which either 60% v/v acetone or 60% v/v 1,3-propane-diol was added to attain a final concentration of 4 mol/L.

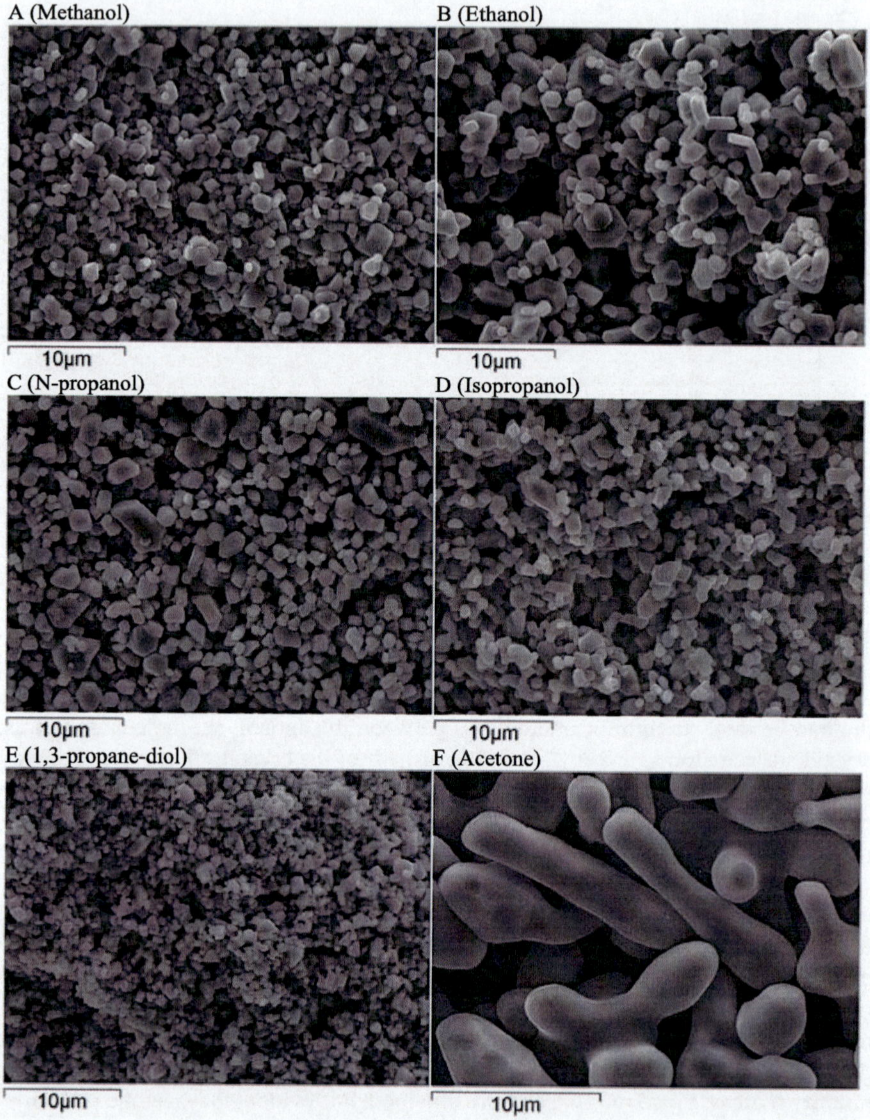

Fig. 2 SEM micrographs for the solid product obtained from experiments where the final concentration of the undiluted antisolvent was 8 mol/L. Note the differences in the scale bars

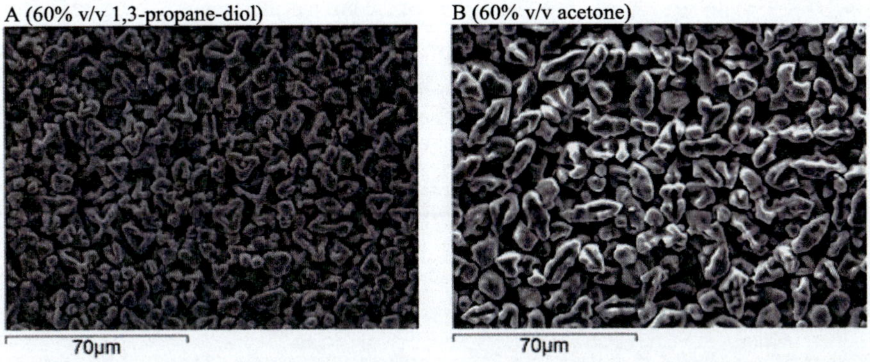

Fig. 3 SEM micrographs for the solid product obtained from experiments conducted with 60% v/v acetone or 1,3-propane-diol where the final ingredient concentration was 4 mol/L

The morphology of $(NH_4)_3ScF_6$ crystals shown in Fig. 3 is compared to the cases where the undiluted antisolvents were used (Fig. 2e, f, respectively). Reducing the concentration of 1,3-propane diol to 60% v/v induced a morphological change on the crystals to triangular, X-shaped, and rod-shaped crystals. The triangular crystals in this case are chipped in the middle. The change in morphology can be attributed partly to an increase in the polarity index of the antisolvent since water with a higher polarity index is used for dilution. In literature, antisolvents with higher polarity indices were also observed to cause elongation of crystals. It could also be attributed partly to the fact that when the initial supersaturation is high as in the case when undiluted antisolvents were used, the difference in the growth rates of specific crystal faces diminishes as desupersaturation occurs, resulting in isodimensional crystals shown in Fig. 2e. There was no significant change observed in the morphology of $(NH_4)_3ScF_6$ crystals when 60% v/v acetone was used as the antisolvent, except that the aspect ratio of the crystals appeared to increase in comparison to the case where undiluted antisolvent was added. Notably, in the case of diluted acetone, the rod-shaped crystals appear to be rough and chipped in the middle.

Figure 4 shows the micrographs for the solid product obtained from experiments in which either undiluted MEK or DMSO were added to attain final concentration of 4 mol/L, undiluted DMSO was added to attain a final concentration of 2 mol/L and 60% v/v DMSO was added to attain a final concentration of 4 mol/L.

The morphology of crystals produced when MEK was added as an antisolvent to attain a concentration of 4 mol/L were isodimensional comparable to the case where alcohol antisolvents were used to attain a final concentration of 8 mol/L (Fig. 2a–e) and contrary to acetone with the same functional group as MEK (Fig. 2f). The experiment with MEK was also conducted to attain a final concentration of 8 mol/L and the crystals obtained (not presented in Fig. 4) had similar morphology and size to those obtained at 4 mol/L. However, bigger crystals were obtained when MEK was used as the antisolvent. The reason for the larger crystals is not immediately clear since the polarity index of MEK is intermediate between that of ethanol and

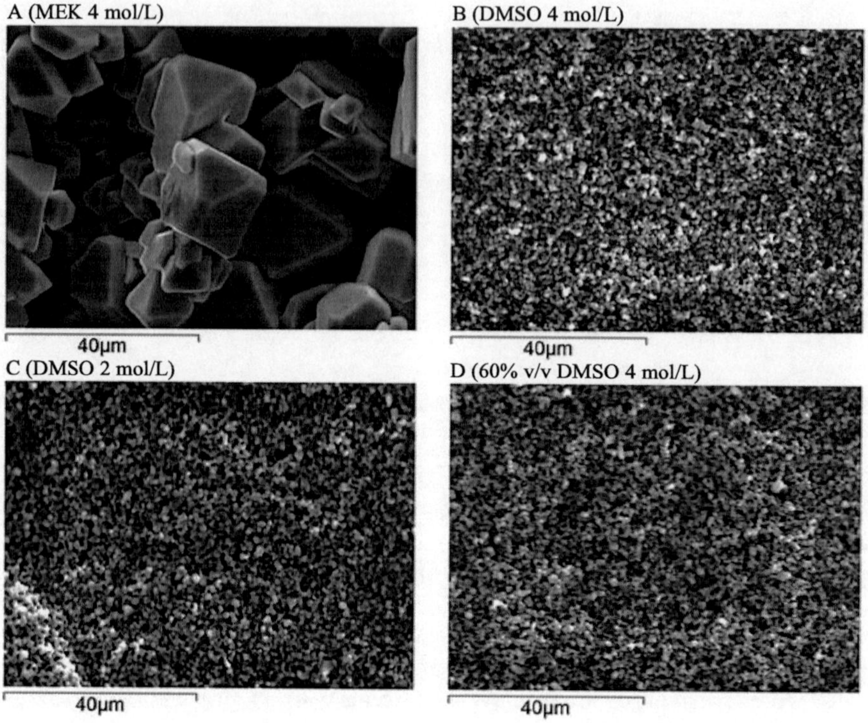

Fig. 4 SEM micrographs for the solid product obtained from experiments conducted with MEK and different DMSO concentrations. The final ingredient concentration was 2 or 4 mol/L

n-propanol, and since MEK possesses the ketone functional group found in acetone. The differences observed between acetone and MEK could be due to steric effects since they have different alkyl groups. However, lower concentrations of MEK were not investigated since it is only miscible with water at about 25% v/v, and this concentration was not sufficient to yield crystals. At higher concentrations, liquid–liquid phase separation occurred. In the case of DMSO, the change in final concentration of the undiluted antisolvent or the change in antisolvent concentration did not have a significant change on morphology and crystal size. Very small crystals were obtained in all cases. This is possibly due to the higher polarity index of DMSO which implies stronger adsorption on crystal faces hindering growth of the crystals and promoting nucleation and/or the sulfoxide functional group which could induce a different adsorption behavior from the hydroxyl or ketone functional group on the crystal faces. Overall, an in-depth understanding of the interactions between the antisolvent molecules and solvent or solute molecules helps to clarify the changes in morphology and crystal size observed.

Conclusions

The changes in morphology and/or crystal size of $(NH_4)_3ScF_6$ induced by different antisolvent types and changes in antisolvent concentration have been demonstrated in an antisolvent crystallization process. All pure alcohol antisolvents employed yielded $(NH_4)_3ScF_6$ product with mostly isodimensional crystals and few slightly elongated crystals although the dihydric alcohol, 1,3-propane-diol, caused a reduction in the crystal size. This was attributed to the presence of two hydroxyl groups which strengthens the interactions between the antisolvent molecules and the atoms on the crystal faces, thereby hindering growth and promoting nucleation. When undiluted or diluted acetone was used as antisolvent, rod-shaped crystals were obtained implying that acetone preferentially adsorbs on specific crystal faces causing crystal elongation. However, MEK produced large isodimensional crystals despite possessing a ketone functional group like acetone. The difference in crystal morphology when acetone and MEK were used as antisolvents is likely due to steric effects which could impact the degree of interactions between the antisolvent molecules and the crystal surface atoms. In the case of DMSO, changing the antisolvent concentration or the final concentration in solution did not alter the crystal morphology but caused a significant reduction in crystal size. This is attributed to its higher polarity index and/or the sulfoxide functional group which possibly increases the degree of adsorption on crystal faces, thereby suppressing growth and promoting nucleation.

References

1. U.S. Geological Survey (2021) Mineral commodity summaries 2021. Reston, Virginia
2. Duyvesteyn WPC, Putnam GF (2014) Scandium: a review of the element, its characteristics and current and emerging commercial applications. Nevada, EMC Metal Corporation White Paper
3. Krishnamurthy N, Gupta CK (2016) Extractive metallurgy of rare earths, 2nd edn. CRC Press Taylor and Francis Group, New York
4. Shaoquan X, Suqing L (1996) Review of the extractive metallurgy of scandium in China (1978–1991). Hydrometallurgy 42:337–343. https://doi.org/10.1016/0304-386X(95)00086-V
5. Wang W, Pranolo Y, Cheng CY (2011) Metallurgical processes for scandium recovery from various resources: a review. Hydrometallurgy 108:100–108. https://doi.org/10.1016/j.hydromet.2011.03.001
6. Liu Y, Naidu R (2014) Hidden values in bauxite residue (red mud): Recovery of metals. Waste Manag 34:2662–2673. https://doi.org/10.1016/j.wasman.2014.09.003
7. Liu Z, Li H (2015) Metallurgical process for valuable elements recovery from red mud—a review. Hydrometallurgy 155:29–43. https://doi.org/10.1016/j.hydromet.2015.03.018
8. Onghena B, Borra CR, van Gerven T, Binnemans K (2015) Selective recovery of scandium (III) from bauxite residue leachates by Sslvent extraction with a carboxyl-functionalised ionic liquid. In: Bauxite residue valorization and best practices. Leuven, pp 331–338
9. Harata M, Nakamura T, Yakushiji H, Okabe TH (2008) Production of scandium and Al–Sc alloy by metallothermic reduction. Miner Process Extr Metall 117:95–99. https://doi.org/10.1179/174328508X290876
10. Yagmurlu B, Zhang W, Avdibegovic D, et al (2018) Advances on scandium recovery beyond state of the art. In: ALTA 2018 uranium-REE-lithium proceedings. Perth, pp 85–93

11. Yagmurlu B, Dittrich C, Friedrich B (2018) Effect of aqueous media on the recovery of scandium by selective precipitation. Metals (Basel) 8:314. https://doi.org/10.3390/met8050314
12. Kaya Ş, Peters EM, Forsberg K et al (2018) Scandium recovery from an ammonium fluoride strip liquor by anti-solvent crystallization. Metals (Basel) 8:767. https://doi.org/10.3390/met8100767
13. Peters EM, Kaya Ş, Dittrich C, Forsberg K (2019) Recovery of scandium by crystallization techniques. J Sustain Metall 5:48–56. https://doi.org/10.1007/s40831-019-00210-4
14. Peters EM, Kaya Ş, Dittrich C, et al (2020) Recovery of alcohol after anti-solvent precipitation of (NH4)3ScF6 from NH4F strip liquors. In: XXX international mineral processing congress 2020. Cape Town
15. Haleblian JK (1975) Characterization of habits and crystalline modification of solids and their pharmaceutical applications. J Pharm Sci 64:1269–1288. https://doi.org/10.1002/jps.2600640805
16. Nokhodchi A, Bolourtchian N, Dinarvand R (2003) Crystal modification of phenytoin using different solvents and crystallization conditions. Int J Pharm 250:85–97. https://doi.org/10.1016/S0378-5173(02)00488-X
17. Tran TTD, Tran PHL, Park JB, Lee BJ (2012) Effects of solvents and crystallization conditions on the polymorphic behaviors and dissolution rates of valsartan. Arch Pharm Res 35:1223–1230. https://doi.org/10.1007/s12272-012-0713-7
18. Khamskii EV (1976) Some problems of crystal habit modification. In: Mullin JW (ed) Industrial crystallization. Plenum Press, New York
19. Zhou X, Shan J, Chen D, Li H (2019) Tuning the crystal habits of organic explosives by antisolvent crystallization: the case study of 2,6-dimaino-3,5-dinitropyrazine-1-oxid (LLM-105). Crystals 9:392. https://doi.org/10.3390/cryst9080392
20. Di Martino P, Censi R, Malaj L et al (2007) Influence of solvent and crystallization method on the crystal habit of metronidazole. Cryst Res Technol 42:800–806. https://doi.org/10.1002/crat.200710908
21. Larsen KL, Nielsen R, Do TT (2013) Production of channel type cyclodextrin crystals. US Patent Application No. 14/383,891
22. Nti-Gyabaah J, Gbewonyo K, Chiew YC (2010) Solubility of artemisinin in different single and binary solvent mixtures between (284.15 and 323.15) K and NRTL interaction parameters. J Chem Eng Data 55:3356–3363. https://doi.org/10.1021/je100125x
23. Abarca-Vargas R, Peña Malacara CF, Petricevich VL (2016) Characterization of chemical compounds with antioxidant and cytotoxic activities in bougainvillea x buttiana holttum and standl, (Var. rose) extracts. Antioxidants 5:45. https://doi.org/10.3390/antiox5040045
24. Gupta MN, Batra R, Tyagi R, Sharma A (1997) Polarity index: the guiding solvent parameter for enzyme stability in aqueous-organic cosolvent mixtures. Biotechnol Prog 13:284–288. https://doi.org/10.1021/bp9700263
25. Sudha C, Srinivasan K (2014) Understanding the effect of solvent polarity on the habit modification of monoclinic paracetamol in terms of molecular recognition at the solvent crystal/interface. Cryst Res Technol 49:865–872. https://doi.org/10.1002/crat.201400200
26. Boistelle R (1976) Survey of crystal modification in solution. In: Mullin JW (ed) Industrial crystallization. Plenum Press, New York
27. Gouthami KS, Kumar D, Thipparaboina R et al (2015) Can crystal engineering be as beneficial as micronisation and overcome its pitfalls?: A case study with cilostazol. Int J Pharm 491:26–34. https://doi.org/10.1016/j.ijpharm.2015.06.009
28. Peters EM, Svärd M, Forsberg K (2020) Phase equilibria of ammonium scandium fluoride phases in aqueous alcohol mixtures for metal recovery by anti-solvent crystallization. Sep Purif Technol 252:117449. https://doi.org/10.1016/j.seppur.2020.117449

Purification of a Low-Grade Molybdenite Ore for Industrial Steel Production

Alafara A. Baba, Christianah O. Adeyemi, Mamata Mohapatra, Mustapha A. Raji, Fausat T. Akanji, and Abdul G. F. Alabi

Abstract Continuous demands for industrial molybdenum compounds such as MoO_3 for steel coating applications cannot be over emphasized, as it's highly characterized by improved pitting corrosion and heat resistance properties. In this study, leaching of molybdenum from a Nigerian low-grade molybdenite ore was investigated by sulphuric acid solution. The raw ore and selected products at optimal conditions were accordingly characterized. At optimal leaching conditions: temperature = 75 °C, [H_2SO4] = 2.5 mol/L, particle size <50 μm, S/L ratio = 10 g/L at 300 rpm, 64.7% molybdenum dissolution was achieved within 120 min. The calculated activation energy of 35.46 kJ/mol and a reaction order 1.02 established suggests a first-order diffusion-controlled reaction mechanism. The residual products analyzed by X-ray diffraction containing co-mixtures of siliceous materials including *quartz* (SiO_2: 96-900-5020) and *haleniusite* ($La_{1.84}Ce_{1.80}Nd_{0.28}Pr_{0.08}O_{4.48}F_{3.52:}$ 96-901-0570) could be further re-processed as a precursor for stainless steel production.

Keywords Molybdenum · Nigeria · Sulphuric acid solution · Leaching · Industrial steel

---

A. A. Baba (✉) · C. O. Adeyemi (✉) · M. A. Raji · F. T. Akanji
Department of Industrial Chemistry, University of Ilorin, P.M.B. 1515, Ilorin 240003, Nigeria
e-mail: alafara@unilorin.edu.ng

C. O. Adeyemi
e-mail: christianaholawale@ymail.com

C. O. Adeyemi
Department of Science Laboratory Technology, The Federal Polytechnic, P.M.B. 420, Offa, Nigeria

C. O. Adeyemi · M. Mohapatra
Department of Hydro & Electrometallurgy, CSIR-Institute of Minerals and Materials Technology, Bhubaneswar 751013, India

F. T. Akanji
Sheda Science and Technology Complex, Sheda, Abuja, Nigeria

A. G. F. Alabi
Department of Materials Science and Engineering, Kwara State University, Malete, Nigeria

T. Ouchi et al. (eds), *Rare Metal Technology 2022*, The Minerals, Metals & Materials Series, https://doi.org/10.1007/978-3-030-92662-5_13

Introduction

Molybdenum (Mo) recovery from primary and secondary resources is of great importance, due to the rapid depletion of rich minerals along with increasing industrial growth demands [1]. Molybdenum (Mo) is a critical and highly valuable metal that exhibits many industrial applications [2–4], particularly when alloyed with other metals [5, 6]. The world's molybdenum reserves are concentrated in nine countries (China, United States, Chile, Canada, Russia, Armenia, Peru, Iran, and Kazakhstan) [7–9]. There are different techniques for Mo (VI) extraction, including leaching, solvent extraction, ion exchange resin, precipitation electro-winning, or adsorption onto activated matrices [10] among others. In recent years, molybdenum extraction from its parent ore for industrial value addition has been investigated [11, 12]. Therefore, this work is aimed at establishing a hydrometallurgical method for direct treatment of a Nigerian low-grade molybdenite ore [13].

Materials and Method

Materials

The molybdenite ore for this work was sourced from Kigom village $(9^{\circ}45'50'' \text{ N}, 8^{\circ}37'31''\text{E})$, the south-west axis of Jos city, Plateau State, Nigeria. The ore was crushed, pulverized, and sieved into three particle sizes: <50 μm, <63 μm, and <75 μm. The finest particle size of <50 μm was used for all the experiments unless otherwise stated.

Leaching Procedure

The leaching experiment was performed in a 250 cm^3 glass reactor. The reaction was carried out at temperature 75 °C, particle size <50 μm, S/L ratio 10 g/L, stirring speed 300 rpm in 120 min with varying sulphuric acid concentrations (0.1–3 mol/L). After the required period of leaching, the residual product was collected by filtration and dried at 80 °C in an oven overnight and weighed. The extent of ore dissolution over time was accordingly evaluated. The undissolved species at optimal conditions were assessed for better understanding of the kinetic mechanism for the ore dissolution reaction *cum* ore beneficiation activities [6, 8].

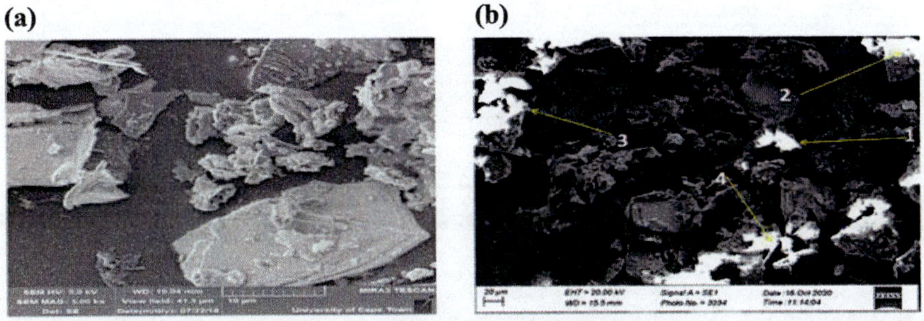

Fig. 1 **a** SEM morphology of raw molybdenite ore; **b** SEM morphology of the H_2SO_4 leached residual product at optimal condition

Results and Discussion

Characterization of Raw Ore and Residual Product

The chemical analysis of the raw molybdenite ore, using ED-XRF gave 47.46 wt% MoS_2, 13.13 wt% SiO_2, 3.06 wt% Al_2O_3, 18.42 wt% SO_2, 1.11 wt% FeO_2, 1.10 wt% K_2O_2, and 0.5 wt% WO_2. Tshe XRD results showed molybdenite ore contained MoS_2 ($Mo_{2.00}S_{4.00}$: 96-900-9145) and quartz ($Si_{3.00}O_{6.00}$: 96-901-3322) and the SEM examined at 5000 magnification confirmed different topographies of heterogeneous phase joined with homogeneous deposits on the ore surfaces as shown in Fig. 1a. According to Fig. 1b, the unreacted product exhibits a homogenous phase with the identified trace of whitish surface in a ring shape and a spot at the center labelled (1–4), suggesting that the molybdenum was separated into different homogenous phases as compared to the raw sample in Fig. 1a. The whitish surface could be attributed to the presence of unreacted molybdenum in the residue. The EDS gave MoL (30.84%), SiK (53.61 &), AlK (5.89%), and KK (7.41%), while the XRD contain siliceous impurities: *Quartz* (96-900-5020), *Mo* (96-900-7661)) with traces of *Haleniusite* (96-901-0570) was identified. These siliceous impurities could serve as a by-product for some defined industrial applications.

Discussion

Kinetic Studies

In this study, rate controlling step and kinetic parameter examinations were analyzed using the shrinking core model (SCM), with diffusion controlled mechanism reactions fitted perfectly to the dissolution data with an average correlation coefficient $R^2 = 0.998$ as shown in Fig. 2

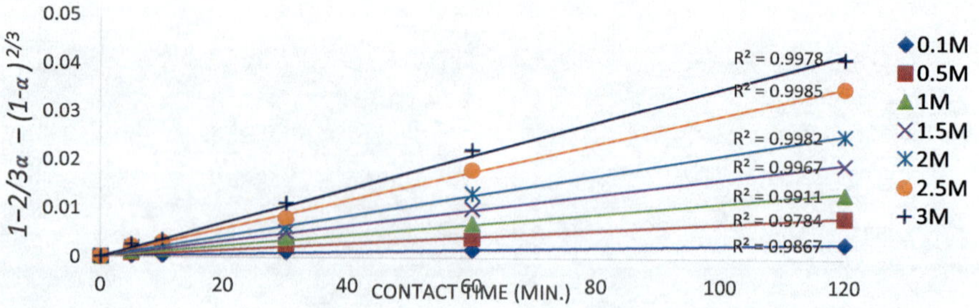

Fig. 2 Plots of $1-\frac{2}{3}\alpha-(1-\alpha)^{\frac{2}{3}}$ versus contact time (t) as a function of leaching time

$$1-\frac{2}{3}\alpha-(1-\alpha)^{\frac{2}{3}}=k_d t \tag{1}$$

where k_d is the rate constant for diffusion controlled leaching. The fraction of ore α (%), k is the specific rate constant (min^{-1}), and t is the contact time (min) [14, 15]. Also, the temperature of the specific rate constant can be estimated from the Arrhenius expression dependence of Eq. (2)

$$k = A^{e-Ea/RT} \tag{2}$$

where A is the frequency factor, k is the specific rate constant, Ea is the activation energy (kJ/mol), R is the universal gas constant (8.314 J/mol/K), and T is the absolute temperature (K).

Thus, Eq. (2) was linearized to obtain Eq. (3)

$$\text{Ink} = \text{In A} - \frac{Ea}{RT} \tag{3}$$

The activation energy (Ea) from the necessary Arrhenius plot and reaction order were calculated as 35.46 kJ/mol and 1.02 (Fig. 3), supporting the proposed first-order diffusion control mechanism for the dissolution process.

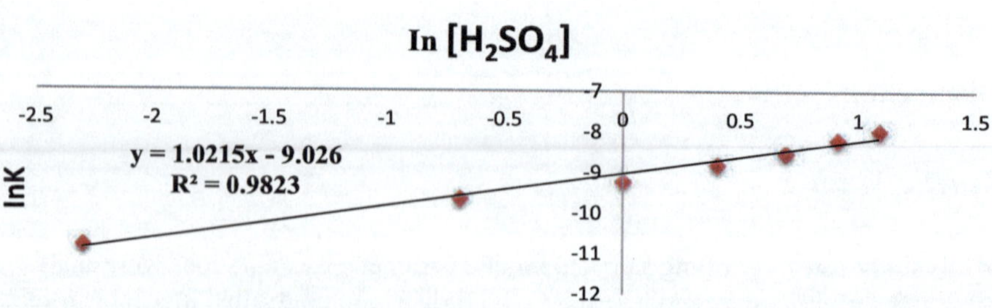

Fig. 3 Plot of lnk against ln $[H_2SO_4]$

Conclusions

In this study, an investigation of molybdenite ore dissolution behavior was carried out using H_2SO_4 solution. The effect of reaction temperature, leachant concentrations at a fixed particle size diameter (<50 μm), on the extent of the ore dissolution was investigated. The results obtained suggest that the dissolution rate increased significantly with leachant concentration and reaction temperature. At optimum conditions (2.5 mol/L H_2SO_4, 75 °C, <50 μm, 300 rpm, 10 g/L and 120 min), 64.7% extraction of the initial 10 g/L ore was reacted within 120 min, and the unleached product showed by XRD and EDS analyses contained siliceous impurities that could serve as a by-product for some defined industrial applications. The dissolution data were also subjected into the shrinking core model (SCM) and found to follow the diffusion-controlled reaction. The calculated activation energy, *Ea* was 35.46 kJ/mol supported the proposed reaction mechanism, whose reactions order 1.02 predicts the first-order dissolution relationship.

References

1. Tran TT, Liu Y, Lee MS (2021) Recovery of pure molybdenum and vanadium compounds from spent petroleum catalysts by treatment with ionic liquid solution in the presence of oxidizing agent. Sep Purif Technol 255:117734. https://doi.org/10.1016/j.seppur.2020.117734
2. Frascoli F, Hudson-Edwards K (2018) Geochemistry, mineralogy and microbiology of molybdenum in mining-affected environments. Minerals 8(2):42. https://doi.org/10.3390/min8020042
3. Gao S, Cui X, Kang S, Ding Y (2020) Sustainable applications for utilizing molybdenum tailings in concrete. Elsevier. https://www.sciencedirect.com/science/article/pii/S095965262032069. Last Accessed 22 Apr 2021
4. Ali E, Dariush A, Pavel AN, Faïçal L, Louis-Cesar P (2020) Selective recovery of molybdenum over rhenium from molybdenite flue dust leaching solution using PC88A extractant. Metals 10(11):1423
5. Liu Z, Yin Z, Chen Y, Xiong L (2015) Leaching kinetics of molybdenum from Ni-Mo ore in sulfuric acid solution with sodium peroxodisulfate as oxidant. J Cent South Univ 22:874–879
6. Lasheen TA, El-Ahmady ME, Hassib HB, Helal AS (2013) Oxidative leaching kinetics of molybdenum-uranium ore in H_2SO_4 using H_2O_2 as an oxidizing agent. Front Chem Sci Eng 7:95–102
7. Xie K, Wang H, Wang S (2019) Direct leaching of molybdenum and lead from lean wulfenite raw ore. Trans Nonferrous Metals Soc China 29:2638–2645
8. Kumar M, Mankhand TR, Murthy DSR, Mukhopadhyay R, Prasad PM (2007) Refining of a low-grade molybdenite concentrate. Hydrometallurgy 86(1):56–62
9. Cao ZF, Zhong H, Jiang T, Wang S, Liu G-Y, Xia L-Y (2012) A novel hydrometallurgy of molybdenite concentrate and its kinetics. J Chem Technol Biotechnol 87(7):938–942
10. Pradhan D, Kim D, Sukla LB, Pattanaik A, Lee W (2020) Evaluation of molybdenum recovery from sulfur removed spent catalyst using leaching and solvent extraction. Sci Rep 10:1–14. https://doi.org/10.1038/s41598-020-58972-x
11. Mehdi MG, Bahram R, Anahita S (2014) Leaching and kinetic modelling of molybdenite concentrate using hydrogen peroxide in sulfuric acid solution. Iran J Chem Chem Eng 33(2):73–85

12. Medvedev AS, Aleksandrov PV (2009) Investigations on processing low-grade molybdenum concentrate by the nitric-acid method. Russ J Nonferrous Metals 50:353–358
13. Jha MK, Kumari A, Panda R, Kumar JR, Yoo K, Lee JY (2016) Review on hydrometallurgical recovery of rare earth metals. Hydrometallurgy 165:2–26
14. Ilhan S, Akgun D (2021) Leaching kinetics of Mo, Ni, and Al oxides from spent nickel-molybdenum hydrodesulphurization catalyst in H_2SO_4 solution. J Sustain Metall. 7(2):470–480. https://doi.org/10.1007/s40831-021-00351-5
15. Baba AA, Girigisu S, Raji MA, Ibrahim AS, Ayinla KI, Adeyemi CO, Abdulkareem AY, Abdul MJ, Alabi AG (2021) Preparation of high grade ammonium metatungstate (AMT) as precursor for industrial tungsten catalyst. In: Rare metal technology, part II. Springer International Publishing, New York, pp 73–78

Hydrometallurgical Extraction of Rhenium from Mo-Bearing Leach Liquor of Molybdenite Flue-Dusts

Sadia Ilyas, Hyunjung Kim, Ijaz Ahmad Bhatti, Humma Akram Cheema, and Rajiv R. Srivastava

Abstract In the present study, the hydrometallurgical separation of refractory metals molybdenum and rhenium from the molybdenite flue-dusts processed leach liquor was investigated using the liquid–liquid extraction technique. Tri-butyl phosphate (TBP) applied as the potential organic extractant for separation of refractory metals from leach liquor that containing 11.8 g/L Mo, 208 mg/L Re, 1.1 g/L Fe, 443 mg/L Cu, 675 mg/L Na, and 43.5 mg/L Ca. The extraction behavior of refractory metal ions as a function of equilibrium pH showed that the maximum rhenium could be separated at a pH_{eq} below zero. Extraction equilibrium between immiscible phases was observed to be ($HReO_4 \cdot 3TBP$) species into the organic phase that exhibiting the exothermic nature of spontaneous process. Further, Re-stripping from loaded TBP was quantitatively yielded > 99% recovery at a stripping pH $\geq$3.0.

Keywords Rare metals · Rhenium · Molybdenum · Roasting flue-dust · TBP · Solvent extraction

Introduction

Rhenium is a less common rare metal with its crustal abundance <7 parts per million in the earth crust. Owing to the incomparable characteristics with other refractory

S. Ilyas (✉) · H. Kim
Department of Mineral Resources and Energy Engineering, Jeonbuk National University, Jeonju, Jeonbuk 54896, Republic of Korea
e-mail: sadiailyas1@yahoo.com

H. Kim
Department of Environment and Energy, Jeonbuk National University, Jeonju, Jeonbuk 54896, Republic of Korea

I. A. Bhatti · H. A. Cheema
Department of Chemistry, University of Agriculture, Faisalabad 38040, Pakistan

R. R. Srivastava
Center for Advanced Chemistry, Institute of Research and Development, Duy Tan University, Da Nang 550000, Vietnam

T. Ouchi et al. (eds), *Rare Metal Technology 2022*, The Minerals, Metals & Materials Series, https://doi.org/10.1007/978-3-030-92662-5_14

metals, rhenium is prominently used in petrochemicals, aerospace, and chemical industries [1]. More than 75% rhenium is alone used in the manufacturing of single-crystal (Ni-based) superalloy. Recently, the progress in aerospace industries has soared its demand that resulted in keeping rhenium in list of energy-critical metals [2, 3]. In contrast, rhenium is yet to be recovered from its natural ore rheniite and molybdenite roasting flue-dust is the only source of its commercial extraction as by-product of molybdenum metallurgy. MoO_3 and Re_2O_7 remained in the flue-dusts gets readily hydrolyzed in aqueous solution to form different species, as the reactions given below [4, 5]:

$$Re_2O_7 + H_2O \rightarrow 2ReO_4^- + 2H^+ \quad \Delta G^\circ = -64.4\,\text{kJ/mol} \tag{1}$$

$$ReO_4^- + H_3O^+ \rightarrow HReO_4 + H_2O \quad \Delta G^\circ = -1465.8\,\text{kJ/mol} \tag{2}$$

$$MoO_3 + H_2O \rightarrow H_2MoO_4 \quad \Delta G^\circ = -31.9\,\text{kJ/mol} \tag{3}$$

$$H_2MoO_4 + 2H^+ \rightarrow MoO_2^{2+} + 2H_2O \quad \Delta G^\circ = -2.79\,\text{kJ/mol} \tag{4}$$

Rhenium concentration in scrubbed liquor remains significantly less (30–200 mg/L) than Mo(VI) concentration [6]; however, the high economic value of rhenium over molybdenum is more attractive to the metallurgists in recent time.

Flue-dusts' leaching in water/dil. H_2SO_4 followed by ion exchange and solvent extraction are the major techniques employed to separation of both metals. The higher efficiency, selectivity, and recyclability of the organic extractants have been found to be advantageous over the slow ion exchange process [7, 8]. The solvating extractants (tri-butyl phosphate, TBP; tri-n-octyl phosphine oxide, TOPO; and Cyanex 923) and anionic (amine) extractants (Alamine 336; tri-octyl/alkyl amine; and Aliquat 336) have been widely employed to separate the anionic rhenium (ReO_4^-) over multi-species molybdenum [9, 10]. TBP can be appropriately used to extract both anionic and neutral species through ion-pairing or solvation mechanism, respectively. Less cost, easy-to-handle, and absence of interfacial substance are other associated advantages to use TBP [11–13].

Literature survey reveals that several research works have been reported for liquid–liquid separation of rhenium and molybdenum using TBP albeit from the single metal system. Alamdari et al. [13] performed the separation studies on the basis of pH by selective extraction of molybdenum at 1.2 pH value. Hosseinzadeh et al. [14] reported above 95% extraction in three stages of counter-current contact than the efficiency (87%) obtained at two stages of contact in another study [15]. The analysis of available literature indicated that the extraction efficiency varied with change in impurity profile and the nature of flue-dusts solution. Hence, systematic study from a real leach liquor has been performed in the present study for separating rhenium and molybdenum. The parameters like contact time, equilibrium pH, TBP concentration, organic-to-aqueous volume phase ratio, and temperature were varied to optimize the extraction conditions for their maximum separation. The results presented in this study can be supportive in the future metallurgy of rhenium.

Experimental

The leach liquor obtained after sulfuric acid leaching of molybdenite roasting flue-dust (of composition leach liquor that containing 11.8 g/L Mo, 208 mg/L Re, 1.1 g/L Fe, 443 mg/L Cu, 675 mg/L Na, and 43.5 mg/L Ca) was used as the feed aqueous, while tri-butyl phosphate (TBP, supplied by Sigma Aldrich of purity 99%) was used as the organic extractant and diluted in distilled kerosene. All the extraction studies were performed by equilibrating the aqueous and organic phases for a predetermined time (10 min) and at an O/A ratio 1 at room temperature (30 °C), unless mentioned specifically. After extraction and settling of both phases, the raffinate was properly diluted in 5 wt.% HCl solution to analyze the metals contents using ICP. The concentration of metal ions into organic phase was estimated by mass balance between two phases for before and after extraction, and the extraction efficiency (*%E*) was calculated as follows:

$$\%E = \frac{D\text{x}100}{D + (V_{aq.}/V_{org.})} \quad (5)$$

where $V_{aq.}$ and $V_{org.}$ denote the volumes of aqueous and organic phases, respectively. *D* is the distribution of metal between organic and aqueous phases, can be obtained as

$$D = \frac{\text{Metal concentration in organic phase}}{\text{Metal concentration in aqueous phase}} \quad (6)$$

Results and Discussion

Effect of Extractant Concentration

The extraction efficiency of Re(VII) and Mo(VI) were investigated in the range of 0.2–1.0 mol/L TBP in kerosene. The other parameters were invariably maintained as O/A, 1; pH_{eq}, ~0; temperature, 30 °C; and time, 10 min. As can be seen from Fig. 1a that the extraction efficiency of Re(VII) increases sharply with increasing TBP concentration from 0.2 mol/L to 0.65 mol/L, where Re(VII) extraction get reached to 86.8% with <1% co-extraction of Mo(VI). Above that, the extraction of Re(VII) goes up to 94.2% with 1.0 mol/L TBP though the co-extraction of Mo(VI) was

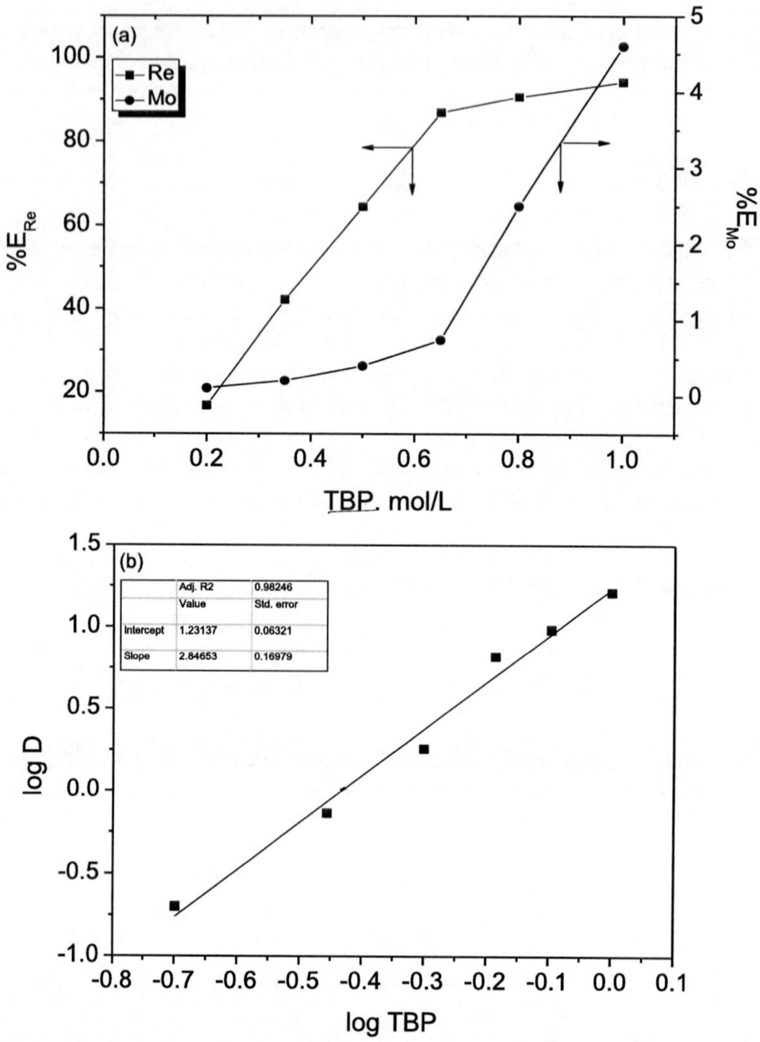

Fig. 1 Extraction of rhenium and molybdenum as a function of TBP concentration in organic phase (**a**). Logarithmic plot for distribution coefficient versus TPB concentration (**b**)

also increased to approximately 5%. Such an extraction trend may cause due to the shift in distribution curve with increasing TBP concentration [16]. It is noteworthy to mention that the co-extraction of other metal impurities in leach liquor was negligible in all the concentration range of TBP used in this study.

For seeking the selective extraction of Re(VII) from Mo(VI), the use of 0.65 mol/L TBP was invariably maintained in further experiments. Furthermore, the distribution of Re(VII) in between the organic and aqueous phases were calculated at each TBP concentration and a graph of log D vs. log TBP was plotted to determine the stoichiometry of the extracted species. The slope value (2.846) of the linear plot shown in Fig. 1b, indicates the association of three moles TBP is required for each mole of Re(VII) extraction in the organic phase.

Effect of pH$_{eq}$

To investigate the effect of equilibrium pH on extraction behavior of Re(VII) and Mo(VI), the experiments were performed in the pH range of (−0.3–3.0). The results shown in Fig. 2a clearly depicts that the separation of Re(VII) and Mo(VI) as a function of pH_{eq} is possible by using TBP as an extractant. It depicts a decreasing trend in Re(VII) extraction with respective increase in pH_{eq}, which declined from 86.7% to 53.4% when pH_{eq} increased from −0.3 to 1.0 and was negligible (<0.5%) at $pH_{eq} = 3.0$. However, the extraction curve of Mo(VI) showed the maximum 84.3% efficiency at $pH_{eq} = 1.5$. At $pH_{eq} > 1.5$ or, < 1.5, the extraction of Mo(VI) declined and found to be 0.7 and 6.4% at $pH_{eq} = -3.0$ and 3.0, respectively.

The different extraction behavior shown by both metals as a function of pH_{eq} can largely be attributed to the changes occur in metal speciation at respective solution pH [5, 9]. At lower pH < 1.0, Mo(VI) exists in cationic form (MoO_2^{2+}) form, whereas Re(VII) exists as anionic ReO_4^- [9]. When the solution pH increased to > 1.0, Re(VII) remains unchanged but Mo(VI) forms neutral H_2MoO_4 species. Hence, the ReO_4^- gets solvated via the ion pair assembly with protonated oxygen atom of TBP under high acid environment, giving separation over Mo(VI) extraction [9]. In contrast at high pH, the neutral H_2MoO_4 competitively gets solvated with TBP, leaving ReO_4^- in the aqueous. The low extraction exhibited by both metals at $pH_{eq} \geq 2.0$ is due to the insufficient acid in solution that of required for protonating the oxygen atom of TBPC to be paired with anionic Re(VII) and Mo(VI) [5]. Henceforth, to achieve a selective extraction of Re(VII) from Mo(VI), the extraction at $pH_{eq} = -0.3$ was found to be the optimal value.

The results of this study was found to be quite similar with using other molybdenite flue-dusts [14], also indicating that vary in metals composition of flue-dusts has a negligible effect on the extraction behavior of Re(VII) and Mo(VI). Furthermore, by calculating the distribution of Re(VII) in between two phases, a graph of log D vs. pH_{eq} was plotted to elucidate the extraction equilibrium. The obtained slope (−1.032) of the linear plot shown in Fig. 2b, indicates the association of proton-to-metal ion on a mole ratio of 1:1.

Effect of Temperature and Extraction Thermodynamics

The extraction reaction of Re(VII) by considering the linear log–log plot of D vs. TBP and pH_{eq} (as shown in Figs. 1b and 2b, respectively) can commonly be given

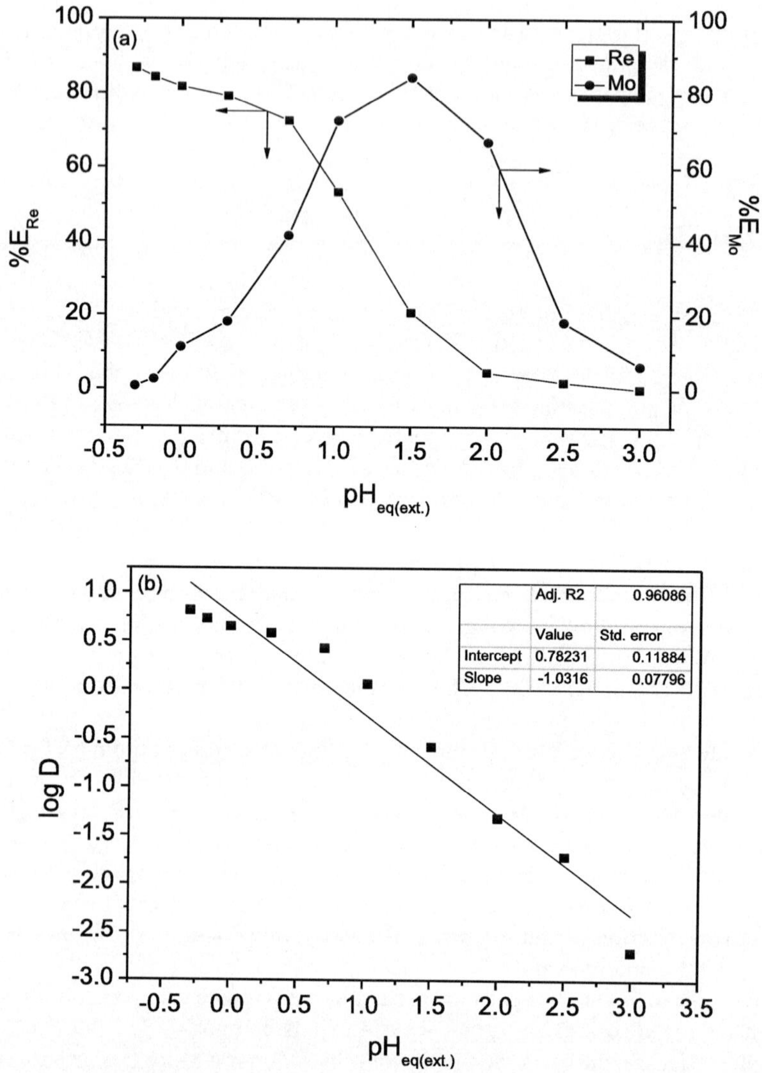

Fig. 2 Extraction behaviour of rhenium and molybdenum as a function of equilibrium pH (**a**). Logarithmic plot for distribution coefficient vs. equilibrium pH of extraction (**b**)

as

$$H^{+}{}_{aq.} + ReO_{4}^{-}{}_{aq.} + 3TBP_{org.} \rightarrow [HReO_4 \cdot 3TBP]_{org.} \tag{7}$$

The equilibrium constant (K) for the above equation can be determined as

$$K = \frac{[\mathrm{HReO_4} \cdot 3\mathrm{TBP}]_{org.}}{[\mathrm{H^+}]_{aq.} \cdot [\mathrm{ReO_4^-}]_{aq.} \cdot [\mathrm{TBP}]^3_{org.}} \tag{8}$$

By substituting D value from Eq. (6) into Eq. (8), the above equation can be written as

$$K = \frac{D}{[\mathrm{H^+}]_{aq.} \cdot [\mathrm{TBP}]^3_{org.}} \tag{9}$$

Rearranging and taking logarithm of Eq. (9) comes as

$$\log D = \log K + 3\log[\mathrm{TBP}] - \mathrm{pH} \tag{10}$$

Furthermore, considering the van't Hoff equation, the equilibrium constant as a function of thermodynamic factors can be written as [12]

$$\log k = -\frac{\Delta H^\circ}{2.303\mathrm{R}T} + \frac{\Delta S^\circ}{2.303\mathrm{R}} \tag{11}$$

By substituting Eq. (11) into Eq. (10), it comes as

$$\log D = -\frac{\Delta H^\circ}{2.303\mathrm{R}T} + \frac{\Delta S^\circ}{2.303\mathrm{R}} + 3\log[\mathrm{TBP}]_{org.} - \mathrm{pH} \tag{12}$$

Hence, at a constant TBP concentration and pH_{eq}, the distribution of Re(VII) as a function of extraction temperature, temperature can be obtained and a graph for log D vs. $1/T$ can be plotted to determine the extraction thermodynamics. For this, a set of experiment was performed by varying temperature in the range of 10–50 °C, while keeping other parameters constant as: 0.65 mol/L TBP in kerosene, O/A = 1, pH_{eq} = −0.3, and time = 10 min. It was observed that an increase in temperature from 10 to 50 °C led to a decreased Re(VII) extraction from 90.2% to ~77%. The slope and intercept values obtained from the linear plot of log D versus ($1/T \times 10^{-3}$) in Fig. 3 was used for calculating the thermodynamic factors, yielding $\Delta H^\circ = -18.15$ kJ/mol and $\Delta S^\circ = -45.41$ J/K/mol. The negative values of entropy and enthalpy indicated for an increase in order of extracted species during the exothermic extraction process. Using the values of entropy and enthalpy, the change in standard Gibbs free energy (ΔG°) can be calculated as

$$\Delta G^\circ = \Delta H^\circ - T\Delta S^\circ \tag{13}$$

The obtained negative values of $\Delta G^\circ_{(298\,\mathrm{K})}$ = 4.62 kJ/mol clearly shows the spontaneity of Re(VII) extraction with TBP.

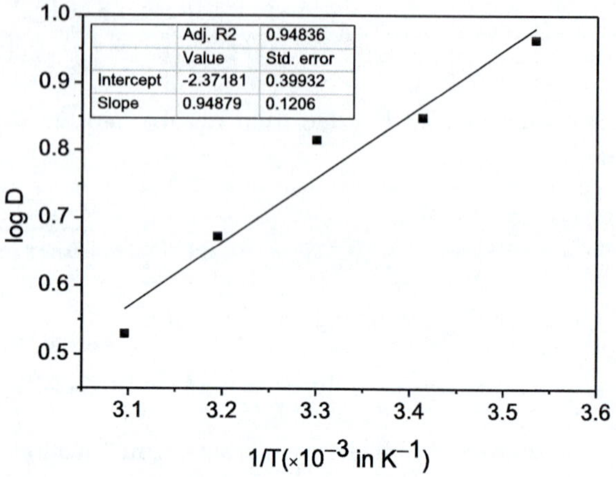

Fig. 3 Plot for rhenium distribution as a function of temperature (log *D* versus 1/*T*)

Effect of Volume Phase Ratio and Extraction Isotherm

The effect of phase ratio (O/A) on Re(VII) extraction from aqueous solution with 0.65 mol/L TBP was investigated by varying O/A in the range of 1/3–3/1. Results showed an increased extraction of Re(VII) with increasing phase ratios. The extraction efficiency could be achieved from 33.4 to 99.6% when the corresponding aqueous feeds were contacted by varying O/A from 1/3 to 3/1. Further, the theoretical number of stages for maximum Re(VII) extraction at a chosen phase ratio was determined by plotting the McCabe–Thiele diagram (shown in Fig. 4a). A vertical line drawn from Re(VII) concentration in aqueous feed at a chosen operating line, the obtained slope value represents the O/A phase ratio to be used in extraction. Figure 4a revealed that three theoretical stages in counter-current are needed for quantitative extraction of Re(VII) using 0.65 mol/L TBP in kerosene at an O/A = 1. A batch simulation of three stages counter-current extraction carried out under this condition is shown in Fig. 4b. As can be seen that the distribution of Re(VII) in both phases are very close to the theoretical values, validating the predicted isotherm given by the McCabe–Thiele plot in Fig. 4a.

Stripping of Rhenium from Loaded Organic

To reuse the organic phase and recovery of Re(VII) from loaded organic, the stripping behaviour of rhenium was investigated as a function of pH. Thus, a loaded organic containing 205.6 mg/L Re(VII) (obtained from the counter-current extraction stage) was used for testing the stripping behaviour as a function of pH. The results shown in Fig. 5 depicts that the stripping efficiency of Re(VII) was lower than 18.2% when

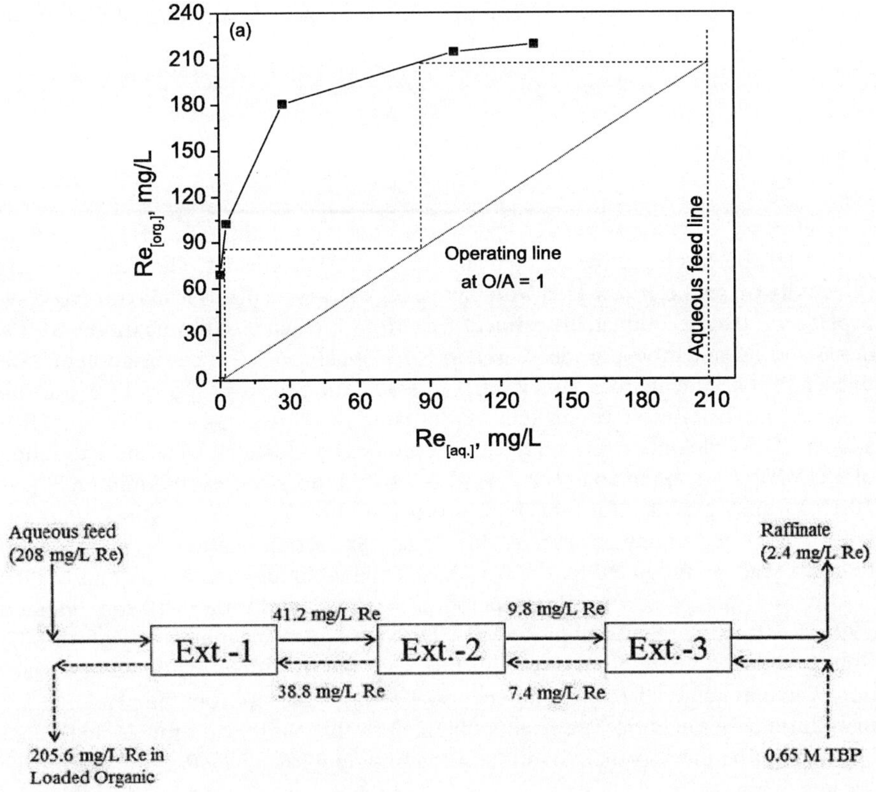

Fig. 4 McCabe–Thiele plot for rhenium extraction as a function of phase ratio variation (**a**). Counter-current batch simulation of rhenium extraction in steady state (**b**)

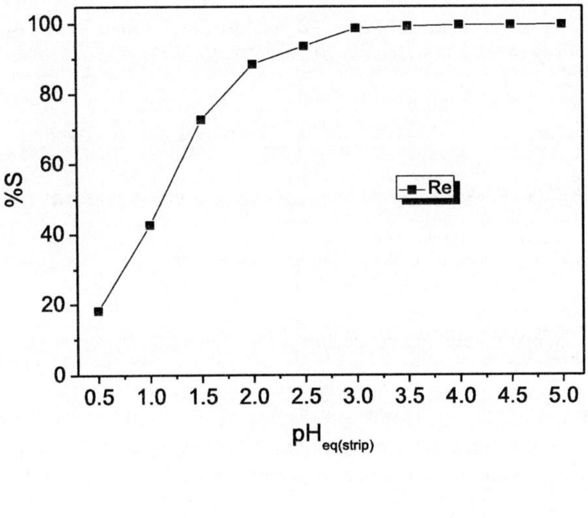

Fig. 5 Stripping behaviour of rhenium from loaded TBP as a function of equilibrium pH

pH was 0.5, which surpassed 50% at pH >1.0 and reached more than 99% with pH 3.0–5.0.

Conclusions

The selective recovery of Re(VII) from the wet-scrub solution of molybdenite flue-dust with TBP has been investigated and found to be promising. The extraction and selectivity of Re(VII) decreased with increasing extraction pH, was maximum ~87% at $pH_{eq} = -0.3$. In contrast, the extraction of Mo(VI) reached to the maximum 84.3% at $pH_{eq} = 1.5$ and thereafter decreased to 6.4% at $pH_{eq} = 3.0$. The variation of TBP concentration indicated for stoichiometric association of three moles of extractant with each mole of Re(VII) extracted in the organic phase, forming the $[HReO_4 \cdot 3TBP]$ adduct. The calculated thermodynamic parameters indicated that the extraction of Re(VII) is a spontaneous ($\Delta G°_{298\ K}$, −4.62 kJ/mol) and exothermic ($\Delta H°$, −18.15 kJ/mol) process, forming highly ordered complex ($\Delta S°$, −45.41 J/K/mol) with TBP in the organic phase. A three-stage extraction isotherm was simulated in batch scale as predicted by the McCabe–Thiele diagram using 0.65 mol/L TBP at O/A = 1, $pH_{eq(ext.)} = -0.3$, temperature = 30 °C, and time = 10 min, yielded >99.6% extraction efficiency of Re(VII). The subsequent stripping study of Re(VII) from the loaded organic phase was in-line with the pH effect exhibited in extraction. The maximum stripping efficiency >99% was achieved from the pH 3.0 to 5.0 in sulfuric acid medium. The results obtained in this study can provide additional information on the solvent extraction of rhenium employing TBP as the prominent organic extractant.

Acknowledgements This work was supported by the Brain Pool Program through the National Research Foundation of Korea (NRF) funded by the Ministry of Science and ICT (Grant No. 2019H1D3A2A02101993 and 2021H1D3A2A01100016) and Basic Science Research Program through the National Research Foundation of Korea (NRF) funded by the Ministry of Education (Project no. 2020R1I1A1A01074249).

References

1. Anderson CD, Taylor PR, Anderson CG (2013) Extractive metallurgy of rhenium: a review. Miner Metall Proc 30:59–73
2. Srivastava RR, Kim M, Lee J, Jha MK, Kim BS (2014) Resource recycling of superalloys and hydrometallurgical challenges. J Mater Sci 49:4671–4686
3. Ilyas S, Kim M, Lee J, Jabeen A, Bhatti HN (2017) Bio-reclamation of strategic and energy critical metals from secondary resources. Metals 7:207. https://doi.org/10.3390/met7060207
4. Habashi F (2010) Rhenium: chemistry, metallurgy, and history. Invest Apl Innov 4:121–127
5. Srivastava RR, Kim M, Lee J (2016) Novel aqueous processing of the reverted turbine-blade superalloy for rhenium recovery. Ind Eng Chem Res 55:8191–8199

6. Abisheva ZS, Zagorodnyaya AN, Bekturganov NS (2011) Review of technologies for rhenium recovery from mineral raw materials in Kazakhstan. Hydrometallurgy 109:1–8
7. Zhan-fang C, Hong Z, Zhao-hui Q (2009) Solvent extraction of rhenium from molybdenum in alkaline solution. Hydrometallurgy 97:153–157
8. Srivastava RR, Kim M, Lee J, Ilyas S (2015) Liquid-liquid extraction of rhenium(VII) from an acidic chloride solution using Cyanex 923. Hydrometallurgy 157:33–38
9. Srivastava RR, Lee J, Kim M (2015) Complexation chemistry in liquid-liquid extraction of rhenium. J Chem Technol Biotechnol 90:1752–1764
10. Srivastava RR et al (2015) Rhenium separation using solvent extraction techniques. In: Paper presented at 104th annual meeting of The Korean society of mineral and energy resource, Daejeon, South Korea
11. Muhsan MA, Ilyas S, Cheema HA, Masud S, Shabbir N (2017) Recovery of nitric acid from effluent streams using solvent extraction with TBP: a comparative study in absence and presence of metal nitrates. Sep Purif Technol 186:90–95
12. Zagorodnyaya AN, Abisheva ZS, Sadykanova SE, Bobrova VV, Sharipova AS (2010) The characterization and origins of interphase substances (cruds) in the rhenium solvent extraction circuit of a copper smelter. Hydrometallurgy 104:308–312
13. Alamdari EK, Darvishi D, Haghshenas DF, Yousefi N, Sadrnezhaad SK (2012) Separation of Re and Mo from roasting-dust leach liquor using solvent extraction technique by TBP. Sep Purif Technol 86:143–148
14. Hosseinzadeh M, Alizadeh M, Ranjbar M (2014) Optimization of Re solvent extraction from molybdenite roasting dust leaching solution and the performance evaluation of extraction in a multi-stage mixer-settler. Int J Miner Process 130:88–94
15. Hosseinzadeh M, Ranjbar M, Alizadeh M (2014) Effect of operational parameters and internal recycle on rhenium solvent extraction from leach liquors using a mixer-settler. ESTIJ 17:45–49
16. Ritcey GM, Ashbrook AW (1984) Process metallurgy 1: solvent extraction Part I. Elsevier, Amsterdam

Extraction for Neodymium from NdFeB Magnet Using Supercritical Carbon Dioxide and Organophosphorus Ligands

Jiakai Zhang, Nattanai Kunanusont, Yusuke Shimoyama, and Gisele Azimi

Abstract Supercritical carbon dioxide (sc-CO_2) is an emerging green solvent for recovery of metals from secondary resources. The sc-CO_2 is a non-polar solvent; thus, a suitable chelating agent is required to accommodate the polarity difference between the solute and solvent during the extraction of metal ions with organometallic compounds. This study investigates four different organophosphorus ligands including triethyl phosphate (TEP), tri-n-butyl phosphate (TBP), tributyl phosphine oxide (TBPO), and trioctyl phosphine oxide (TOPO), and their effect on the extraction of neodymium from NdFeB magnet in sc-CO_2. A COSMO-vac model is developed to estimate solubility of the four agents in sc-CO_2. The results show that the order of solubility is TEP > TBPO ~ TBP > TOPO. The coordination environment of neodymium was determined by UV–Vis spectroscopy, resulting a TEP:Nd = 1, TBP:Nd = 3, TBPO:Nd = 3, and TOPO:Nd = 4 coordination chemistry. The highest extraction for neodymium is achieved with TEP, because of low coordination number which results in less hydrophobic interactions between aliphatic functionalities. Also, the smaller micellar assemblies have a higher solubility in sc-CO_2; thus, a higher extraction efficiency is achieved.

Keywords NdFeB magnet · Supercritical fluid extraction · Rare earth element · Neodymium · COSMO-vac (vacancy) model

Introduction

Over the past few years, there has been an increasing interest on the application of supercritical fluids (SCFs) in a wide variety of industries including chemical, biochemical, food, petroleum, extraction, separation, pharmaceutical, and materials preparation because of their roles as green solvents.

J. Zhang · G. Azimi (✉)
University of Toronto, 200 College Street, Toronto, ON M5S35, Canada
e-mail: g.azimi@utoronto.ca

N. Kunanusont · Y. Shimoyama
Tokyo Institute of Technology, S1-33, 2-12-1, Ookayama, Meguro-ku, Tokyo 152-8550, Japan

T. Ouchi et al. (eds), *Rare Metal Technology 2022*, The Minerals, Metals & Materials Series, https://doi.org/10.1007/978-3-030-92662-5_15

Supercritical carbon dioxide (sc-CO_2) is among the most widely used SCFs, [1–4] as it offers the advantages of moderate critical points, safety, abundance, low cost, and ease of recycling. The sc-CO_2 is a non-polar solvent, so there is a polarity difference across solute (metal ions, adducts) and solvent (sc-CO_2). For narrowing the difference and satisfying charge neutrality, complexing agents (chelating agents) with high solubility in sc-CO_2 in both their complexed and non-complexed form are required for improvement of solvent–solute interactions. Hence, the solubility data of complexing agents and organometallic compounds in sc-CO_2 are essential for examining their feasibility, establishing optimum operating conditions, and designing of supercritical fluid extraction (SCFE) processes. Moreover, mathematical modeling of complexing agents and organometallic compounds can provide a better understanding of the dissolution phenomenon. Solubility modelling can also provide a method to estimate solubility under different operating conditions experimental data is not available, which can facilitate the development and progress rate of an SCFE process.

Many previous studies have used equation of state, solution models, and empirical models to estimate or correlate the solubility of solutes in supercritical CO_2, and they optimized the model parameters using experimental data [5]. This limits the application of the model to the systems for which solubility data are available. In addition, using equation of state method requires the critical properties and the sublimation pressure of solutes. The solubility and critical properties data of organometallic compounds are limited. Activity coefficient models based on conductor-like screening model (COSMO) can be used to model thermodynamics of mixtures without data fitting [5]. A previous study has successfully developed a COSMO-based model for a system composed of the surface segments on the solvent molecule and vacancy. This model can predict the solubility of solutes in sc-CO_2 at high pressure (15 MPa) [5].

There are two main factors that impact solubility of compounds in sc-CO_2, so it can also affect extraction efficiency of metals from feed materials. First, the density of sc-CO_2 can affect solubility since it determines the free volume difference between the solute and the solvent. With increasing density of sc-CO_2, the solubility of solute increases according to Chrastil model [6]. Second, the nature and structure of the metal and ligand species can also affect the solubility of solute. Previous studies reported that the solubility of the metal–ligand complex in sc-CO_2 increases with oxidation state of the central metal ion and shorter metal–ligand distances [7, 8]. In addition, the solubility of free ligands affects physical and chemical interactions between the solute and the solvent, which also affect the solubility of the metal–ligand complex in sc-CO_2. Hence, the structure and property of free ligand is important in predicting solubility of metal–ligand complexes in sc-CO_2 and extraction efficiency. Four organophosphorus compounds including triethyl phosphate (TEP), tributyl phosphate (TBP), tributyl phosphine oxide (TBPO), and trioctyl phosphine oxide (TOPO) in terms of solubility and extraction were tested. These adducts are widely used in sc-CO_2 extraction of REEs from various sources [2–4, 9–15].

In this study, the COSMO-vac (vacancy) model is used to predict the solubility of organophosphorus adducts in sc-CO_2. The effects of structure and property of organophosphorus ligands on solubility in sc-CO_2 and extraction efficiency are investigated. The study is focused on the extraction of neodymium (Nd) from neodymium–iron–boron (NdFeB) magnet using organophosphorus adducts in sc-CO_2. Moreover, the coordination chemistry of Nd and ligands are determined using UV–Vis spectroscopy. The synergistic effect of organophosphorus ligand solubility in sc-CO_2 and Nd–ligand complex chemistry on the extraction efficiency are discussed.

Materials and Methods

Materials

Chemicals used in this study are following: tri-n-butyl phosphate (TBP, 98%, Alfa Aesar), tributyl phosphine oxide (TBPO, 90%, Millipore Sigma), triethyl phosphate (TEP, 99.8%, Sigma Aldrich), trioctyl phosphine oxide (TOPO, 98%, Alfa Aesar), neodymium (III) nitrate hexahydrate ($Nd(NO_3)_3 \cdot 6H_2O$, 99.9%, Alfa Aesar), acetone (ACS grade, Fisher Chemical), hexane (ACS grade, Fisher Chemical), nitric acid (HNO_3, 70 wt%, VWR), hydrochloric acid (HCl, 37 wt%, VWR), and carbon dioxide (CO_2, grade 4.0, Linde Canada). The NdFeB magnet was acquired from CMS magnetics. The magnet was first demagnetized in a box furnace manufactured by CARBOLITE® at 350 °C for 1 h. The surface coating (nickel–copper–nickel triple layers) of the magnet was removed with a rotary tool (Dremel 4000). The magnet was crushed and grinded in a roller mill into powder.

Preparation of Ligand–HNO_3 Adduct

Each one of organophosphorus compounds was mixed with 70 wt% nitric acid. For TEP and TBP, the mixing ratios by volume were TEP/HNO_3 = 2 and TBP/HNO_3 = 1. For TBPO and TOPO, the mixing ratios were corresponding to weight (g) to volume (mL), which both were one. The mixtures were agitated for 5 min, then the top liquid layer was collected after settling for 10 min.

SCFE Process

Experiments were conducted in a high-pressure reactor system by Supercritical Fluid Technology Inc., USA. The magnet powder and adducts were added first, then liquid CO_2 was pumped into the system until desired pressure and temperature reached. After extraction, metal–ligand complexes (organometallic compounds) carried by sc-CO_2 to a collection vessel, where CO_2 becomes a gas due to depressurization and organometallic compounds can be separated and collected. More details can be found in a previous study [2].

Calculation for Extraction Efficiency

Elemental composition of magnet particles before and after extraction was measured by dissolving them in aqua-regia solution in a microwave (MARS6 Xpress) system at 200 °C and followed by Inductively Coupled Plasma Optical Emission Spectrometry (ICP-OES Perkin Elmer Optima 8000). Detailed characterization of the magnet can be found in a previous study [2]. The extraction efficiency, *E*, can be calculated as following:

$$E(\%) = \frac{\Delta C}{C_i} 100\% \tag{1}$$

where ΔC is the concentration difference between the sample before and after extraction, and C_i is the concentration of unextracted sample.

UV–Vis Spectroscopy Measurement for Coordination Chemistry

For measuring the coordination number within a neodymium–ligand complex, a series of solutions with various mole ratios of ligands (TEP, TBP, TBPO, TOPO) to neodymium nitrate was prepared, and liquid hexane was the chosen solvent to mimic the non-polar environment in sc-CO_2 system. Mole ratios varies from 0.1 to 9. Each sample was analyzed by a UV–Vis spectroscopy (Cary 300), and UV–Vis spectra were collected in the range of 500–700 nm with a scan rate of 2 nm s^{-1}.

Calculation for Solubility of Organophosphorus Compounds in sc-CO_2

As extraction temperature of the process is at 55 °C, TEP, TBP, and TOPO are liquid and TBPO is solid at this temperature according to their melting points. Hence, liquid–liquid equilibrium can be used for calculating solubility of TEP, TBP, and TOPO in sc-CO_2. Solid–liquid equilibrium can be used for calculating solubility of TBPO in sc-CO_2.

For liquid–liquid equilibrium, the two phases are the mixture of ligand and sc-CO_2 and the ligand-rich phase, which can be represent as following:

$$x_2^m y_2^m = x_2^l y_2^l \tag{2}$$

where x_2^m and x_2^l are the mole fraction of the mixture (m) and the ligand-rich phase (l), and y_2^m and y_2^l are the activity coefficient of the ligand in the mixture (m) and the ligand-rich phase (l), respectively. Assume the mole fraction of sc-CO_2 in the ligand-rich phase is nearly zero, so $x_2^m y_2^m$ is about one. Therefore, the mole fraction of the ligand in sc-CO_2 can be calculated as following:

$$x_2^l = \frac{1}{y_2^l} \tag{3}$$

Solid–liquid equilibrium can be calculated as following:

$$x_2 = \frac{1}{\gamma_2}\left(\frac{f_2^{\circ S}}{f_2^{\circ L}}\right) \tag{4}$$

where x_2 and γ_2 are the mole fraction and the activity coefficient of the solid solute in the liquid phase, and $f_2^{\circ S}$ and $f_2^{\circ L}$ are the fugacity of the pure solute in the solid and liquid phases at reference state, respectively.

The ratio of the fugacity can be determined by

$$\ln\frac{f_2^{\circ S}}{f_2^{\circ L}} = \frac{\Delta H_2^m}{RT}\left(\frac{T}{T_2^m} - 1\right) \tag{5}$$

where R is the universal gas constant, T is the absolute temperature, ΔH_2^m is enthalpy of fusion of solid solute, and T_2^m is the melting temperature of solid solute. In this work, ΔH_2^m of TBPO was estimated from group contribution method, [16] and it was estimated to be 32.02 kJ mol^{-1}.

COSMO-vac Model

In this work, we used a COSMO-vac model developed in a previous study [5] to calculated activity coefficient of ligands in sc-CO_2. As shown in Fig. 1, the geometries of ligand and CO_2 were first calculated using TmoleX software based on density functional (DFT) theory. Then, surface charge density was used to derive the sigma profile, molecular volume, and molecular surface area of CO_2. The density of sc-CO_2 was calculated based on temperature and pressure by an equation of state method [16]. By entering all parameters, solubility of the ligand in sc-CO_2 was calculated via a loop calculation using Visual Basic in MS Excel.

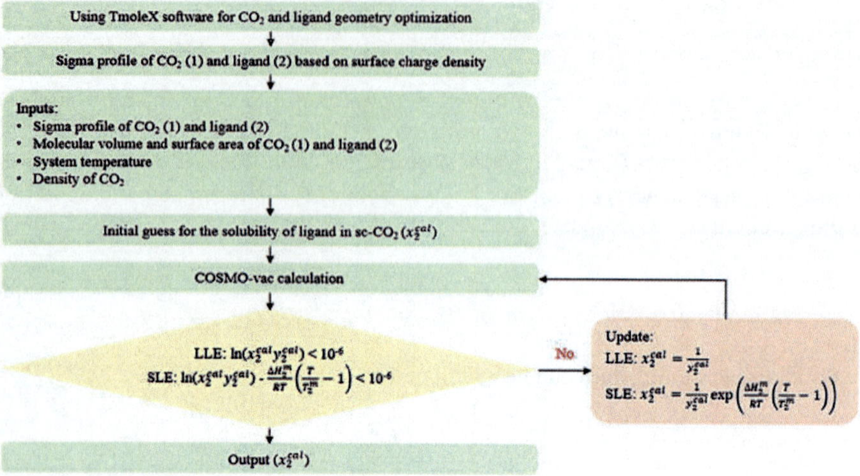

Fig. 1 Flowchart of solubility calculation process for the ligand in sc-CO_2

Results and Discussion

Solubility of Organophosphorus Ligands in sc-CO_2

The solubility of TEP, TBP, TBPO, and TOPO in sc-CO_2 was estimated using COSMO-vac model. The results were compared against literature values showing in Table 1, which summarizes the average deviation of logarithms of the ligand solubilities between the experimental and estimated results from the COSMO-vac model. The average deviation was calculated as following:

$$\Delta \log(x) = \left|\log x_2^{\text{exp}} - \log x_2^{\text{calc}}\right| \tag{6}$$

where x_2^{exp} and x_2^{calc} are the solubilities of literature values and estimated values from the COSMO-vac model, respectively.

As can be seen, all the average deviations are less than unity, which means the predicted results are within the order of solubility to that of the experimental results. The deviation is reasonable, and literature indicates that the COSMO-vac model is capable of estimating the solubility of these ligands in sc-CO_2 [16].

After validation of constructed COSMO-vac model, solubilities of TEP, TBP, TBPO, and TOPO in sc-CO_2 were estimated at conditions (55 °C and 10.0–30.0 MPa) of SCFE for Nd from NdFeB magnet. The estimation results show that the order of solubility is TEP > TBP ~ TBPO > TOPO. The high solubility of TEP is because of its short alkyl chains which facilitates boning with sc-CO_2 solvent [17]. Ligands can interact not only with sc-CO_2 but also can interact with each other via van der Waals

Table 1 The difference in predicted results by COSMO-vac model and literature values of solubilities of ligands in sc-CO_2

Solute	Temperature (°C)	Pressure (MPa)	$\Delta\log(x)$	Data source
Triethyl phosphate (TEP)	40	10.3	0.39	[18]
	40	12.3	0.42	
	40	15.0	0.44	
	40	20.0	0.41	
	40	25.0	0.53	
	60	14.0	0.41	
	60	16.0	0.41	
	60	18.0	0.42	
	60	20.0	0.43	
	60	25.0	0.45	
Tributyl phosphate (TBP)	60	15.0	0.37	[19, 18]
	60	20.0	0.01	
	60	25.0	0.13	
	40	10.3	0.35	
	40	12.3	0.18	
	40	15.0	0.07	
	40	20.0	0.05	
	40	25.0	0.13	
	60	14.0	0.61	
	60	16.0	0.39	
	60	18.0	0.27	
	60	20.0	0.19	
	60	25.0	0.06	
Tributyl phosphine oxide (TBPO)	60	20.0	0.10	[20]
Trioctylphosphine oxide (TOPO)	40	15.0	0.49	[20, 21]
	45	20.0	0.58	
	37	11.1	0.30	
	37	14.0	0.31	
	37	20.3	0.26	
	37	11.1	0.10	
	47	14.0	0.17	
	47	20.3	0.20	

forces; the longer alkyl chain and the larger the size, the stronger the van der Waals forces; thus, their interactions with sc-CO_2 decreases. Hence, TBP and TBPO have similar solubilities as they both have three butyl chains. The TOPO has the lowest solubility due its three long octyl chains.

Coordination Number Measurement by UV–Vis Spectroscopy

According to a previous study, complexation stoichiometry for REEs and organophosphorus ligands are similar in both hexane solvent and sc-CO_2 solvent [22]. Neodymium nitrate and organophosphorus ligands were added to hexane, a chemical reaction [7] can occur and form Nd-ligand complex which is soluble in non-polar solvent such as hexane and sc-CO_2. The complex has the *f-f* electronic transition in the ultraviolet, visible and near infrared regions of the electromagnetic spectrum. Since free neodymium metal and neodymium nitrate are not soluble in non-polar solvents, the absorbance measured by UV–Vis spectrophotometer is because of the absorption by the complex dissolved in hexane. The higher the mole ratios of ligand to neodymium, the higher the concentration of the corresponding complex which increases the absorbance at 585 nm (peak position) [22]. However, once the mole ratio of ligand to neodymium is higher than the coordination number, no additional complexes can be formed. Therefore, the absorbance reaches a plateau, and coordination number of the ligand to neodymium can be determined.

$$\mathrm{REE}^{3+} + 3(\mathrm{NO}_3)^- + \mathrm{xTBP} = \mathrm{REE}(\mathrm{NO}_3)_3\mathrm{TBP}_x \tag{7}$$

where x is the number of TBP molecules.

According to the UV–Vis results, coordination numbers of TEP to Nd, TBP to Nd, TBPO to Nd, and TOPO to Nd are one, three, three, and four, respectively. The results match with other studies [2].

Extraction of Neodymium from NdFeB Magnet

All supercritical fluid extraction experiments were conducted under following conditions: 55 °C, 20.7 MPa, 1 h duration, adduct to solid (CA/S) ratio of 10 mL/g, and 1500 rpm agitation. The difference between experiments was the adduct type. The experimental conditions were chosen based on the previous work [2], which was set as the baseline in this work. The extraction results are illustrated in Fig. 2, the extraction efficiency from the highest to the lowest for REEs were achieved using TEP–HNO_3, TBP–HNO_3, TBPO–HNO_3, and TOPO–HNO_3 adducts, respectively.

This trend can be explained by the following mechanism. In order to dissolve a metal in sc-CO_2, it must be charge neutral and coordinatively satisfied. Charge

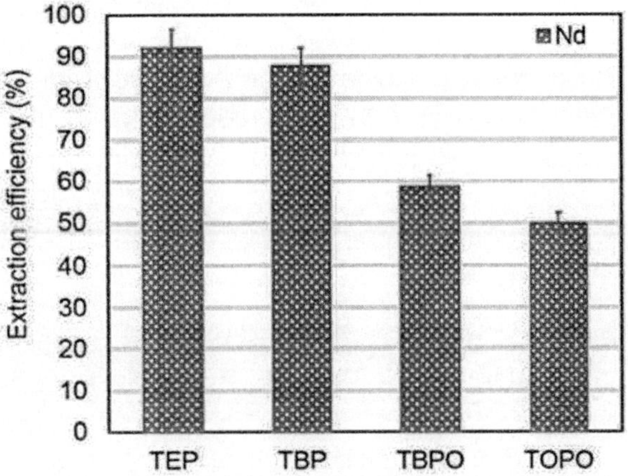

Fig. 2 Extraction efficiency of neodymium from NdFeB magnet

neutrality can be satisfied by having three bidentate nitrate anions from ligand–HNO_3 adduct to surround one REE^{3+} ion, forming neutral lanthanide nitrate, $Ln(NO_3)_3$. Phosphate or phosphine oxide groups of the ligand are more polar than aliphatic tails, so the groups point inward and coordinate with neutral lanthanide nitrates, and the aliphatic tails point outward forming non-polar clusters in non-polar sc-CO_2. The higher the coordination number, the more ligands in the complex which can cause more hydrophobic interactions between aliphatic functional groups, leading to a larger micellar assembly with lower solubility in sc-CO_2. Hence, by considering the combination effect of both solubility of ligands and structure of complexes, the solubility of complexes can be expected to follow the order of $Nd(NO_3)_3(TEP)_1$ > $Nd(NO_3)_3(TBP)_3$ > $Nd(NO_3)_3(TBPO)_3$ > $Nd\ (NO_3)_3(TOPO)_4$, which is consistent with the observed extraction results.

Conclusions

The COSMO-vac model was developed to predict the solubility of organophosphorus ligands including TEP, TBP, TBPO, and TOPO in sc-CO_2, and the deviation of estimated results from experimental results are acceptable. The predicted solubility follows the order of TEP > TBP ~ TBPO > TOPO. This trend is because of the length of alkyl chains; the longer the chain, the lower the solubility of the ligand. Coordination numbers were determined using UV–Vis spectroscopy, which are TEP/Nd = 1, TBP/Nd = 3, TBPO/Nd = 3, and TOPO/Nd = 4. Moreover, the effect of the four ligands on the extraction for neodymium from NdFeB magnet using sc-CO_2 as a solvent was studied by conducting extraction at 55 °C, 20.7 MPa, 1 h, 1500 rpm, and adduct to solid ratio of 10 mL/g. The results show that TEP–HNO_3 adduct results in the highest extraction, followed by TBP–HNO_3, TBPO–HNO_3, and TOPO–HNO_3. This is because of attraction between aliphatic functional groups formed between hydrophobic aliphatic tails and hydrophilic neodymium nitrate polar core, leading to highly ordered structures. The highly ordered structures increase the number of ligands resulting larger micellar assemblies with lower solubility in sc-CO_2. The results of this study provide a method for selecting ligands for SCFE of REEs by considering their solubility and metal–ligand complex structures.

References

1. Zhang J, Zhang F, Azimi G (2018) Efficient recovery of neodymium from Neodymium–Iron–Boron Magnet BT. In: Davis BR et al (eds) Extraction 2018, 2609–2617. Springer International Publishing
2. Zhang J, Anawati J, Yao Y, Azimi G (2018) Aeriometallurgical extraction of rare earth elements from a NdFeB magnet utilizing supercritical fluids. ACS Sustain Chem Eng 6:16713–16725
3. Yao Y, Nina F, Farac GA (2017) Supercritical fluid extraction of rare earth elements from nickel metal hydride battery. ACS Sustain Chem Eng. https://doi.org/10.1021/acssuschemeng.7b03803
4. Shimizu R, Sawada K, Enokida Y, Yamamoto I (2005) Supercritical fluid extraction of rare earth elements from luminescent material in waste fluorescent lamps. J Supercrit Fluids 33:235–241
5. Shimoyama Y, Iwai Y (2009) Development of activity coefficient model based on COSMO method for prediction of solubilities of solid solutes in supercritical carbon dioxide. J Supercrit Fluids 50:210–217
6. Chrastill, J (1982) Solubility of solids and liquids in supercritical fluids. J Phys Chem, 3016–3021, Wiley. https://doi.org/10.1002/0470867833.ch13
7. Aschenbrenner O, Kemper S, Dahmen N, Schaber K, Dinjus E (2007) Solubility of β-diketonates, cyclopentadienyls, and cyclooctadiene complexes with various metals in supercritical carbon dioxide. J Supercrit Fluids 41:179–186
8. Saito N, Ikushima Y, Goto T (1990) Liquid/solid extraction of acetylacetone chelates with supercritical carbon dioxide. Bull Chem Soc Jpn 63:1532–1534
9. Tomioka O, Enokida Y, Yamamoto I (2013) Solvent extraction of lanthanides from their oxides with TBP in supercritical carbon dioxide. 日本畜産学会報 **84**, 487–492
10. Fox RV, Ball RD, Harrington PDB, Rollins HW, Wai CM (2005) Holmium nitrate complexation with tri-n-butyl phosphate in supercritical carbon dioxide. J Supercrit Fluids 36:137–144
11. Vincent T, Mukhopadhyay M, Wattal PK (2009) Direct in situ supercritical fluid extraction of neodymium ion from its oxide using thenoyl tri fluoro acetone-tri butyl phosphate-methanol in carbon dioxide. J Supercrit Fluids 48:230–237
12. Duan W, Cao P, Zhu Y (2010) Extraction of rare earth elements from their oxides using organophosphorus reagent complexes with HNO_3 and H_2O in supercritical CO_2. J Rare Earths 28:221–226
13. Baek DL et al (2016) Extraction of rare earth oxides using supercritical carbon dioxide modified with Tri-n-Butyl phosphate-nitric acid adducts. Ind Eng Chem Res 55:7154–7163
14. Batorshin GS et al (2015) Recovery of rare earth elements, uranium, and thorium from monazite concentrate by supercritical fluid extraction. Radiochemistry 57:343–347
15. Sinclair LK (2019) Supercritical extraction of lanthanide tributyl phosphate complexes : current status and future directions
16. Jain A, Yang G, Yalkowsky SH (2004) Estimation of melting points of organic compounds. Ind Eng Chem Res 43:7618–7621
17. Bouali S et al (2019) Amidophosphonate ligands as cerium extractants in supercritical CO_2. J Supercrit Fluids 149:64–69
18. Pitchaiah KC, Sivaraman N, Lamba N, Madras G (2016) Experimental determination and model correlation for the solubilities of trialkyl phosphates in supercritical carbon dioxide. RSC Adv 6:51286–51295

19. Meguro Y, Iso S, Sasaki T, Yoshida Z (1998) Solubility of organophosphorus metal extractants in supercritical carbon dioxide. Anal Chem 70:774–779
20. Smart NG et al (1997) Solubility of chelating agents and metal-containing compounds in supercritical fluid carbon dioxide. Talanta 44:137–150
21. Schmitt WJ, Reid RC (1988) The solubility of paraffinic hydrocarbons and their derivatives in supercritical carbon dioxide. Chem Eng Commun 64:155–176
22. Fox RV et al (2004) Praseodymium nitrate and neodymium nitrate complexation with organophosphorus reagents in supercritical carbon dioxide solvent. J Supercrit Fluids 31:273–286

Part III
New Extractant and Biometallurgical Processing

Development of New Extractants for Platinum Group Metal Ions in Hydrochloric Acid Solutions

Hirokazu Narita

Abstract Solvent extraction is widely used for the separation and purification of platinum group metal ions. This paper describes our development of two series of novel extractants for Pd(II) and Rh(III) chloridometalates. We have found that sulfide-diamide-type thiodiglycolamides (TDGAs) extract Pd(II) chloridometalates much faster than the conventional extractant di-*n*-hexyl sulfide (DHS). The TDGAs also show high stability against strongly acidic solutions, unlike DHS. The Pd(II) extracted with the TDGAs can be stripped by using aqueous ammonia. We have also found that amide-containing tertiary amines (ACTAs), especially the diamide-type ACTA (BisAA: *N*–*n*-hexyl-bis(*N*-methyl-*N*–*n*-octyl-ethylamide)amine) and the triamide-type ACTA (TrisAA: tris(*N*-methyl-*N*–*n*-octyl-ethylamide)amine), efficiently extract Rh (III) from 1–5 mol/L HCl solutions, even though a typically unextractable pentachlororhodate complex, $[RhCl_5(H_2O)]^{2-}$, is dominantly present in these solutions. This high extraction efficiency is considered to be the result of the formation of unique outer-sphere coordination assemblies between the $[RhCl_5(H_2O)]^{2-}$ and protonated BisAA or TrisAA molecules. The loaded Rh (III) can be stripped with either a highly concentrated HCl solution or a NaOH aqueous solution.

Keywords Solvent extraction · Extractant · Palladium · Rhodium · Extraction mechanism

Introduction

In hydrometallurgical processes, solvent extraction (liquid–liquid extraction) is a widely used separation and purification method for metal ions [1]. An important key to the building of economical hydrometallurgical processes is the use of highly efficient extractants. To develop such extractants, a detailed understanding of the

H. Narita (✉)
Global Zero Emission Research Center, National Institute of Advanced Industrial Science and Technology (AIST), 16-1 Onogawa, Tsukuba 305-8569, Japan
e-mail: hirokazu-narita@aist.go.jp

T. Ouchi et al. (eds), *Rare Metal Technology 2022*, The Minerals, Metals & Materials Series, https://doi.org/10.1007/978-3-030-92662-5_16

extraction mechanism is important, which requires knowledge of the basic characteristics of the target metal ion and candidate extractants, structural properties of the metal complex, and distribution behavior of the target metal ion between organic and aqueous phases.

Platinum group metals (PGMs) are indispensable industrial materials, especially for their use as exhaust gas catalysts. However, due to their cost and rarity, refining technologies for PGMs from both natural ores and urban mines must be improved to ensure that the supply of PGMs remains stable. At some PGM refineries, PGMs are dissolved with HCl solution containing Cl_2 [2]. All PGM ions are dominantly present as chloridometalates in relatively concentrated HCl solutions (>1 mol/L) [3]. Therefore, to develop improved PGM extractants, the chemical properties of PGM chloridometalates must also be understood.

In Japan, PGM refining processes based on the Vale (Acton) process are widely used, where the mutual separation of PGM ions is performed mainly by solvent extraction [2, 4]. However, there are some problems related to solvent extraction with this process, which are discussed later. Our group has focused on Pd (II) and Rh (III), and we have carried out extraction studies examining the extraction of these ions from HCl solutions using newly synthesized extractants. Also, we have elucidated the extraction mechanisms by Fourier-transform infrared (FT-IR), nuclear magnetic resonance (NMR), extended X-ray absorption fine structure (EXAFS), and small angle neutron scattering (SANS) measurements, and density functional theory (DFT) calculations. This paper discusses the extraction of Pd (II) and Rh (III) chloridometalates using two series of novel extractants developed by our group.

Palladium Extraction

Di-alkyl sulfide (DAS) compounds, such as di-*n*-hexyl sulfide (DHS, Fig. 1A) and di-*n*-octyl sulfide (DOS), used in the Vale process are well-known extractants of Pd (II) from HCl solutions [1, 2]. DAS compounds have a *soft* donor S atom and therefore have a high affinity for *soft* ions such as Pd (II), in accordance with the HSAB (hard and soft acids and bases) concept [5]. However, DAS compounds show low extraction rates for Pd (II) and have low oxidation resistance [1, 3]. Therefore, our group began the development of a novel extractant with an improved extraction rate and

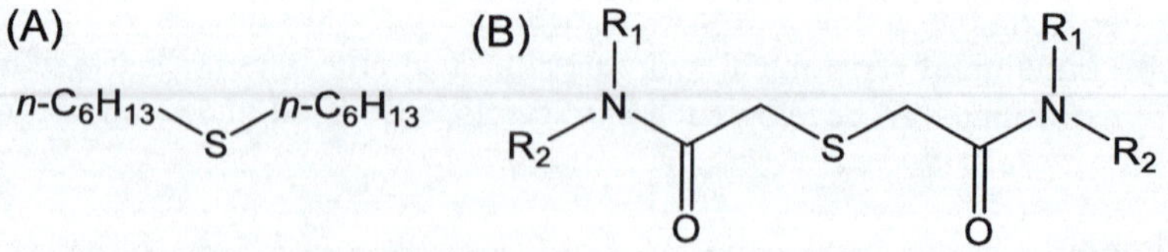

Fig. 1 Structures of extractants **A** DHS, **B** TDGA

oxidation resistance as well as high selectivity for Pd (II) over Pt(IV). To have selectivity, we used an extractant that could directly coordinate to Pd (II). To improve the oxidation resistance, we used a functional group with high durability against strong acids. To improve the extraction rate, we used both hydrophobic and hydrophilic groups to enhance interfacial activity. As a result, we identified thiodiglycolamides (TDGAs), which are characterized by a sulfide atom sandwiched between two *N,N*-disubstituted amide groups, as candidate extractants (Fig. 1B) [6–8]. TDGAs can be readily synthesized and soluble in most organic solvents by introducing long chain alkyl groups to the *N*-substitutes.

Figure 2 shows the extraction efficiency of Pd (II) and Pt(IV) with DHS and TDGA (R_1 = methyl, R_2 = *n*-octyl) as a function of extraction (shaking) time [6]. The TDGA extracted Pd (II) much faster than DHS and showed good selectivity for Pd (II) over Pt(IV). The higher interfacial tension afforded by the TDGA compared with that afforded by the DHS may explain the faster Pd (II) extraction. FT-IR and EXAFS analyses of the Pd (II) complex extracted into the organic phase revealed that the Pd (II) was bound to the extractant via the sulfur atom, not the oxygen atoms [8]. The extraction reaction in the TDGA systems was the same as that in the DHS system (org = organic phase):

$$[PdCl_4]^{2-} + 2TDGA_{(org)} \rightleftharpoons \left[PdCl_2(TDGA)_2\right]_{(org)} + 2Cl^- \quad (1)$$

A durability test under strongly acidic conditions was also performed [7]. A TDGA ($R_1 = R_2$ = *n*-octyl) and DHS were stirred with 0.75 mol/L HNO_3 + 2.25 mol/L HCl for one week, after which extraction efficiency of Pd (II) was determined. The extraction efficiency of DHS decreased from around 100% at the start of the experiment to around 40% after one week of stirring under acidic conditions. In contrast, the extraction efficiency of the TDGA remained at around 100% throughout the experiment. At the end of the experiment, FT-IR analyses of the extractants were

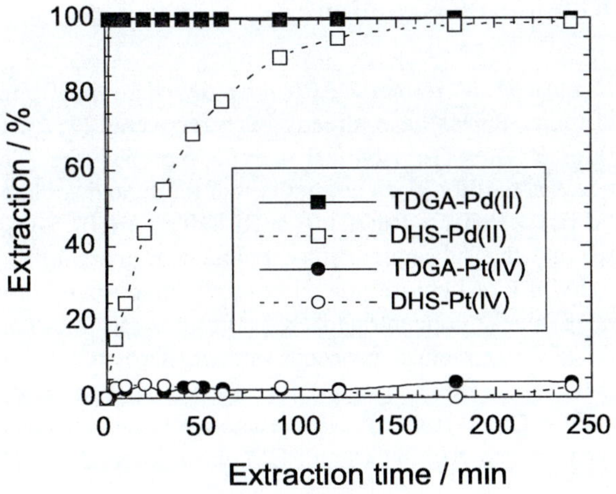

Fig. 2 Extraction of Pd (II) and Pt(IV) with DHS and TDGA (R_1 = methyl, R_2 = *n*-octyl) as a function of the extraction time. Organic phase: 0.05 mol/L DHS or TDGA; Aqueous phase: 3.0 mol/L HCl. Reproduced with permission from Ref. [6]

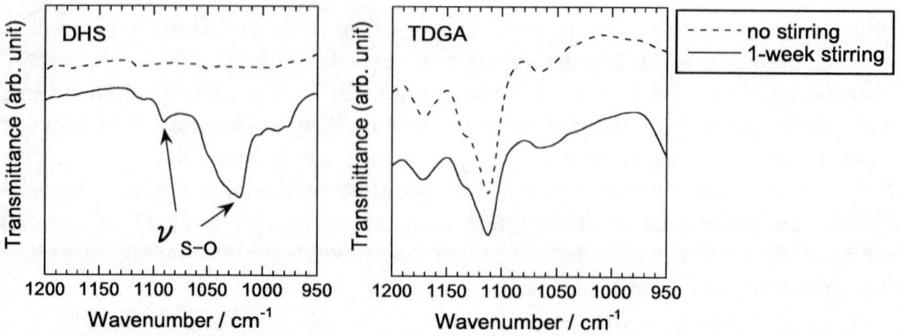

Fig. 3 FT-IR spectra of DHS and TDGA ($R_1 = R_2 =$ *n*-octyl). Dotted: free TDGA or DHS. Solid: TDGA or DHS after the one-week stirring with 0.75 mol/L HNO_3–2.25 mol/L HCl solutions. Reproduced with permission from Ref. [7]

conducted (Fig. 3). In the FT-IR spectrum of DHS, the stretching frequency of the S–O bond ($\nu_{S–O}$) was observed at around 1020 and 1090 cm^{-1}, indicating that the sulfide group had been oxidized to sulfoxide and/or sulfone. This deterioration of the sulfide group likely explains the decreased extraction efficiency of DHS. In contrast, no peaks assigned to $\nu_{S–O}$ appeared in the TDGA, indicating that the TDGA had high oxidation resistance.

Finally, we investigated the effects of different *N*-substitutions for the optimization of TDGA with respect to considerations for industrial use such as third phase formation, solubility in organic solvents, and synthesis cost. As a result, we obtained *N,N,N′,N′*-tetra-2-ethylhexyl-TDGA as an extractant with excellent extraction properties. Currently, the TDGA is commercially available and used at PGM refineries in Japan.

Rhodium Extraction

In most PGM refineries, Rh is retrieved by precipitation from raffinate in which all other PGMs have already been recovered [1, 2]; this is because Rh (III) chloridometalates are a typical unextractable species. There are two reasons for this unextractability [3]. The first is that the ligand exchange rate proceeds in the order Pd (II) > Pt(II) > Ru(III) > > Rh (III) > Ir(III) > Os(III) > > Ir(IV) Pt(IV), with Rh and the subsequent ions regarded as inert ions. The second is that the reactivity of PGM chloridometalates to form ion-pair complexes with cationic reagents in relatively concentrated HCl solutions (>1 mol/L), which are extensively used for PGM extraction, proceeds in the order $[MCl_4]^-$ >> $[MCl_6]^{2-}$ > $[MCl_4]^{2-}$ >> $[MCl_6]^{3-}$ > $[MCl_{n-x}(H_2O)_x]^{m-}$, and Rh (III) forms two complexes, $[RhCl_5(H_2O)]^{2-}$ and $[RhCl_6]^{3-}$, that have low reactivity. Usually, a basic extractant such as an amine-type compound can extract PGM chloridometalates. However, a typical amine-type extractant such as tri-*n*-octyl amine (TOA), which is also an ion-pair-type extractant, is unable to efficiently extract Rh (III) from relatively concentrated HCl solutions. We, therefore, sought to develop an improved ion-pair-type extractant.

We began by synthesizing three tertiary amine compounds possessing 1–3 *N,N*-disubstituted amides (i.e., amide-containing tertiary amines (ACTAs) (Fig. 4)) and then examined their performance for the extraction of Rh (III) from 2 mol/L HCl solution [9–13]. Each of the three ACTAs extracted Rh (III) more efficiently than TOA (Fig. 5). BisAA and TrisAA showed much higher extraction efficiency than either MonoAA or TOA. We also found that the dependency of extraction efficiency on the concentration of HCl in the aqueous phase was similar for BisAA and TrisAA and that the maximum extraction efficiency was obtained at around 2 mol/L HCl. The loaded Rh (III) was stripped with a highly concentrated HCl (>6 M) or a NaOH aqueous solution. We also investigated the specificity of these extractants for PGMs

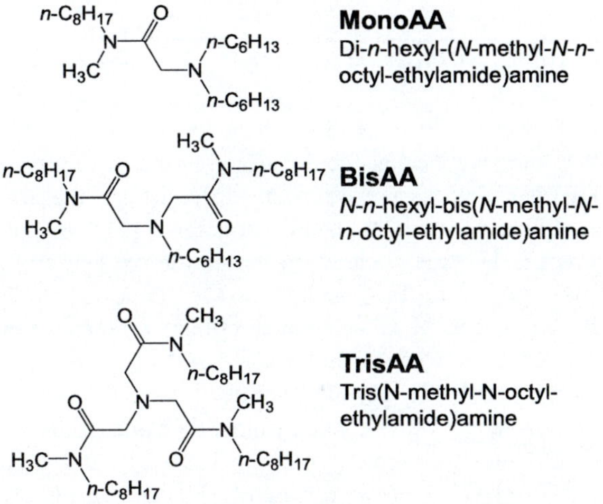

Fig. 4 Structures of ACTA compounds

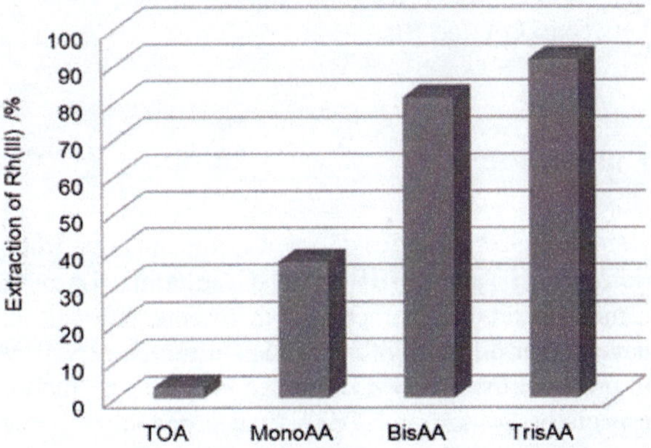

Fig. 5 Extraction of Rh(III) from 2 mol/L HCl with 0.5 mol/L extractants in $CHCl_3$. Reproduced with permission from Ref. [12]

(i.e., Pd(II), Pt(IV), and Rh (III)) over base metals (i.e., Fe(III), Cu(II), and Ni(II)). Appropriate combinations of extraction, scrubbing, and stripping enabled specific recovery of the individual PGMs [13].

Overall, extraction efficiency proceeded in the order TrisAA ≥ BisAA > MonoAA > > TOA, suggesting that the amide group plays an important role in enhancing extraction efficiency. We, therefore, examined the extraction mechanisms for the three ACTAs. Slope analyses of the distribution ratios of Rh (III) and uptake of HCl to the organic phase from 2 mol/L HCl revealed that each Rh(III) ion was extracted by two protonated ACTA molecules. In addition, FT-IR, NMR, and EXAFS analyses revealed that $[RhCl_5(H_2O)]^{2-}$ was the predominant species in both the aqueous and organic phases and that the amide oxygen atoms were involved in the protonation but not in the Rh(III) coordination [11]. Thus, the extraction reactions can be expressed as follows (E = ACTA):

$$H^+Cl^- + E_{(org)} \rightleftharpoons E \cdot H^+Cl^-_{(org)} \tag{2}$$

$$[RhCl_5(H_2O)]^{2-} + 2E \cdot H^+Cl^-_{(org)} \rightleftharpoons [RhCl_5(H_2O)] \cdot (E \cdot H)_{2(org)} + 2Cl^- \tag{3}$$

However, the same reactions were also observed in the TOA system, suggesting that there were other factors affecting the extraction efficiency. We, therefore, used DFT calculations to examine the differences in the structural properties of the Rh(III) complexes formed with the extractants [11]. In the complexes with BisAA and TrisAA, H^+ was chelated to the amide oxygen and amine nitrogen atoms. In the complexes with TOA and MonoAA, chloride ion directly interacted with the $N–H^+$ group (i.e., $N–H^+–Cl^-$ interactions), whereas in the complexes with BisAA and TrisAA, there were several weak $C–H^{\delta+}–Cl^-$ interactions. In the lowest energy structures of the $[RhCl_5(H_2O)]^{2-}$ complexes with BisAA and TrisAA, but not with TOA or MonoAA, H^+ was chelated by more than one amide oxygen atom, and they used only $C–H^{\delta+}$ sites to interact with $[RhCl_5(H_2O)]^{2-}$. The charge-diffuse $C–H^{\delta+}$ sites and $[RhCl_5(H_2O)]^{2-}$ are *softer* than $N–H^+$ and Cl^-, likely leading to the higher extraction efficiency of BisAA and TrisAA. Furthermore, SANS measurements showed that the outer-sphere assembly during the extraction of $[RhCl_5(H_2O)]^{2-}$ was different between TOA and BisAA.

Conclusions

Here, I report two series of novel extractants our group has developed for the extraction of Pd (II) and Rh(III) from HCl solutions. Extraction of Pd (II) by using TDGAs is much faster than extraction with the conventional extractant DHS, and TDGAs also have higher durability than DHS. Currently, *N,N,N′,N′*-tetra-2-ethylhexyl-TDGA is in practical use at PGM refineries in Japan. Extraction of Rh(III) by using ACTAs, especially BisAA and TrisAA, afforded higher extraction efficiencies than that afforded by the typical ion-pair-type extractant TOA. We attribute these improved extraction efficiencies to the unique interaction between $[RhCl_5(H_2O)]^{2-}$ and C–$H^{\delta+}$ sites in BisAA and TrisAA.

Acknowledgements The author thanks the following people for their fruitful collaboration and helpful discussions: Dr. Mikiya Tanaka (National Institute of Advanced Industrial Science and Technology); Dr. Tsuyoshi Yaita, Dr. Ryuhei Motokawa, and Dr. Hideaki Shiwaku (Japan Atomic Energy Agency); and Dr. Rebecca M. Nicolson, Prof. Jason B. Love, Prof. Peter A. Tasker, and Prof. Carole A. Morrison (University of Edinburgh). Part of this work was financially supported by the Industrial Technology Research Grant Program (03A44001) of the New Energy and Industrial Technology Development Organization (NEDO) of Japan and by the Japan Society for the Promotion of Science KAKENHI grants (20686060, 23360407, 26289283, 17H03438, and 20H02497).

References

1. Cox M (2004) Solvent extraction in hydrometallurgy. In: Rydberg et al (eds) Principles and practices of solvent extraction, 2nd edn. Marcell Dekker Inc., New York, pp 455–505
2. Crundwell FK et al (2011) Extractive metallurgy of nickel, cobalt and platinum-group metals. Elsevier Science BV, Amsterdam
3. Narita H et al (2020) Precious metal separations. In: Encyclopedia of inorganic and bioinorganic chemistry. Hoboken, NJ, Wiley & Sons Ltd. https://doi.org/10.1002/9781119951438.eibc2756
4. Shibata J, Okuda A (2002) Recycling technology of precious metals. J MMIJ 118:1–8
5. Pearson RG (1963) Hard and soft acids and bases. J Am Chem Soc 85:3533–3539
6. Narita H et al (2004) Rapid separation of palladium(II) from platinum(IV) in hydrochloric acid solution with thiodiglycolamide. Chem Lett 33:1144–1145
7. Narita H, Tanaka M, Morisaku K (2008) Palladium extraction with N,N,N',N'-tetra-n-octyl-thiodiglycolamide. Miner Eng 21:483–488
8. Narita H et al (2004) Comparison of the extractabilities of tetrachloro- and tetrabromo-palladate(II) ions with a thiodiglycolamide compound. Anal Sci 33:1305–1309
9. Narita H, Tanaka M, Morisaku K (2008) The first effective extractant for trivalent rhodium in hydrochloric acid solution. Chem Commun 45:5921–5923
10. Narita H, Tanaka M, Morisaku K (2015) Highly efficient extraction of rhodium(III) from hydrochloric acid solution with amide-containing tertiary amine compounds. Solvent Extr Ion Exch 33:407–417
11. Narita H et al (2019) Proton chelating ligands drive improved chemical separations for rhodium. Inorg Chem 58:8720–8734
12. Narita H, Suzuki T, Motokawa R (2017) Recent research in solvent extraction of platinum group metals. J Jpn Inst Met Mater 81:157–167
13. Narita H et al (2014) Separation of rhodium from palladium, platinum, and base metals with amide-containing tertiary amine compounds. Proc ISEC 2014:1058–1063

Development of Antagonistic Solvent Extraction Systems for Selective Separation of Copper, Cobalt, and Nickel in Ammoniacal Solution

Kurniawan Kurniawan, Jae-chun Lee, Jonghyun Kim, Ha Bich Trinh, and Sookyung Kim

Abstract Selective separation of copper, cobalt, and nickel from ammoniacal solution using antagonistic solvent extraction (SX) systems has been developed. Copper was first selectively extracted using a combination of LIX 84-I and TBP. Subsequently, cobalt was separated from nickel using a combination of PC88A and Alamine 336. Effects of parameters that were investigated include the total ammonia, initial metal concentration, extractants molar ratio, total extractant concentration and phase ratio. The results show that antagonistic interaction of each combination was suitable for selective separation of copper and cobalt, leaving nickel in the ammoniacal solution for further processing. The two-stage antagonistic SX method is simple and promising for further appropriate treatment steps, and can be regarded as the basis for a sustainable metal separation process.

Keywords Ammoniacal solution · Copper · Nickel · Cobalt · Antagonistic solvent extraction

Introduction

Ammoniacal leaching is recognized as an effective hydrometallurgical method for recovering valuable and rare metals, including copper, cobalt, and nickel from various sources, such as sulfide deposits, manganese sea nodules, or secondary resources [1–5]. The leaching brings the metals into solutions as metal ammine complexes. Despite the capability of quantitative leaching of the metals, the subsequent separation and

K. Kurniawan (✉) · J. Lee · J. Kim · S. Kim
Resource Recycling, Korea University of Science and Technology, Daejeon 34113, Republic of Korea
e-mail: k.kurniawan@kigam.re.kr

Mineral Resources Research Division, Korea Institute of Geoscience and Mineral Resources (KIGAM), Daejeon 34132, Republic of Korea

H. B. Trinh
Department of Integrated Energy and Infra System, Kangwon National University, Chuncheon, Gangwon-do 24341, Republic of Korea

T. Ouchi et al. (eds), *Rare Metal Technology 2022*, The Minerals, Metals & Materials Series, https://doi.org/10.1007/978-3-030-92662-5_17

purification process of copper, cobalt, and nickel ions in ammoniacal solutions is significantly difficult because of their similar physicochemical properties. Thus, the separation of these metals has gained particular interest due to the demand for high-purity metals. The hydrometallurgical method through solvent extraction (SX) is commonly practiced for separating these metals. A number of commercial organic extractants from β-diketones (DK-16, LIX 54) [6, 7], oximes (ACORGA M5640, LIX 84, LIX 84-I, LIX 984 N, LIX 64 N, SME 529) [4, 8, 9], phosphatic ones (D2EHPA, PC88A, Cyanex 272, TOPS-99) [9, 10], etc., were particularly used separation of these metals, exhibiting high extraction efficiency. However, these extractants lacked selectivity as they could extract the three metals altogether. Although separation of the co-extracted metals can be further achieved by selective stripping, the process suffered from high consumptions of organic extractants and stripping solution. As a result, the separation of copper, cobalt, and nickel is still considered challenging.

In the present study, a strategy based on a two-stage antagonistic SX process for selective separation of copper, cobalt, and nickel from ammoniacal solution has been applied. Selective separation of copper was firstly conducted using a combination of LIX 84-I and TBP. Subsequently, cobalt was separated from nickel using a combination of PC88A and Alamine 336. The effects of different parameters were studied, particularly the total ammonia, initial metal concentration, extractants molar ratio, total extractant concentration and phase ratio. These combinations are practical and efficient, and can be successfully adapted for selective separation of respective copper and cobalt, leaving nickel in the solution for further processing.

Experimental

Synthetic solutions containing 1.5 g/L of each Cu^{2+}, Co^{2+}, and Ni^{2+} were prepared by dissolving their sulfate salts in the ammonia/ammonium sulfate (NH_4OH/ammonia salt ratio of 4/1) with a total ammonia concentration of 4.0 mol/L (until specified for investigating the effect of total ammonia and initial metal concentrations). Organic extractants were prepared by dissolving the combined extractants (LIX 84-I + TBP and PC88A + Alamine 336) in kerosene according to the desired concentration. Particularly 0.1 M TBP was used in the second stage as a modifier. All the chemicals and reagents (NH_4OH (Daejung, South Korea); $CuSO_4$ (Kanto Chemical, Japan); $(NH_4)_2SO_4$, $CoSO4._7H_2O$, $NiSO_4.6H_2O$, TBP, kerosene (Junsei, Japan); LIX 84-I, Alamine 336 (BASF, Germany); PC88A (Daihachi, Japan)) were used as received. Solvent extraction experiments were performed by shaking a total volume of 20 mL in a 50-mL separating funnel for 5 min. After extraction, the phases were allowed to separate and the metal contents in the raffinates were determined using an inductively coupled plasma optical emission spectrometer (ICP-OES). The distribution ratio (D) and extraction percentage ($\%E$) were calculated using the expressions (1) and (2):

$$D = \frac{[M]_{org}}{[M]_{raff}} \tag{1}$$

$$\%E = \frac{D}{D+1} \times 100 \tag{2}$$

where $[M]_{org}$ and $[M]_{raff}$ are the metal concentration in the organic and raffinate solutions.

Results and Discussion

Selective Separation of Copper

Extraction reaction of metal ammine complexes with organic extractant of LIX 84-I can be commonly written as reaction (3):

$$m(NH_3)_q^{p+} + p\overline{(RH)_2} = \overline{m(R.RH)_2} + p(NH_4)^+ + (q-p)NH_3 \tag{3}$$

where m is metal and $\overline{(RH)_2}$ is organic extractant.

Selectivity of the LIX 84-I toward copper is enhanced by combining with TBP. Effects of several parameters were studied.

Antagonism Interaction in a Mixed System of LIX 84-I and TBP

A series of experiments using LIX 84-I, TBP and their mixture was conducted to gain insight into the interaction of both extractants. Results and experimental conditions are presented in Table 1. If there is no interaction between the two extractants, the total distribution ratio of the LIX 84-I and TBP alone, i.e., (D_1+D_2) should be equal to that of mixed extractant system (D_3). However, it was observed that $D_1+D_2 \gg D_3$ for the three metals, confirming that there was an antagonistic interaction of the mixed extractants towards the metals which was more pronounced for Co and Ni. At a proper molar ratio, complete separation of Cu from Co and Ni could be achieved. The mixed LIX 84-I and TBP is thus worth exploiting for the selective separation of Cu from Co and Ni.

Table 1 Effect of organic phase composition on the distribution ratio of Cu, Co, and Ni

Concentration (M)		Cu		Co		Ni	
LIX 84-I	TBP	D_1+D_2	D_3	D_1+D_2	D_3	D_1+D_2	D_3
0.1	0.9	1.88	0.07	0.11	0.00	0.07	0.00
0.2	0.8	7.16	0.32	1.89	0.00	0.16	0.00
0.3	0.7	13.28	0.97	7.15	0.00	0.50	0.00
0.4	0.6	47.29	2.09	9.97	0.00	2.22	0.00
0.5	0.5	67.83	6.74	12.48	0.10	4.28	0.00
0.6	0.4	121.55	40.54	14.91	0.54	7.00	0.02
0.7	0.3	629.17	203.58	17.87	1.97	12.17	0.04
0.8	0.2	1653.51	1372.74	20.10	3.46	25.70	1.01
0.9	0.1	9234.34	7981.74	25.60	4.20	72.54	6.58

(Experimental conditions: Aqueous phase, 1.5 g/L of each Cu^{2+}, Co^{2+}, and Ni^{2+}, 4.0 total ammonia; organic phase, 0.1–0.9 M LIX 84-I and 0.9–0.1 M TBP both in separate (D_1+D_2) and in mixed systems (D_3); 1/1 A/O phase ratio)

Effect of SX Parameters

Effects of total ammonia concentration, initial metal concentration, and total extractant concentration at a different molar ratio of LIX 84-I and TBP were studied, and results were presented in Fig. 1. It was observed from Fig. 1a that extraction of the three metals decreased with the increase of total ammonia concentration from 2.0 to 5.0 M. With the increase of ammonia concentration, the metals tended to form more stable metal–ammine complexes [11]; as a result, the metal extraction was limited. Even so, a sufficient amount of ammonia in an aqueous solution is crucial to ensure the formation of metal–ammine complexes. While varying the initial concentration of Cu^{2+}, Co^{2+}, and Ni^{2+} (Fig. 1b), the extraction behavior can be ascribed as follows. At a fixed Cu^{2+} concentration of 1.5 g/L, the Co^{2+} and Ni^{2+} extraction decreased as a result of increasing their concentration from 0.25 to 1.5 g/L. At the initial concentration of 1.5/1.5/1.5 g/L, Ni^{2+} extraction was almost zero. In the case where Co^{2+} and Ni^{2+} concentrations were fixed at 1.5 g/L, increasing the Cu^{2+} concentration decreased the Co^{2+} and Ni^{2+} extraction, showing that there was a crowding effect of Cu^{2+} over the extraction of Co^{2+} and Ni^{2+}.

Further, the extraction behavior was examined as a function of total extractant concentration, ([LIX 84-I] + [TBP]) at different LIX 84-I and TBP ratios. Results shown in Fig. 1c depicted that controlling the extractant ratio and total extractant concentration is vital to obtain a complete separation of Cu^{2+} from Co^{2+} and Ni^{2+}. For an instance, using 0.35 M LIX 84-I + 0.15 M TBP (or 7/3 molar fraction of LIX 84-I/TBP at total concentration of 0.5 M), Cu^{2+} extraction could reach 95.4%, while no Co^{2+} and Ni^{2+} extraction was observed. This ratio was thus applied to conduct experiments at varying A/O phase ratios (from 1/4 to 4/1), and data were used to construct the McCabe–Thiele diagram for Cu^{2+} (Fig. 1d). This plot clearly shows that

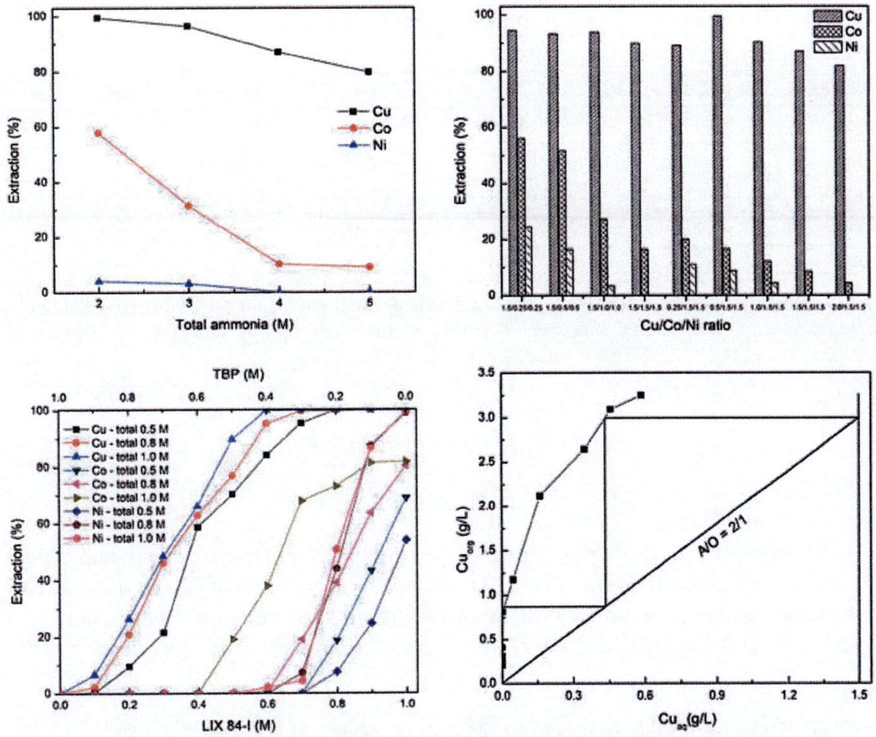

Fig. 1 **a** Effect of total ammonia concentration (aqueous: 1.5 g/L of each Cu^{2+}, Co^{2+}, and Ni^{2+}, 2.0–5.0 M total ammonia; organic: 0.5 M LIX 84-I + 0.5 M TBP. A/O phase ratio, 1/1), **b** effect of initial metal concentration (aqueous: 0.5–2.0 g/L of each Cu^{2+}, Co^{2+}, and Ni^{2+} 4.0 M total ammonia; organic: 0.5 M LIX 84-I + 0.5 M TBP. A/O phase ratio, 1/1), **c** effect of total extractant concentration at different LIX 84-I and TBP molar ratio (aqueous: 1.5 g/L of each Cu^{2+}, Co^{2+}, and Ni^{2+}, 4.0 M total ammonia; organic: 0.5–1.0 M [LIX84-I] + [TBP] at different [LIX84-I]/[TBP] ratio. A/O phase ratio, 1/1), and **d** McCabe–Thiele diagram of Cu^{2+} extraction (aqueous: 1.5 g/L of each Cu^{2+}, Co^{2+}, and Ni^{2+}, 4.0 M total ammonia; organic: 0.35 M LIX 84-I + 0.15 M TBP)

two theoretical extraction stages are needed to completely extract and enrich Cu^{2+} from the feed solution to the extract (organic) at an A/O ratio of 2/1. The Cu-loaded organic solution could be subsequently stripped by a H_2SO_4 solution and recovered separately.

Selective Extraction of Cobalt

PC88A is usually used for extraction of metal cations. The extraction reaction can be expressed as (Eq. (4)):

Table 2 Effect of organic phase composition on the distribution ratio of Co and Ni

Concentration (M)		Co		Ni	
PC88A	Alamine 336	D_1+D_2	D_3	D_1+D_2	D_3
0.1	0.4	1.29	1.10	0.02	0
0.2	0.3	2.58	2.35	0.04	0
0.25	0.25	3.29	2.91	0.11	0.007
0.3	0.2	3.71	3.36	0.14	0.02
0.4	0.1	4.20	3.75	0.16	0.06

(Experimental conditions: Aqueous phase, 1.5 g/L of each Co^{2+} and Ni^{2+}, 4.0 total ammonia; organic phase, 0.1–0.4 M PC88A and 0.4–0.1 M Alamine 336 both in separate (D_1+D_2) and in mixed system (D_3), 0.1 M TBP as modifier; 1/1 A/O phase ratio)

$$m^{p+} + p\overline{HA} = \overline{mA_p} + pH^+ \quad (4)$$

where $\overline{HA}$ is PC88A.

After separating copper, experiments were continued to separate cobalt and nickel. Here, the solution containing 1.5 g/L of each Co^{2+} and Ni^{2+} was used. For the selective extraction of Co^{2+}, combined PC88A and Alamine 336 extractants were used.

Antagonism Interaction in a Mixed System of PC88A and Alamine 336

A similar procedure was followed to investigate the antagonistic characteristics of mixed PC88A and Alamine 336. Results presented in Table 2 show that the mixture of PC88A and Alamine 336 has antagonistic effects on both Co and Ni extraction, as also reported in SX studies of other metals using the combination of acid organophosphorus extractants and amine-based extractants [12–14]. The effect is stronger for nickel; at a certain ratio of PC88A and Alamine 336, Ni^{2+} extraction was completely depressed while still providing Co^{2+} extraction. Thus, a mixture of PC88A and Alamine 336 could be used for selective extraction of cobalt.

Effect of Other SX Parameters

Extraction behavior as a function of total ammonia concentration showed a decreasing trend as the ammonia concentration increased (Fig. 2a). Investigation on the effect of initial concentration of Co^{2+} and Ni^{2+} shows the crowding out effect of Co^{2+} over the Ni^{2+} extraction (Fig. 2b). A significance to control the total extractant concentration ([PC88A] + [Alamine 336]) and a molar ratio of the extractants

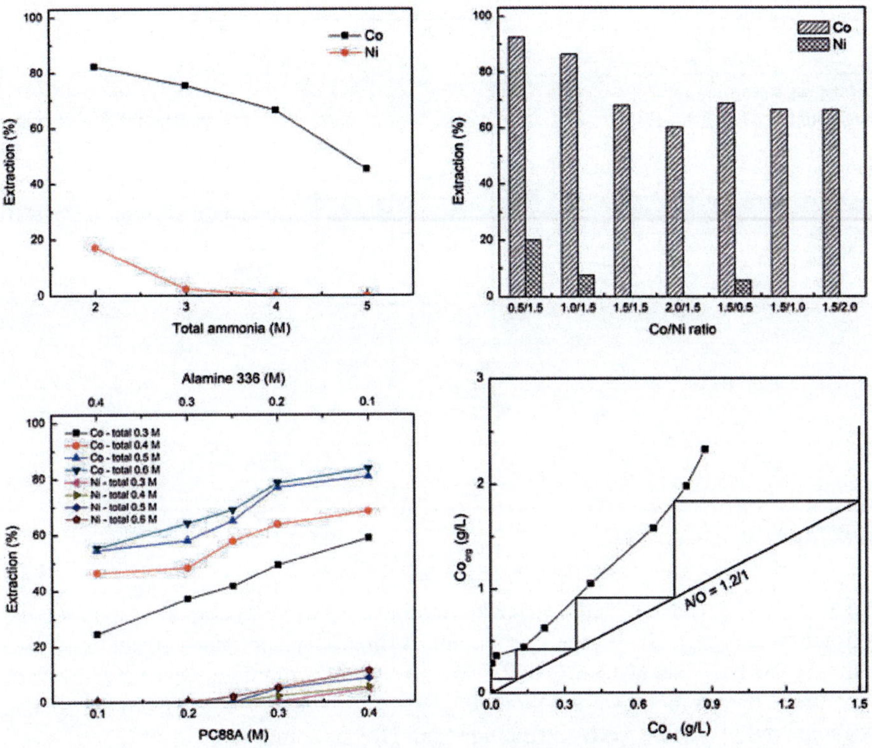

Fig. 2 **a** Effect of total ammonia concentration (aqueous: 1.5 g/L of each Co^{2+} and Ni^{2+}, 2.0–5.0 M total ammonia; organic: 0.25 M PC88A + 0.25 M Alamine 336, 0.1 M TBP as modifier. A/O phase ratio, 1/1), **b** effect of initial metal concentration (aqueous: 0.5–2.0 g/L of each Co^{2+} and Ni^{2+} 4.0 M total ammonia; organic: 0.25 M PC88A + 0.25 M Alamine 336, 0.1 M TBP as modifier. A/O phase ratio, 1/1), **c** effect of total extractant concentration at different PC88A and Alamine 336 molar ratio (aqueous: 1.5 g/L of each Co^{2+} and Ni^{2+}, 4.0 M total ammonia; organic: 0.3–0.6 M [PC88A] + [Alamine 336] at different [PC88A]/[Alamine 336] ratio, 0.1 M TBP. A/O phase ratio, 1/1), and **d** McCabe–Thiele diagram of Cu^{2+} extraction (aqueous: 1.5 g/L of each Co^{2+} and Ni^{2+}, 4.0 M total ammonia; organic: 0.25 M PC88A + 0.25 M Alamine 336, 0.1 M TBP as modifier)

can specifically be found in order to obtain a complete separation of the two metals while still maintaining the Co^{2+} extraction (Fig. 2c). A condition of 0.25 M PC88A and 0.25 M Alamine 336 (or 1/1 molar fraction of PC88A/Alamine 336 at a total concentration of 0.5 M) was used in the subsequent experiments. As per the phase ratio variations at different A/O (4/1–1/5), the Ni^{2+} co-extraction was observed at A/O $\geq$ 2/1. Using the experimental data, the McCabe–Thiele diagram of Co^{2+} was drawn (Fig. 2d), which shows that the quantitative extraction of Co^{2+} can be achieved in four counter-current stages at an A/O ratio of 1.2/1. The loaded organic could be stripped by H_2SO_4 solution and re-used after adequate washing with an ammoniacal solution. The stripped metal can be recovered separately in the desired form.

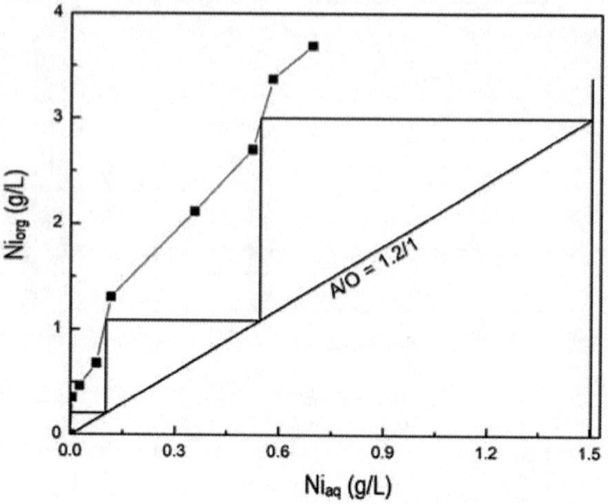

Fig. 3 McCabe–Thiele diagram of Ni^{2+} extraction (aqueous: 1.5 g/L of Ni^{2+}, 4.0 M total ammonia; organic: 0.3 M LIX 84-I)

Extraction of Nickel

Recovery of Ni from Cu- and Co-free ammoniacal solution was carried out by solvent extraction using LIX 84-I as the extractant. A three-stage counter-current extraction using 0.3 M LIX 84-I at an A/O of 2/1 was required to completely extract and enrich Ni^{2+} from the ammoniacal solution (Fig. 3). And recovery of Ni from the loaded organic could be achieved by stripping using H_2SO_4 solution.

Conclusions

Separation of Cu, Co, and Ni from an ammoniacal solution containing 1.5 g/L of each metal was studied using antagonistic solvent extraction (SX) systems. Copper was separated from cobalt and nickel using an antagonistic system involving the LIX 84-I and TBP and the ammoniacal solution containing cobalt and nickel was made Co^{2+}-free by solvent extraction using an antagonistic system comprising of PC88A and Alamine 336. Effects of SX parameters, such as total ammonia-, initial metal-, total extractant concentrations, and molar ratio of the extractants were investigated. The significant effects of these parameters on the extraction and separation behavior of the metals were established. As a consequence of treating these metals from the solutions, the separation and enrichment of copper could be achieved in two-stage counter-current extraction using 0.35 M LIX 84-I + 0.15 M TBP at an A/O phase ratio of 2/1. And a four-stage counter-current extraction was theoretically required for the complete extraction and separation of $Co2^{+}$ from Ni^{2+} using 0.25 M PC88A + 0.25 M Alamine 336. These results established that an antagonistic solvent extraction system is a useful tool for selective separation of copper, cobalt, and nickel present in ammoniacal solutions.

Acknowledgements This paper was supported by the Energy Technology Development Program of Korea Institute of Energy Technology Evaluation and Planning (KETEP) grant funded by the Ministry of Trade, Industry and Energy (Project no. 20217510100080).

References

1. Ilyas S, Srivastava RR, Kim H, Ilyas N, Sattar R (2020) Extraction of nickel and cobalt from a laterite ore using the carbothermic reduction roasting-ammoniacal leaching process. Sep Purif Technol 232:115971
2. Ku H, Jung Y, Jo M, Park S, Kim S, Yang D, Rhee K, An E-M, Sohn J, Kwon K (2016) Recycling of spent lithium-ion battery cathode materials by ammoniacal leaching. J Hazard Mater 313:138–146
3. Liu ZX, Yin ZL, Chen YG, Xiong LZ (2014) Leaching of calcareous bornite ore in ammoniacal solution containing ammonium persulfate. Metall Mater Trans B 45(6):2027–2032
4. Sridhar V, Verma JK (2011) Extraction of copper, nickel and cobalt from the leach liquor of manganese-bearing sea nodules using LIX 984N and ACORGA M5640. Miner Eng 24(8):959–962
5. Kim TG, Srivastava RR, Jun M, Kim MS, Lee J-C (2018) Hydrometallurgical recycling of surface-coated metals from automobile-discarded ABS plastic waste. Waste Manage 80:414–422
6. Fu W, Chen Q, Hu H, Niu C, Zhu Q (2011) Solvent extraction of copper from ammoniacal chloride solutions by sterically hindered β-diketone extractants. Sep Purif Technol 80(1):52–58
7. Przeszlakowski S, Wydra H (1982) Extraction of nickel, cobalt and other metals [Cu, Zn, Fe (III)] with a commercial β-diketone extractant. Hydrometallurgy 8(1):49–64
8. Duan H, Wang S, Yang X, Yuan X, Zhang Q, Huang Z, Guo H (2017) Simultaneous separation of copper from nickel in ammoniacal solutions using supported liquid membrane containing synergistic mixture of M5640 and TRPO. Chem Eng Res Des 117:460–471
9. Parhi PK, Sarangi K (2008) Separation of copper, zinc, cobalt and nickel ions by supported liquid membrane technique using LIX 84I, TOPS-99 and Cyanex 272. Sep Purif Technol 59(2):169–174
10. Reddy BR, Rao SV, Park KH (2009) Solvent extraction separation and recovery of cobalt and nickel from sulphate medium using mixtures of TOPS 99 and TIBPS extractants. Miner Eng 22(5):500–505
11. Kurniawan Kim MS, Chung KW, Kim R, Lee J-C (2021) Simple and complete separation of copper from nickel in the ammoniacal leach solutions of metal coated ABS plastic waste by antagonistic extraction using a mixture of LIX 84-I and TBP. Sep Purif Technol 255:117712
12. Gmar S, Mutelet F, Chagnes A (2020) Effect of the addition of amine in organophosphorus compounds on molecular structuration of ionic liquids–application to solvent extraction. Molecules 25(11):2584
13. Lee JY, Kumar JR, Kim JS, Kim DJ, Yoon HS (2009) Extraction and separation of Pt (IV)/Rh (III) from acidic chloride solutions using Aliquat 336. J Ind Eng Chem 15(3):359–364
14. Oh CG, Son SH, Lee MS (2019) Solvent extraction of Tb (III) from chloride solution using organophosphorous extractant, its mixture and ionic liquids with amines. J Korean Inst Resour Recycl 28(1):40–46

Study on Pre-removal Antimony from Antimony–Gold Concentrate Using Slurry Electrolysis

Yonglu Zhang, Chengyan Wang, Xiaowu Jie, Wei Gao, and Shufeng Ruan

Abstract Antimony has a negative effect on the gold extraction from antimony–gold concentrate, and it is necessary to remove antimony in advance. For a high calcium low antimony–gold concentrate (Ca–Sb–Au ore), this study proposed to pre-remove antimony using slurry electrolysis in HCl–$CaCl_2$ system. The results indicated that the optimal conditions were $CaCl_2$ 300–350 g/L, electric quantity 1.2 times, HCl 30 g/L, time 3.5 h, and temperature 60 °C. The leaching efficiency of antimony and calcium was higher than 96% and 98%, respectively, and gold was less than 1%. The dissolved antimony can be directly electrodeposited on the cathode to produce metallic antimony, the sulfur present in stibnite was transformed into elemental sulfur remaining in the residue, and the leaching residue can be used as raw material for further gold extraction. Thus, the separation of antimony and gold was achieved.

Keywords Antimony · Gold · Stibnite · Slurry electrolysis

Introduction

Antimony–gold concentrate is a refractory gold ore, and the direct cyanide leaching efficiency is very low [1]. For antimony–gold concentrate with high antimony content (Sb > 30%), the volatilization melting of blast furnace and reduction melting of reverberation furnace is generally adopted at present, and the gold is finally enriched in noble antimony for further recovery [2–5]. However, for the one with lower antimony (Sb < 20%), it is difficult to recover antimony economically by existing fire process.

At present, there are two main processes for the treatment of low antimony–gold concentrate. The first is using the ore as the smelting ingredients of copper or lead smelting, using copper or lead melt to collect gold to realize the separation of gold

Y. Zhang · C. Wang (✉) · X. Jie
School of Metallurgical and Ecological Engineering, University of Science and Technology Beijing, Beijing 100083, China
e-mail: chywang@yeah.net

Y. Zhang · X. Jie · W. Gao · S. Ruan
BGRIMM Technology Group, Beijing 100160, China

T. Ouchi et al. (eds), *Rare Metal Technology 2022*, The Minerals, Metals & Materials Series, https://doi.org/10.1007/978-3-030-92662-5_18

[6]. However, copper and lead pyrometallurgy system has certain requirements on the content of antimony in raw materials, so it can only be used in limited quantities [7]. The second is cyanide leaching, but the inclusion of gold by plenty of sulfide minerals seriously affects the direct cyanide leaching [8, 9]. Therefore, pre-roasting desulphurization treatment is generally required before cyanide leaching, but antimony is easy to form low-melting point compounds in the roasting process and form new inclusion on gold, which affects the leaching of gold [10, 11]. Simultaneously, antimony will cause a large amount of cyanide consumption in the cyanide leaching process. Therefore, antimony is a harmful element and needs to be removed in advance, no matter it is collected by copper or lead smelting system or by cyanide leaching process.

Slurry electrolysis (SE) is characterized by short process flow and can realize the leaching, partial solution purifying, and electrowinning stages in an electrolytic cell [12, 13]. It can simultaneously oxidize leach the sulfide ore on the anode and electro-deposit the metal on the cathode.

Wang et al. [14] studied the treatment of refractory jamesonite using SE in HCl–NH_4Cl system, wherein antimony and lead were separated in a single step, and metal antimony was directly deposit at the cathode. Meanwhile, the lead in jamesonite remained in the residue as $PbCl_2$.

In our previous study, the leaching and electrowinning of antimony from a high arsenic and gold-containing stibnite concentrate ore using SE process in a medium of HCl–NaCl was investigated Zhang et al. [15]. The antimony was selectively leached, while gold and arsenic were retained in the residue as a raw material for further gold extraction.

SE process makes full use of anodic reaction to leach sulfide minerals without adding additional oxidants. For gold concentrate with high calcium and low antimony, the HCl–NaCl or HCl–NH_4Cl system aforementioned used in the antimony ore treatment is not suitable, due to a large amount of calcium leaching and circulation accumulation in the electrolyte. Therefore, this study attempts to use the SE process to pre-remove antimony from high calcium low antimony–gold concentrate (Ca–Sb–Au ore) in HCl–$CaCl_2$ system. We investigated the factors affecting the antimony extraction, such as $CaCl_2$ concentration, HCl concentration, electric quantity, temperature, and duration. And the transformation process of stibnite mineral phase was further revealed.

Materials and Methods

Materials

The Ca–Sb–Au ore used in the test was a flotation concentrate, with a particle size of 97.6% below 0.075 mm, and 89.58% below 0.045 mm.

Table 1 Elemental composition of the Ca–Sb–Au ore

Component	Sb	Fe	S	Mg	Ca	Si	As	Al	K	C	Au
Percentage (wt.%)	19.72	6.62	12.52	1.50	7.95	14.35	1.83	0.84	0.33	4.59	65.3(g/t)

The elemental analysis is shown in Table 1, the main valuable elements in the mineral are antimony and gold. Figure 1 shows the X-ray diffraction (XRD) patterns of the minerals. The results of phase analysis showed that, the major mineral constituents were stibnite (Sb_2S_3), pyrite (FeS_2), arsenopyrite (FeAsS), calcite ($CaCO_3$), dolomite ($CaMg(CO_3)_2$), and quartz (SiO_2), along with a small amount of muscovite ($KAl_2(AlSi_3O_{10})(OH)_2$) and other gangue phases. Based on the results of optical microscopy, analysis of XRD, SEM–EDS detection, and chemical phase analysis, it was concluded that antimony was mainly present in stibnite (81.00%) and metallic antimony phase (14.81%), which were easily oxidized and leached during the SE process. Gold was mainly present in sulfide and silicate minerals, which were stable during the SE and were retained in the leaching residue.

All reagents used in the experiments were of analytical grade, and all the aqueous solutions were prepared using distilled water. The antimony extraction efficiency (α, %) was calculated using the formula (Eq. 1).

$$\alpha = \frac{m_0 \times x_0 - m_1 \times x_1}{m_0 \times x_0} \times 100 \quad (1)$$

where m_0 and m_1 are the mass (g) of samples before and after SE process, respectively; x_0 and x_1 are the Sb content (wt%) of samples before and after SE process, respectively.

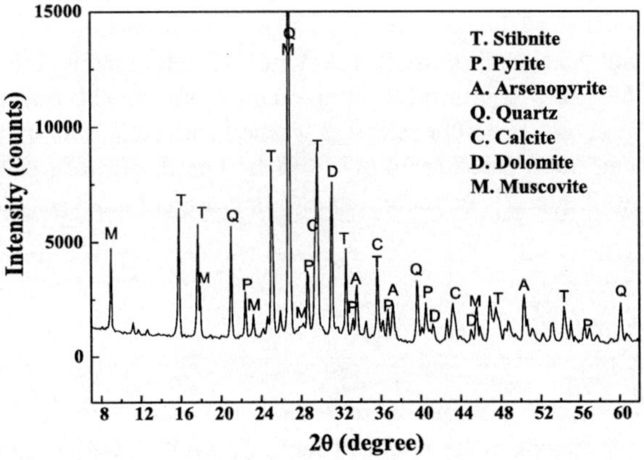

Fig. 1 XRD pattern of the antimony–gold concentrate ore

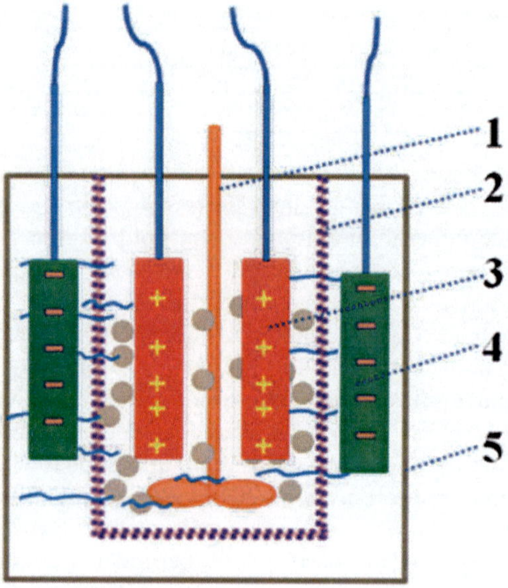

Fig. 2 Schematic diagram of home-made slurry electrolyser (1 Agitator blade, 2 Polypropylene diaphragm bag, 3 Graphite anode, 4 Titanium cathode, 5 Glass beaker)

Experimental Procedure

The SE experiments were performed in a cylindrical home-made electrolyser (Fig. 2). The cathode and anode were made of titanium strip and graphite stick, respectively. The electrolyser was divided into separate concentric anode and cathode departments by a polypropylene permeable membrane. In the anode departments, minerals were kept suspended by stirring. For each experiment, a prepared solution was added first into the anode and then into the cathode department. A thermostatic water bath was used to maintain a constant temperature of the electrolyser. When the temperature reached to the desired value, a quantitative ore was added into the anode department, and the electric current was switched on to conduct the SE process. After SE process, the slurry was filtered by a vacuum pump. The filtrate volume was measured and sampled. The filter cake was washed with HCl (1 mol/L) firstly, and then with distilled water and dried at 70 °C for 8 h. The dried filter cake was weighed, ground, and consequently, elemental and mineralogical analyses were conducted.

Analytical Methods

The metal contents were determined by inductively coupled-plasma atomic emissive spectrometry (IRIS Intrepid II, XRS), and the higher antimony contents (>10% wt.%) were determined by titration with ceric sulfate. The gold contents in the solid

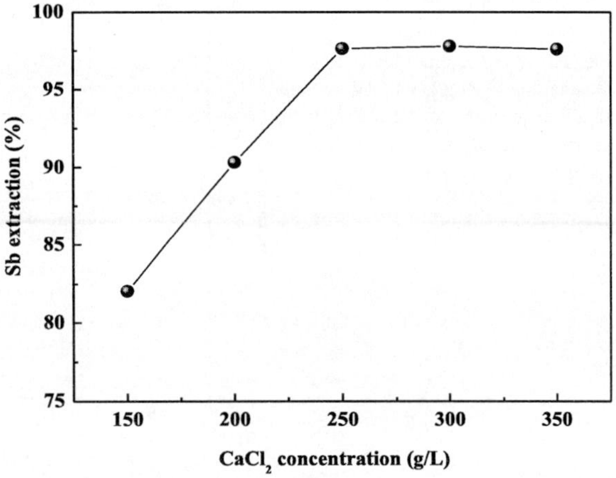

Fig. 3 Effect of $CaCl_2$ concentration on antimony extraction (experimental conditions: 30 g/L HCl, electric quantity of 1.2 times, temperature of 60 °C, duration of 3.5 h)

samples were determined by fire assay. The "silicon molybdenum blue spectrophotometry method" and "combustion-neutralization titrimetric method" were used for the determining the silicon, carbon, and sulfur contents, respectively.

Results and Discussion

Effect of $CaCl_2$ Concentration

The antimony extraction was investigated with the $CaCl_2$ concentrations ranging from 150 to 350 g/L. Figure 3 shows that the antimony leaching efficiency reached 82.03% with 150 g/L $CaCl_2$ and then gradually increased to 97.63% with 250 g/L $CaCl_2$. Further increase in the $CaCl_2$ concentration did not result in any significant increase in the antimony extractions. Since the raw material contains about 8% leachable calcium compounds, the accumulation of calcium chloride is inevitable in the electrolyte recycling process. Meanwhile, the increase of calcium concentration will improve the removal efficiency of calcium in electrolyte. Therefore, a higher concentration of $CaCl_2$, namely 350 g/L, was selected in subsequent tests.

Effect of HCl Concentration

Antimony is easy to hydrolytic precipitation, so it is necessary to maintain a certain acid concentration in the electrolyte. The addition of hydrochloric acid improves the solubility of antimony in the solution and prevents the hydrolysis of antimony.

Fig. 4 Effect of HCl concentration on antimony extraction (experimental conditions: 350 g/L $CaCl_2$, electric quantity of 1.2 times, temperature of 60 °C, duration of 3.5 h)

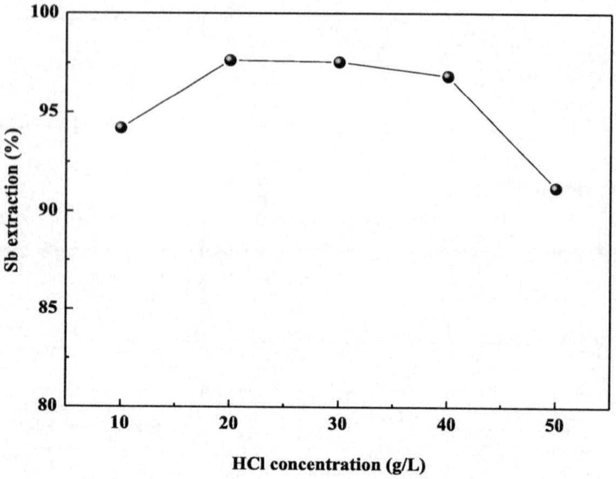

The effect of HCl concentrate on the leaching efficiency of Sb is studied and the result is presented in Fig. 4. The antimony leaching efficiency increased from 94.20 to 97.64% as the HCl concentrate changed from 10 to 20 g/L, and then it remained approximately 97% when the HCl concentrate further increased to 40 g/L. However, the leaching efficiency obviously decreased to 91.21% if the HCl concentrate reached to 50 g/L. This phenomenon needs a further investigation and explanation in the future study. Hence, the HCl concentrate of 20–40 g/L was recommended to the subsequent experiments.

Effect of Electric Quantity (Ah)

Slurry electrolysis is essentially anodic oxidation leaching of sulfide minerals, so the leaching efficiency of antimony is correlated with electric quantity (Ah). Assuming that all the antimony in the ore was present in stibnite. The theoretical electric quantity for the complete oxidative leaching of stibnite (Eq. 2) was considered as the benchmark.

The relation between Sb extraction and the electric quantity was investigated with the electric quantity ranging 0.8–1.4 times of the benchmark value (Fig. 5). The leaching efficiency of Sb increased sharply from 82.35 to 96.86% as the electric quantity ascended from 0.8 to 1.1 times, and then slightly increased to 97.56% as the electric quantity of 1.2 times. However, it decreased to 95.64% when the electric quantity further enhanced to 1.4 times. Therefore, it was concluded that 1.2 times was the optimal electric quantity.

$$Sb_2S_3 = 2Sb^{3+} + 3S^0 + 6e^- \quad (2)$$

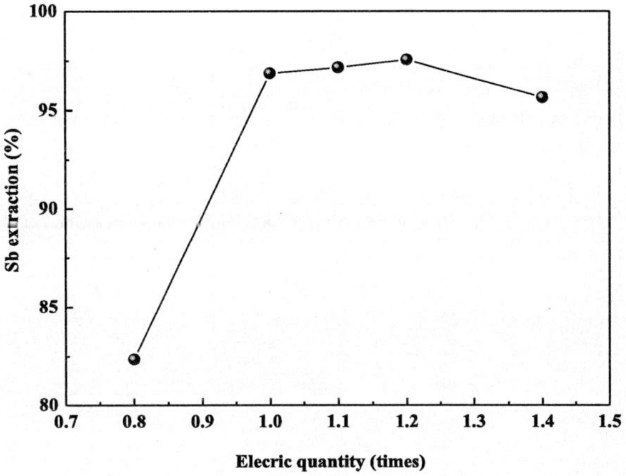

Fig. 5 Effect of electric quantity on antimony extraction (experimental conditions: 30 g/L HCl, 350 g/L $CaCl_2$, temperature of 60 °C, duration of 3.5 h)

Effect of Temperature

Figure 6 shows the effect of temperature on antimony extraction from 30 to 70 °C. An increase in temperature from 30 to 50 °C resulted in a sharp increase in antimony extraction. However, beyond 50 °C, the extraction of antimony did not prominent influence by the increasing of temperature. In addition, an increase in temperature will increase the energy consumption and electrolyte volatilization loss. Therefore, an optimal temperature range was considered to be 50–60 °C.

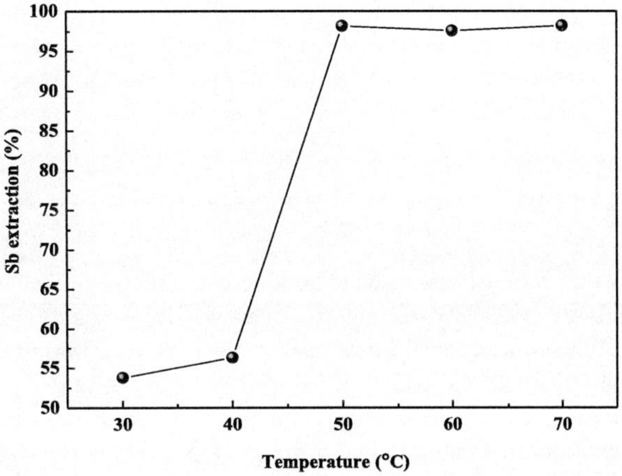

Fig. 6 Effect of temperature on antimony extraction (experimental conditions: 30 g/L HCl, 350 g/L $CaCl_2$, electric quantity of 1.2 times, duration of 3.5 h)

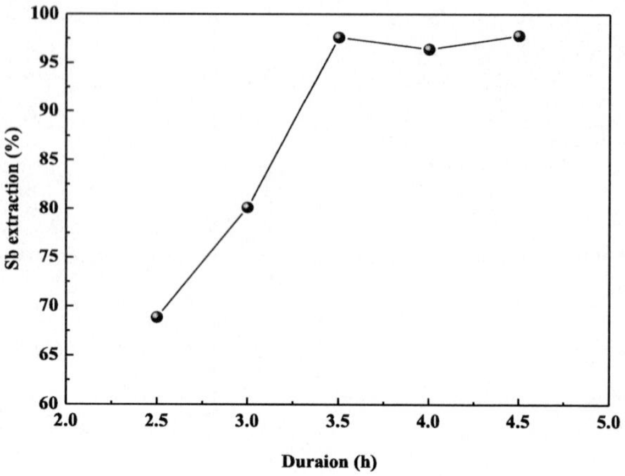

Fig. 7 Effect of duration on antimony extraction (experimental conditions: 30 g/L HCl, 350 g/L $CaCl_2$, electric quantity of 1.2 times, temperature of 60 °C)

Effect of Duration

With the electric quantity of 1.2 times, the effects of duration on antimony extractions are illustrated in Fig. 7. The antimony leaching efficiency increased from 68.89 to 97.59% as the SE duration increased from 2.5 to 3.5 h. Further increase in leaching time from 3.5 to 4.5 h had a slight effect on the extraction of Sb, adversely some other side reactions will occur. Therefore, the appropriate leaching time was 3.5 h. Under the condition of given electricity quantity (Ah), less electrolysis duration (h) means higher electrolysis current (A) was employed in the SE process. The leaching rate of minerals was limited by the kinetics, and had a maximum value. When the current oxidation rate was greater than this maximum value, the excess current will cause side reactions instead of leaching stibnite. Therefore, the leaching efficiency of antimony was lower at a small duration of SE process.

Comprehensive SE Experiments

Comprehensive tests were conducted under the optimized conditions, namely, $CaCl_2$: 30 g/L, HCl: 30 g/L, electric quantity: 1.1 times the benchmark value, temperature: 60 °C, duration: 3.5 h, antimony: 30 g/L. Antimony extraction reached over 96%, the average extraction of Ca and Au were 98.46%, and 0.43%, respectively. Gold remained in the leaching residue, which was further used as a raw material for gold extraction. The metallic antimony deposited on the cathode (Fig. 8) contained more than 99% antimony.

The X-ray diffraction spectra comparison of the SE leaching residue and raw ore is shown in Fig. 9. It indicates that the diffraction peak of stibnite, calcite, and dolomite almost disappeared after the SE, and some elemental sulfur appeared in the leaching residue. Other minerals of pyrite, quartz, arsenopyrite, and muscovite were basically

Fig. 8 Optical photograph of cathode metal antimony

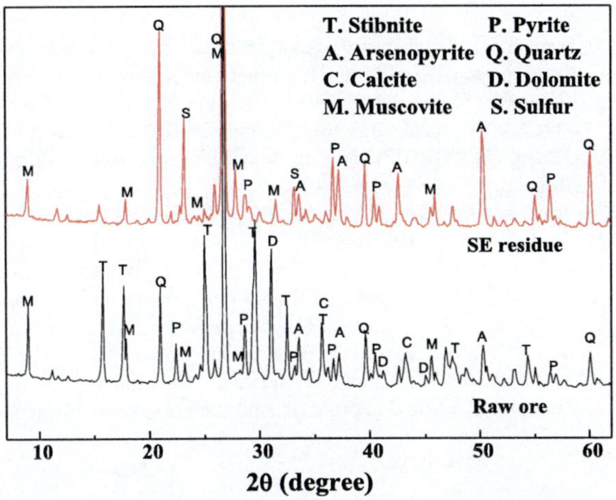

Fig. 9 The XRD patterns comparison of raw ore and SE residue of comprehensive test

consistent with the raw ore. Stibnite was dissolved by oxidation to liberate antimony ion and elemental sulfur was generated in the SE process. Other phases, such as pyrite, quartz, and arsenopyrite were relatively stable under the aforementioned conditions, and still remained in the leaching residue.

Conclusion

The SE process was used to pre-remove antimony from Ca–Sb–Au ore in HCl–$CaCl_2$ system. The stibnite was oxidized to released antimony ion into electrolyte in anode area and then deposited on cathode. The sulfur present in stibnite was transformed into elemental sulfur remaining in the residue. The results show that with the optimum conditions of $CaCl_2$ 300–350 g/L, electric quantity 1.2 times, HCl 30 g/L, time 3.5 h, and temperature 60 °C, the leaching efficiency of antimony and calcium was higher than 96% and 95%, respectively, and gold was less than 1%. The cathode antimony contained more than 99% antimony.

Acknowledgements This work was supported by the National Key R&D Program of China (2019YFC1908301), the Major Science and Technology Projects of Qinghai Province (grant number 2018-GX-A7), and the National Natural Science Foundation of China (51604030).

References

1. Yang TZ, Rao S, Liu WF, Zhang DC, Chen L (2017) A selective process for extracting antimony from refractory gold ore. Hydrometallurgy 169:571–575
2. Wang CY, Shao S, Ma BZ, Li X, Cheng YQ, Xing P (2019) Status and development of antimony and bismuth metallurgy technologies in China. Nonferrous Metals (Extract Metall) 08:11–17
3. Kuang JH (2019) Production practice of antimony sulfide gold concentrate smelting in blast furnace. Gold 40(08):62–65
4. Anderson CG (2012) The metallurgy of antimony. Geochemistry 72:3–8
5. Tian QH, Wang HL, Xin YT, Li D, Guo XY (2016) Ozonation leaching of a complex sulfidic antimony ore in hydrochloric acid solution. Hydrometallurgy 159:126–131
6. Dong ZQ, Li XS, Chen T (2020) Comprehensive recovery and utilization of antimony in copper and lead smelting process. Hunan Nonferrous Metals 36(06):21–24
7. Sun HM (2018) Extraction of antimony from a complex gold concentrate containing antimony in China. World Nonferrous Metals 02:207–208
8. Ubaldini S, Vegliò F, Fornari P, Abbruzzese C (2000) Process flow-sheet for gold and antimony recovery from stibnite. Hydrometallurgy 57:187–199
9. Yang TZ (2005) Metallurgy and product of precious metals. Central South University Press, Changsha, pp 210–248
10. Liu XG (2018) Experimental study on improving the cyanidation leaching rate for a gold concentrate containing arsenic and antimony. Gold 39(04):59–62
11. Xu XY, Xiong M, Cai CK, Ding WT, Guo JY (2018) Research progress of pretreatment technology of refractory gold ores containing arsenic, stibium and sulfur. Gold 39(08):68–72
12. Qiu DF (1999) A clean, energy efficiency extractive metallurgical method-slurry electrolysis process (SEP). Chin Eng Sci 1(01):67–72
13. Zhang S, Li YG, Wang R, Xu ZH, Wang B, Chen S, Chen MJ (2017) Superfine copper powders recycled from concentrated metal scraps of waste printed circuit boards by slurry electrolysis. J Cleaner Prod 152:1–6
14. Wang CY, Qiu DF, Jiang PH, Liang DH (2002) Jamesonite treatment by slurry electrolysis process. Nonferrous Metals 54(03):24–27
15. Zhang YL, Wang CY, Ma BZ, Jie XW, Xing P (2019) Extracting antimony from high arsenic and gold-containing stibnite ore using slurry electrolysis. Hydrometallurgy 186:284–291

Bionanomining: A Revised Insight into Processing of South Africa's Complex Gold Ores

Daniel Ogochukwu Okanigbe, Abimbola Patricia Popoola, Nicholas Malatji, Tsietsi Lesufi, and Gift Sekgobela

Abstract South Africa's economy is partly built on gold (Au) mining; this mining sector is an essential foreign exchange earner, with Au accounting for more than one-third of exports. But due to a decrease in high-grade gold ores, the need, therefore, arises to investigate ground-breaking approaches for sustainable processing of complex gold ores (CGO). Despite CGO mineral composition (MC) assisting in figuring out their responsiveness to conventional processes, the challenge of a low degree of liberation remains. Hence, this paper presented a revised insight into processing methods for energy-efficient recovery of gold from CGO in South Africa. This aim was achieved by reviewing publications on the different processing methods. It was concluded that it is important to develop cost-effective and energy efficient processing methods like bionanomining for environmentally friendly recovery of gold from CGO in order to guarantee a sustainable gold industry in South Africa.

Keywords Complex gold ore · Bionanomining · Energy efficient processing · Gold mining · Gold nanoparticles

Introduction

Gold (Au) mining in South Africa (SA) started more than a century ago in the Witwatersrand goldfields. Most of the high-grade Au ore deposits have been exhausted and mines now exploit lower grades. Hence, the reason for SA's current global ranking with respect to global Au Production [1]. SA was at one time the global leader in gold output, with the capacity of digging up over 1,000 tonnes (i.e. in 1970), however, its capacity has fast dropped, since then. In 2019, it dropped out of the top 10 gold producers' list altogether. Thus prompting an urgent demand for novel approaches

D. O. Okanigbe (✉) · A. P. Popoola · N. Malatji · T. Lesufi · G. Sekgobela
Faculty of Engineering and the Built Environment, Department of Chemical Metallurgical and Materials Engineering, Tshwane University of Technology, Pretoria 0183, Republic of South Africa
e-mail: okanigbdo@tut.ac.za

T. Ouchi et al. (eds), *Rare Metal Technology 2022*, The Minerals, Metals & Materials Series, https://doi.org/10.1007/978-3-030-92662-5_19

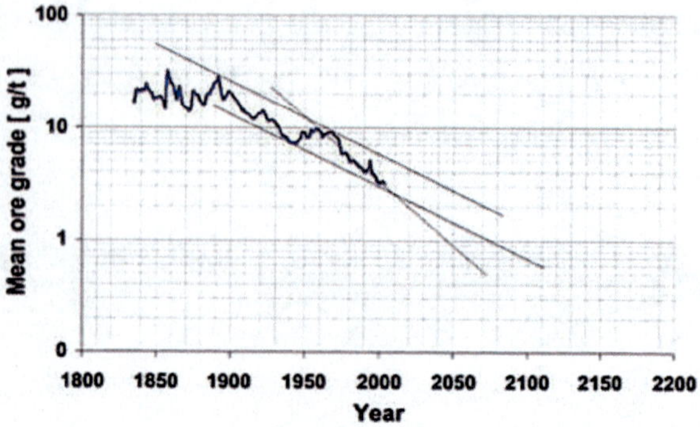

Fig. 1 Projection of ores mined in South Africa, Australia, Canada, Brazil, and the United States from 1830 to 2004 [3]

that can assist to derive best value from the low-grade/complex gold ores [2]. Figure 1 shows the arithmetic mean of the grade of Au for main gold-mining countries.

Based on the information presented in Fig. 1, it can be deduced that Au ore grade (AOG) has constantly dropped over the past 80 years. According to Müller and Frimmel [3], which shared that "should the decrease in AOG be plotted and extrapolated, it can be projected that the average AOG will be as low as 0.9 g/t by 2050. Which presupposes that most current Au operations will be marginal. Corroborating the statement made by Müller and Frimmel [3], while focusing on SA, Musingwini [2] mentioned that the average AOG of SA mines dropped from an estimated 12 g/t in the 1970s to an estimated 5 g/t in 2005 (see Fig. 2).

Based on data related to average recovered AOG for SA mines between 2004 and 2013 [4] in Fig. 2, which shows a drop in AOG from 5.15 g/t in 2005 to 2.91 g/t in the year 2013. This drop in AOG, together with drop in Au price, and increase in production cost, form key value drivers that determine the competitiveness of the SA gold mining industry. Currently, the attractiveness of this mining sector is made to appear irredeemable by the increased depth miners have to cover to explore for Au ores (often leading to low-grade/complex gold ores), a situation that further escalates production cost.

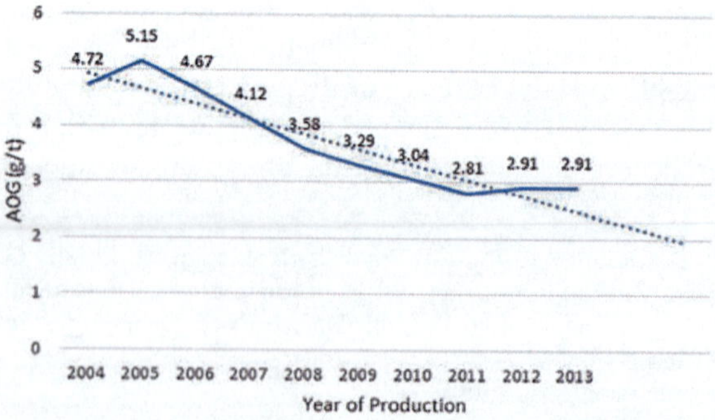

Fig. 2 South African average AOG versus production (adapted from [4])

Complex gold ores (CGO) often contain two or more metals, for example, Au ore with As or Sb minerals, or ore composed almost entirely of several sulphide minerals. Many CGO are difficult or costly to treat by employing the traditional metallurgical unit operations largely because of complex mineralogy [5]. CGO are classified as Au ores that require a lot of cyanide, oxygen or are preg-robbing [6] (a phenomenon whereby the Au cyanide complex ($Au(CN)_2$) is removed from solution by the constituents of the ore) [7]. This explains why treating CGO with traditional processing methods is considered neither economical nor environmentally friendly because of the high concentrations of cyanide required. Hence, the need for an existing processing route to enhance gold recoveries from these ores.

The complexities of CGO present significant mineral processing challenges [8]. Hence, in order to maintain sustained exploration and exploitation of gold resources in SA, an effective processing approach of CGO is urgently required. This paper, therefore, presents a review on techniques currently being employed in SA for gold processing, with the aim of establishing where the gap of knowledge exists for sustainable CGO processing.

Theoretical Background

Gold Ore

The primary mineral of gold is the native gold [9] and electrum [10] (Fig. 3a and b). Some tellurides are also important ore minerals such as calaverite [11], sylvanite [12], and petzite [13] (Fig. 3c). The major ores of gold contain gold in its native form and are either exogenetic, which means formed at the Earth's surface [14] or endogenetic which suggests formed within the Earth [15].

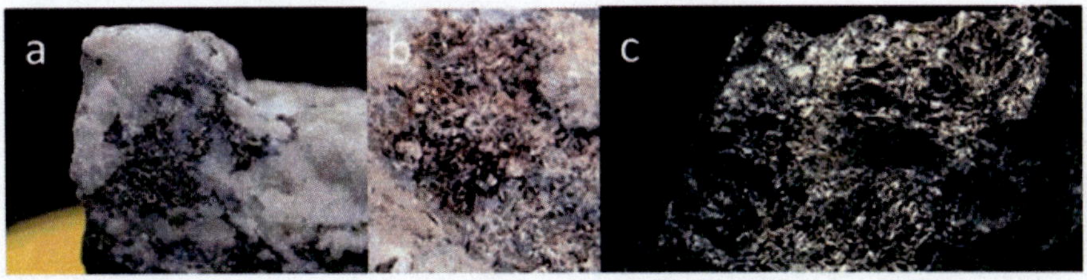

Fig. 3 Images of primary sources of gold ores **a** native gold, **b** electrum, and **c** tellurium (*Source* google image)

Fig. 4 Mponeng gold mine, the deepest operating mine in the world (*Source* google image)

Gold in South Africa

Most SA's Au resources are found in Witwatersrand basin (WB) [16, 17], the world's largest gold placer deposits [18, 19]. The dimension of this geological formation (i.e. WB) is 56 km in length in the direction of the north cliff tapping from east to west, as shown in Fig. 2. The Mponeng gold mine situated in the WB, is the world's deepest operating mine (Fig. 4) and the last underground mine being exploited by AngloGold Ashanti, in SA.

Processing of Gold Ores

The processing of Au can involve the use of either of these unit operations—gravity separation, flotation, microbial oxidation, etc. [20, 21]. As made reference to in the section before, Au occurs often in its native form or hosted in sulphides (especially pyrite, arsenopyrite, and chalcopyrite), silicate, carbonate, and oxide minerals, [20, 22, 23], allowing Au ores to be classified into refractory gold ores or non-refractory gold ores.

Whether a gold ore is refractory and non-refractory is a relative term, which bothers on the degree to which the gold ore is responsive to cyanide extraction, and it ranges from close to 0% through to near 100%. The potential of a mine's sustainability is hugely premised on the gold ore being either refractory gold ores (RGO) or non-refractory gold ores (NRGO). Thus presupposing that processing cost is an integral part of the economics of any gold project.

Au can exist in complex sulphide minerals particularly, pyrite and arsenopyrite [24]. In RGO, the Au is significantly embedded in the sulphide matrix and as such pre-treatment processing becomes a requirement to recover contained Au value from the sulphide minerals. This is because the sulphide minerals often inhibit the recovery of Au during leaching. Often flotation and roasting are processing methods employed to separate gold from its associated gangue minerals. This practice helps to prevent the formation of a passive layer on the mineral surface, which hinders Au recovery from complex sulphide minerals [25]. Additionally, Au can also be found in arsenic-bearing refractory gold ores. These ores commonly contain stones that lock up Au, hence, the need for an initial crushing process to liberate the Au for leaching with an appropriate gold leaching agent [26, 27]. The crushing of rocks is a capital intensive process, which can defeat the whole economics of a gold mining project, particularly if the Au ores are substantially CGO.

Processing CGO in South Africa

High Operating Costs

Although the mining industry (MI) is a leading producer of wealth for SA, it is also a large consumer of electric power, taking about 15% of ESKOM's annual output [28]. Amongst the mining sectors in SA, the Au sector is the largest consumer of the MI's electricity (47%), platinum consumes 33% of the electricity, while the remaining 20% is consumed by other mines (Table 1).

Since expenditure outweighs income, particularly because of the continued declining trend in the AOG in SA (Fig. 2), even though it is well known that most of the Witwatersrand gold ores are highly amenable to gold cyanidation; it becomes necessary for Au mining companies to resort in mining and enhanced processing of CGO, to augment for this short supply of high-grade Au ores created by the current declining AOG (Fig. 2), a situation that threatens the going concern of most, if not all the Au mining companies in SA.

Table 1 Order of electricity usage in the mining industry [28]

S/N	Mining sector	Consumption of electricity provided by ESKOM to the industry (%)
1	Gold	47
2	Platinum	33
3	others	20

Haven't said this, it should also be mentioned that the processing of CGO can be technically challenging, because of low recovery rates, excessive reagent consumption, and high operating costs when compared to free-milling gold ores. It is, therefore, necessary to investigate for novel energy efficient processing methods that can enhance the recovery of Au from CGO.

Proposed Energy Efficiency for Enhanced Processing of CGO in South Africa

As reported by Eskom [28], in SA, the energy consumption by the MI, can be saved by applying efficiency solutions in the following areas in the order of significance: Materials handling, processing, compressed air, pumping, fans, industrial cooling, and lighting (Fig. 5).

With this understanding, which amplifies the need to create a strategic plan, particularly as it relates to the processing of CGO; since the processing aspect of all MI accounts for 19% of the energy efficiency consumption (Fig. 5). It is in this light that the present appraisals of publications as it relates to processing CGO for Au recovery in SA is presented in the sections that follows:

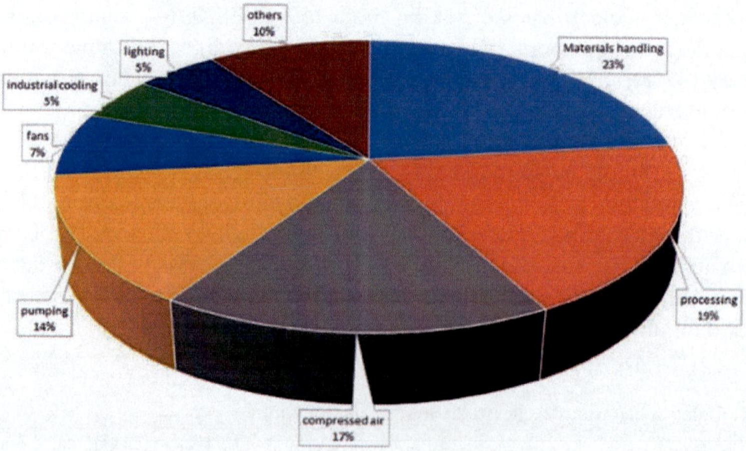

Fig. 5 Energy efficiency in the mining industry (adapted from [28])

Review of Past and Current Publications on Processing of CGO in South Africa

Traditional Methods of Processing

There are basically two main methods recommended for base metal processing and are as follows:

Pyrometallurgy

It involves the use of very high temperature to melt Au ore in a furnace for a certain period of time. The pyrometallurgical recovery process of gold is costly, energy- and time-inefficient, further made complex by releases of toxic gases, such as sulphur dioxide into the atmosphere [29].

Hydrometallurgical Gold Recovery Processes

It involves the use of aqueous solutions to dissolve soluble gold species [30]. Thus, making it the most practiced beneficiation process, for the recovery of gold from GCO using cyanide leaching [31, 32, 33]. The dissolution of gold in cyanide is best described by heterogeneous reactions which occur at the solid–liquid interfaces. This can be outlined as follows:

A. Absorption of oxygen in solution [34].
B. Transportation of dissolved cyanide and oxygen to solid–liquid interface [35].
C. Nernst boundary layer of CN^- and O_2 and adsorption onto the solid surface [36].
D. Electrochemical reaction [37].
E. Production desorption from the surface of solid [38].
F. Transportation of reaction products transportation to the bulk solution [39, 40].

Bionanomining of Gold Nanoparticles (nAu)

In recent years, bionanomining emerged as a cutting-edge scientific–technological development associated with the application of micro-organisms and macro-organisms to generate nanotechnological products by using mining and industrial wastes and wastewaters. It is necessary to use bionanomining to produce low-cost, energy-efficient, and non-toxic metal nanoparticles (NPs) [41]. Biomass of many species of bacteria, plants, etc., have the ability to reduce or oxidize cations, which can physically be deposited as nanometric materials such as nanoparticles [42].

Bacteria

In the works conducted by Shuster et al. [43], it was shared that there are special 'Au nugget-producing' bacteria with the key to more efficient processing of gold ore, mine tailings, and recycled electronics, as well as aid in exploration for new deposits. In another similar work conducted by Bütof et al. [44], the bacterium

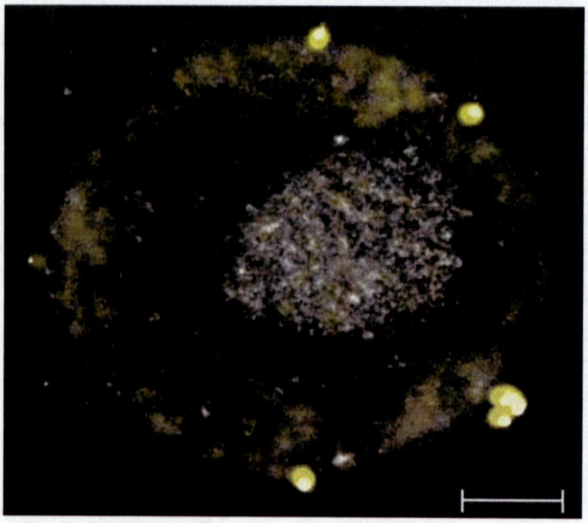

Fig. 6 *C. metallidurans* can produce small gold nuggets. Photo is credited to the American Society for Microbiology [44]

Cupriavidus metallidurans was reported to possess the capacity to reduce toxic Au^{+} and Au^{3+}-complexes into metallic gold (Au) nano-particles, thereby mediating the transformation of Au nuggets (Fig. 6) in Earth surface environments.

Plants

The process of metallic nanoparticles (mNPs) formation in plants is by the reduction of metal ions taken in as dissolvable salts. Even though the actual process underlying this formation is not well understood in plants, it is believed that plants produce mNPs extracellularly. An example is the work conducted by Raju et al. [45], in which the researchers shared how AuNPs are produced extracellularly using living peanut seedlings as shown in Fig. 7a–e. The researchers reported that in the membrane of the cell and in solution, the produced AuNPs were very stable. Characterization of the AuNPs using the transmission electron microscopy, showed particles are of uniform size in the dispersed phase with sizes ranging between 4 and 6 nm. Bali and Harris [46], observed that the process of extracellular production of AuNPs, occurred after the interactions between root exudates (they used *B. juncea* and *M. sativa*) and the aqueous solution of $KAuCl_4$ that resulted in the reduction of Au^{3+} to AuNPs. Particle sizes of the AuNPs were 5–10 nm for *B. juncea* and 10–20 nm for *M. sativa*.

For a better understanding of the mechanism of reduction processes, Beattie and Haverkamp [47] verified that the locations of the most abundant reduction of Au metal salts to AuNPs were the chloroplasts, the areas of high reduction of glucose and fructose. They suggested that glucose and fructose were accountable for the reduction of Au with reduction potentials of over +0.16 V and that the quantity of reducing

Fig. 7 **a** Plant that grew on paper, **b** 24-h, **c** 48-h, **d** 72-h exposed seedlings, and **e** 10 − 4 M $HAuCl_4$ solution [45]

glucose and/or fructose present or produced controls the quantity of AuNPs that is produced. In addition, AuNPs developed using chloroplasts, revealed success [48].

Generally, living plants have been very efficient in the production of AuNPs in a faster and safer approach. It is expected that future investigations discover more suitable living plant species in field application, soil-based systems, and the use of waste streams containing metal ions (bionanomining) [49] to act as a precursor to produce both extracellular and intracellular AuNPs in SA. Hence, within the new production patterns, biological routes are being proposed as an alternative method of mining Au from its resources, an approach that will require less polluting organic solvents and less energy consumption [49].

Conclusions

Added to the global challenges that Au mining faces; SA Au mining operations is faced with its own unique challenges that brutally impact the sector's productivity, and as a result, profitability and sustainability are impaired significantly. Amongst these challenges are environmental issues, high electricity cost, erratic electricity supply, and safety-related stoppages imposed by SA Department of Mineral

Resources. All these international and national drawbacks have led to increased operating costs, squeezed profit margin, and possible future suspension of going concern of Au mining companies.

SA, therefore, requires gold mining companies to develop innovative extraction strategies to ensure the long-term competitiveness and viability of the sector. Mining companies need to move away from traditional mine approaches like pyrometallurgy and hydrometallurgy that are less energy-efficient to biological routes that require less polluting solvents and less energy consumption. It is therefore recommended that future research in the area of Au mining should focus on bionanomining, seeing that this approach is not polluting and energy-efficient. It is therefore recommended that future research should focus on finding efficient ways to concentrate the AuNPs produced using bionanomining as a method of recovering Au from CGO.

Acknowledgements The authors wish to thank Tshwane University of Technology, Pretoria, South Africa for the support given during the preparation of this manuscript.

References

1. Bond P (2013) Debt, uneven development and capitalist crisis in South Africa: from Moody's macroeconomic monitoring to Marikana microfinance mashonisas. Third World Q 34(4):569–592
2. Musingwini C (2014) Introduction of specialization in mine planning and optimization within the Master's degree (MSc) programme at the University of Witwatersrand. In: Proceedings of the 6th international platinum conference, platinum-metal for the future, pp 20–22
3. Müller J, Frimme HE (2010) Numerical analysis of historic gold production cycles and implications for future sub-cycles. Open Geol J 4(1)
4. Chamber of Mines of South Africa (2014) Facts and figures 2014. https://commondatastorage.googleapis.com/comsa/f_f_2014_final.pdf. Accessed 19 May 2015
5. Mabwe NR (2020) Potential pre-treatment and processing routes for recovery of gold from complex (refractory) gold ores. A review
6. Marsden JO, House C (2009) The chemistry of gold extraction, 2nd edn. Society for Mining, Metallurgy, and Exploration Inc., Colorado
7. Ilyas S, Lee JC (2018) Gold metallurgy and the environment. CRC Press
8. Liu G, Wu Y, Tang A, Li B (2020) Recovery of scattered and precious metals from copper anode slime by hydrometallurgy: a review. Hydrometallurgy 197: 105460
9. Chapman RJ, Banks DA, Styles MT, Walshaw RD, Piazolo S, Morgan DJ, Grimshaw MR, Spence-Jones CP, Matthews TJ, Borovinskaya O (2021) Chemical and physical heterogeneity within native gold: implications for the design of gold particle studies. Mineral Depos 1–26
10. Nachlas W, Schranz L, Rogers D (2021) Quantitative X-ray mapping of Au–Ag ratios in native electrum from the Fire Creek epithermal vein deposit, Lander County, Nevada (USA). Microsc Microanal 27(S1):1296–1298
11. Korolev I, Altinkaya P, Haapalainen M, Kolehmainen E, Yliniemi K, Lundström M (2021) Electro-hydrometallurgical chloride process for selective gold recovery from refractory telluride gold ores: a mini-pilot study. Chem Eng J 132283
12. Alexandre P, Fayek M (2021) Formation temperature and ages of the True North orogenic gold deposit in Manitoba. Geological Society, London, Special Publications, Canada, p 516

13. Sorokhtina NV, Zaitsev VA, Petrov SV, Kononkova NN (2021) Estimation of formation temperature of the noble metal mineralization of the Kovdor Alkaline-Ultrabasic Massif (Kola Peninsula). Geochem Int 59(5):474–490
14. Nikiforova Z (2019) New data on microhardness of placer gold. In: International congress on applied mineralogy. Springer, Cham, pp 115–118
15. Kazhenkina AG, Nikiforova ZS (2018) Prediction of possible primary sources of gold-platinoid mineralization within the Lena-Viluy interfluve (East Siberian platform)
16. Ahiakwo NI, Egwuonwu AC, Okeke OC (2018) A review of the geology and mineral resources of South Africa. Int J Adv Acad Res 4:90–121
17. Nwaila GT, Ghorbani Y, Zhang SE, Frimmel HE, Tolmay LC, Rose DH, Nwaila PC, Bourdeau JE (2021) Valorisation of mine waste-Part I: characteristics of, and sampling methodology for, consolidated mineralised tailings by using Witwatersrand gold mines (South Africa) as an example. J Environ Manag 295:113013
18. Frimmel HE (2019) The Witwatersrand basin and its gold deposits. In: The Archaean geology of the Kaapvaal Craton, Southern Africa. Springer, Cham, pp 255–275
19. Nguimatsia FWD, Bolarinwa AT, Yongue RF, de Dieu Ndikumana J, Olajide-Kayode JO, Olisa OG, Abdu-Salam MO, Kamga MA, Djou ES (2017) Diversity of gold deposits, geodynamics and conditions of formation: a perspective view. Open J Geol 7(11):1690
20. Allan GC, Woodcock JT (2001) A review of the flotation of native gold and electrum. Miner Eng 14(9):931–962
21. Asamoah RK, Skinner W, Addai-Mensah J (2018) Leaching behaviour of mechano-chemically activated bio-oxidised refractory flotation gold concentrates. Powder Technol 331:258–269
22. O'Connor CT, Dunne RC (1994) The flotation of gold bearing ores—a review. Miner Eng 7(7):839–849
23. Stewart M, Kappes D (2012) SART for copper control in cyanide heap leaching. J Southern Afr Inst Mining Metall 112(12):1037–1043. Syed S (2012) Recovery of gold from secondary sources—a review. Hydrometallurgy 115–116, pp 30–51
24. Asamoah RK, Zanin M, Skinner W, Addai-Mensah J (2019) Refractory gold ores and concentrates part 2: gold mineralisation and deportment in flotation concentrates and bio-oxidised products. Miner Process Extract Metall 1–14
25. Cho K, Kim H, Myung E, Purev O, Choi N, Park C (2020) Recovery of gold from the refractory gold concentrate using microwave assisted leaching. Metals 10(5):571
26. Yang EG, Li YJ (2008) Northeastern University (Natural Science) 12:1742
27. Zhou ST, Liu L (2009) Metal mine. 2:98
28. Eskom DSM (2010) The energy efficiency series: towards an energy efficient mining sector
29. Syed S (2012) Recovery of gold from secondary sources—a review. Hydrometallurgy 115–116:30–51
30. Oraby EA, Eksteen JJ (2016) Gold dissolution and copper suppression during leaching of copper–gold gravity concentrates in caustic soda-low free cyanide solutions. Miner Eng 87:10–17
31. Brooy SRL, Linge HG, Walker GS (1994) Review of gold extraction from ores. Miner Eng 7:1213–1241
32. Moses LB, Peterson FW (2000) Flotation as a separation technique in the coal gold agglomeration process. Miner Eng 13:255–264
33. Vaughan JP, Kyin A (2004) Refractory gold ores in Archaean greenstones, Western Australia: mineralogy, gold paragenesis, metallurgical characterization and classification. Mineral Mag 68:255–277
34. Sitando O, Dai X, Senanayake G, Nikoloski AN, Breuer P (2020) A fundamental study of gold leaching in a thiosulfate-oxygen-copper system in the presence of activated carbon. Hydrometallurgy 192:105232
35. Burat F, Demirağ A, Şafak MC (2020) Recovery of noble metals from floor sweeping jewelry waste by flotation-cyanide leaching. J Mater Cycles Waste Manag 22(3):907–915
36. Srivastava RR, Ilyas S, Ilyas N, Kim H (2021) Urban mining of precious metals with cyanide as lixiviant. Sustain Urban Mining Precious Metals 91–122

37. Azizitorghabeh A, Wang J, Ramsay JA, Ghahreman A (2021) A review of thiocyanate gold leaching–Chemistry, thermodynamics, kinetics and processing. Miner Eng 160:106689
38. Fritz PA, Bera B, van den Berg J, Visser I, Kleijn JM, Boom RM, Schroën CGPH (2021) Electrode surface potential-driven protein adsorption and desorption through modulation of electrostatic, van der Waals, and Hydration Interactions. Langmuir
39. Ringe S, Morales-Guio CG, Chen LD, Fields M, Jaramillo TF, Hahn C, Chan K (2020) Double layer charging driven carbon dioxide adsorption limits the rate of electrochemical carbon dioxide reduction on Gold. Nat Commun 11(1):1–11
40. Zhang BA, Costentin C, Nocera DG (2020) Driving force dependence of inner-sphere electron transfer for the reduction of CO_2 on a gold electrode. J Chem Phys 153(9):094701.
41. Wong-Pinto LS, Menzies A, Ordóñez JI (2020) Bionanomining: biotechnological synthesis of metal nanoparticles from mining waste—opportunity for sustainable management of mining environmental liabilities. Appl Microbiol Biotechnol 104(5):1859–1869
42. Pan HW, Iizuka A, Shibata E (2020) Gold recovery from dilute aqueous solution by a biosorbent derived from woody biomass. Chem Eng Commun 1–14
43. Shuster J, Reith F, Cornelis G, Parsons JE, Parsons JM, Southam G (2017) Secondary gold structures: relics of past biogeochemical transformations and implications for colloidal gold dispersion in subtropical environments. Chem Geol 450:154–164
44. Bütof L, Wiesemann N, Herzberg M, Altzschner M, Holleitner A, Reith F, Nies DH (2018) Synergistic gold–copper detoxification at the core of gold biomineralisation in Cupriavidus metallidurans. Metallomics 10(2):278–286
45. Raju D, Mehta UJ, Ahmad A (2012) Phytosynthesis of intracellular and extracellular gold nanoparticles by living peanut plant (*Arachis hypogaea* L.). Biotechnol Appl Biochem 59:471–478
46. Bali R, Harris AT (2010) Biogenic synthesis of Au nanoparticles using vascular plants. Ind Eng Chem Res 49:12762–12772. https://doi.org/10.1021/ie101600m
47. Beattie IR, Haverkamp RG (2011) Silver and gold nanoparticles in plants: sites for the reduction to metal. Metallomics 3:628–632
48. Shabnam N, Pardha-Saradhi P, Sharmila P (2014) Phenolics impart Au^{3+} stress tolerance to cowpea by generating nanoparticles. PLoS One 9
49. Wong-Pinto LS, Mercado A, Chong G, Salazar P, Ordóñez JI (2021) Biosynthesis of copper nanoparticles from copper tailings ore—an approach to the 'Bionanomining'. J Cleaner Product 315:128107

Research Progress in Biohydrometallurgy of Rare Metals and Heavy Nonferrous Metals with an Emphasis on China

Jianzhi Sun, Bowei Chen, He Shang, Xiaolan Mo, and Jiankang Wen

Abstract In the twenty-first century, progress in the field of biohydrometallurgy had been significant. Since 2006, a total of 17 novel biomining microorganisms were discovered, and 8 copper heap bioleaching plants and 12 gold biooxidation plants were established or expanded. This review summarized the it was summarized the physiological properties of the newly isolated biomining microorganisms and three novel microbial ecological methods for studying microbial community dynamics and structure. In addition, biohydrometallurgy research on rare metals such as uranium, molybdenum, tellurium, germanium, indium, and secondary rare metal resources, as well as heavy nonferrous metals such as copper, nickel, cobalt, and gold has been reviewed, with an emphasis on China. In future, further studies on bioleaching of low grade chalcopyrite ore, rare metals, secondary resources from waste, and environmental pollution caused by resource utilization are necessary.

Keywords Biohydrometallurgy · Microbial ecology · Rare metals · Practice

Introduction

The industrial application of biohydrometallurgical technology began in the 1980s and subsequently entered the first golden period in the 1990s. During this period, 17 copper heap bioleaching plants and 9 gold biooxidation plants went into operation. From 2000 to 2005, 3 copper biohydrometallurgy plants and 11 gold biooxidation plants were commissioned. Since 2006, the application of biohydrometallurgical technology has entered the second golden period. In this period, 8 copper heap bioleaching plants and 12 gold biooxidation plants had been established or expanded, and the highest copper production capacity of heap bioleaching plants using primary and secondary sulfide copper ore reached 180,000 and 200,000 ton·year^{-1}, respectively. Furthermore, the largest gold biooxidation plant, which treats 2138 ton·day^{-1} gold concentrates, annually produces more than 20 ton gold. In addition, industrial

J. Sun · B. Chen · H. Shang · X. Mo · J. Wen (✉)
GRINM Resources and Environment Tech. Co. Ltd, Beijing, China
e-mail: wjk@grinm.com

T. Ouchi et al. (eds), *Rare Metal Technology 2022*, The Minerals, Metals & Materials Series, https://doi.org/10.1007/978-3-030-92662-5_20

application of nickel–cobalt–copper–zinc–uranium ore heap bioleaching had also been realized with an annual output of more than 16,000 ton nickel and 32,000 ton zinc. Based on its rapid development and enormous potential in terms of refractory resource utilization, biohydrometallurgical technology had been described as one of the 10 "world-changing ideas" by Scientific American in 2011 [1].

Biomining Microorganisms

Newly Isolated Biomining Microorganisms

Biomining microorganisms refer to microorganisms having ore-leaching characteristics that can be used in bioleaching and biooxidation processes. Currently, bioleaching and biooxidation processes are conducted under acidic conditions, hence, biomining microorganisms are usually acidophiles. Since 2016, 17 new biomining microorganisms had been isolated. Some of their physiological properties are given in Table 1.

Key Issues in Breeding of Efficient Biomining Microorganisms

In traditional microbiological methods, the microorganisms are enriched and screened in the culture media, and identified based on physiological and biochemical characteristics and scanning electron microscope (SEM). In contrast, in modern molecular biological methods, the design of the culture media can be guided with the "eyes" to achieve rapid isolation and screening. For example, the nitrogen fixer *Leptospirillum ferrodiazotrophum* was isolated from acid mine drainage using the genome-directed method [2].

The ultimate goal of breeding of biomining microorganisms is utilization in mineral bioleaching/biooxidation; therefore, the target minerals should be used in breeding and domestication processes. As oxidation of sulfide ores usually involves iron and sulfur oxidation, occasionally, ideal results cannot be achieved using microorganisms with single oxidation function. Thus, use of a combination of iron oxidizer, sulfur oxidizer, and heterotroph has been suggested. Biomining microorganisms usually have a broad temperature range and high resistance to toxic ions, and must be gradually adapted to improve their performance. Through years of experience, it is found that the use of native source strains is better than utilization of strains from other mines. Therefore, the initial samples for the breeding of microorganisms should be collected from the target mine [3, 4].

Table 1 Optimum and range of growth for pH and temperature of newly isolated biomining microorganisms

Species	Optimal pH	pH range	Optimal temperature/°C	Temperature range/°C
Acidithiobacillus concretivorus	2.0–4.0	−0.6–6.0	28	10–37
Acidithiobacillus ferrianus	2.2	1.3–3.5	28–30	20–32
Acidithiobacillus ferriphilus	2.0	1.5–3.0	30	10–30
Acidithiobacillus sulfuriphilus	3.0	1.8–7.0	25–28	15–30
Thiomicrospira aerophila	9.8–10.0	7.5–10.6	ND	25–41
Thiomicrospira cyclica	9.5	7.5–10.5	ND	ND
Thiomicrospira hydrogeniphila	6.0	5.0–8.0	30	2–40
Thiomicrospira microaerophila	9.0	8.0–10.0	25–28	ND
Thiomicrospira sibirica	9.8–10.0	7.5–10.6	ND	20–41
Sulfurovum denitrificans	7.0	6.5–7.5	30	10–35
Sulfurovum indicum	6.0	5.0–8.6	37	4–50
Sulfurovum riftiae	6.0	5.0–8.0	35	25–40
Thermithiobacillus plumbiphilus	6.4–7.1	5.8–8.7	28–32	5–37
Sulfurimonas crateris	8.0	5.5–9.5	30	5–40
Alicyclobacillus montanus	~3.0	1.5–4.5	45	25–55

Application of Biohydrometallurgy

Currently, about 20% of the world's mined copper and 3% of the world's mined gold are produced by biological processing [5]. In addition, this technology is now being applied to rare metals such as uranium, molybdenum. Recently, BHP Billiton accomplished a 6-m column bioleaching of chalcopyrite [6]. An average temperature of 70 °C was attained in the column through heat generated from bioleaching, and the copper recovery was 75% in a leach cycle of 283 days. For high acid-consuming ores, bioleaching is performed at elevated pH [7]. In the "Ferredox" process employed for nickel laterite treatment, the bacterium *At. ferrooxidans* is used to combine oxidation of elemental sulfur (S^0) with reduction of ferric iron contained in the mineral goethite (or limonite ore), releasing nickel, cobalt, and manganese [8]. The following are the recent applications of biohydrometallurgy on rare metals and heavy nonferrous metals.

High-Fluorine Uranium Ore

High-fluorine uranium ore accounts for a large proportion of the world's uranium resources, with more than 60% found in China. The fluorine concentration in the leachate of high-fluorine uranium ore can reach 2–4 g·L^{-1}. The lowest inhibition concentration of fluoride for bacterial acid production is 1–2 mg·L^{-1}, and fluorine concentrations of 100–250 mg·L^{-1} can be lethal to microorganisms. The pH and concentrations of some soluble components such as manganese, aluminum, iron is determined by the parameters controlling the bioleaching process (ore particle size, acidification process, spraying conditions, etc.), which influence the form of fluorine, thus affecting the bacterial oxidation activity. Mo et al. [9] isolated an *At. ferrivorans* strain that could completely oxidize 9 g·L^{-1} Fe^{2+} in 24 h in a medium containing 1.0 g·L^{-1} fluorine (Fig. 1). Recently, a high-fluorine uranium ore heap bioleaching experiment was conducted in Southern China by the East China University of Technology and General Research Institute for Nonferrous Metals, China (Fig. 2). When compared with acid leaching, the leaching cycle in the bioleaching experiment was reduced by 18–28%, acid consumption was decreased by 10–13%, and uranium leaching efficiency was more than 90% [10].

The Bioleaching of Rare Metals

Chen et al. [11, 12] studied bioleaching of nickel–molybdenum sulfide ore with *Sulfolobus metallicus* and compared the difference in leaching between column reactor and membrane biological reactor (MBR). Their results showed that the leaching efficiency of nickel and molybdenum in the reactor with *S. metallicus* were 93.17% and 73.52%, whereas those in the reactor with *Thiobacillus ferrooxidans* were 67.34% and 38.36%, respectively. Lei and Xie [13] investigated the leaching of tellurium from low-grade sulfide mineral in shake flasks by using mixed bacteria enriched from the acid mine drainage. Their results showed that the tellurium extraction efficiency reached 76.3% after 30 days of bioleaching. Xie [14] studied bioleaching of indium from zinc concentrate using bacteria and noted that the leaching efficiency of zinc and indium with bacteria reached 98.70% and 72.46% at 33 °C after 40 days, respectively, whereas the leaching efficiency of indium without bacteria was only 26.93%.

Secondary Rare Metal Resources

With social and economic development, as well as owing to human activity and production, a large amount of wastes are produced every year, which contain a considerable amount of rare metals. Xin et al. [15] investigated the bioleaching mechanism of cobalt and lithium from spent lithium-ion batteries using mixed culture of sulfur- and iron-oxidizing bacteria. Their results revealed that the dissolution concentration of cobalt in the S + FeS system (1068 $mg{\cdot}L^{-1}$) was higher than that in the S system (866 $mg{\cdot}L^{-1}$) or FeS_2 system (989 $mg{\cdot}L^{-1}$) after 5 days of bioleaching, whereas lithium exhibited the highest and lowest bioleaching efficiency in the S system (441 $mg{\cdot}L^{-1}$) and FeS_2 system (353 $mg{\cdot}L^{-1}$), respectively. Furthermore, Qu and Lian [16] investigated biological leaching of rare earth elements and radioactive elements from red mud and showed that the recoveries of Y, Sc, La, Lu, Yb, Ho, Tm, Gd, Sm, and Eu were more than 75%, 70%, 28%, 65%, 65%, 65%, 65%, 55%, 40%, and 40%, respectively.

Copper Ore

In December 2005, a heap bioleaching plant with a capacity of 10,000 $ton{\cdot}year^{-1}$ was constructed in Zijinshan Copper Mine, China. In 2010, the capacity of the plant was expanded to 30,000 $ton{\cdot}year^{-1}$, and in 2012, production was restarted (Fig. 3). The average grade of copper added into the heap is 0.308%. The ore is crushed to a coarse particle size (40 mm accounting for 80%), and the crushed ore is simply stacked on the heap without acid agglomeration. The heap is formed with three lifts, each with a height of 8 m. The solution pH is controlled at 1.5–1.7, and redox potential is maintained at 730–760 mV (vs. standard hydrogen electrode); thus, the heap temperature could reach above 45 °C. During the 185-day leaching cycle, copper leaching efficiency of more than 76% and iron leaching efficiency of less than 6% are achieved, indicating a high copper sulfide leaching efficiency and inhibition of pyrite dissolution [17].

Nickel and Cobalt Ore

The Baishan nickel–cobalt–copper Mine is located in Baishan, Jilin, China, 15 km away from the Yalu River. The average temperature of the mine is below 0 °C from the end of November to the end of April, with the lowest temperature of −30 °C. The ore contains 0.180% nickel, 0.169% copper, 0.058% cobalt, 0.520% sulfur, and 0.070% arsenic. In 2007, a microbial consortia with high ferrous ion oxidation ability at 15 °C was isolated from the acid mine drainage of the mine. Flask and column bioleaching experiments showed that the nickel leaching efficiency could reach more than 70%. In 2014, a 60,000-ton ore heap bioleaching test was completed (Fig. 4). The ore particle size was −20 mm accounting for 80%. The nickel, cobalt, and copper leaching efficiency reached 71.29%, 70.55%, and 41.61% in 12 months, and 79.18%, 77.25%, and 46.93% in 16 months, respectively. The leachate containing nickel, cobalt, and copper was recovered by sulfide precipitation, and the recovery rate was 98.02% [18].

Talvivaara mining company owns a low-grade nickel sulfide mine (Ni 0.23%, Zn 0.51%, Cu 0.13%, Co 0.02%) in eastern Finland. Due to the low grade of raw ore, it is difficult to realize economic utilization by conventional beneficiation and metallurgy. In 2005, heap building was deployed for bioleaching. After crushing, sulfuric acid was used to acidify the raw ore to pH = 1.8, double heap leaching was used, and nickel and zinc were leached in the first heap. After leaching for one and a half years, the ore moved to the second reactor for leaching copper and cobalt. The temperature in the reactor is 30–90 °C and the bacterial concentration of the leaching solution is $1 \times 10^5 \sim 1 \times 10^8$/mL. At the end of 2006, the leaching efficiency of nickel, zinc, cobalt, and copper in the first pile reached 94%, 83%, 14%, and 3%, respectively. After moving the pile, the final leaching efficiency of nickel, zinc, and copper reached 100%, and the leaching efficiency of cobalt and copper were 35% and 22%, respectively. In 2020, the storage yard have 30,000 tons of nickel and 60,000 tons of zinc.

Gold Ore

Currently, the number of refractory gold ore biooxidation plants around the world is more than 30. Since 2006, 11 plants have been established, of which 6 are located in China [19]. Table 2 shows a list of the biooxidation plants operated.

Conclusion and Outlook

Although pyrometallurgy has been conventionally used for the recovery of rare metals, with the draining of resources and increasing environmental issues arising from the use of secondary resources, green and efficient extraction techniques are gaining increasing attention, among which biohydrometallurgy is one of the most effective techniques. While the mechanisms for the extraction of rare metals using biohydrometallurgy are similar to those of traditional biohydrometallurgy, the type of microorganisms, extractants, and ion-exchange resin used for solution separation may differ. The field of rare metals biohydrometallurgy is broader and is still in the early stages of development, with leaching–separation being the key factor.

Currently, the biohydrometallurgy techniques employed for secondary copper sulfide ores, copper–nickel–cobalt ores, refractory gold ores, and uranium ores have advanced. However, further research is necessary for developing methods that can

Table 2 New built/expanded refractory gold ore biooxidation plants from 2006

Plant	Country	Capacity/(ton·day^{-1})	Process	Operational status
Bogoso	Ghana	750	BIOX	2006–present
Sanhe	China	80	CCGRI	2006–present
Olimpiada	Russia	1000	BIONORD	2001–present (expansion in 2007)
Jinfeng	China	790	BIOX	2007–present
Hatu	China	100	Tank reactors	2008–present
Kokpatas	Uzbekistan	2138	BIOX	2009–present
Maud Creek	Australia	109	Geobiotics	2010–present
Wulaga	China	100	CCGRI	2010–present
Agnes	South Africa	20	BIOX	2010–present
Yili	China	80	Tank reactors	2012–present
Huibao	China	100	Tank reactors	2012–present
Runruno	Philippine	–	BIOX	Construction

[a]CCGRI, Changchun Gold Research Institute

be used in extreme environments (low temperature, high altitude, arid/rainstorm conditions) and on low-grade ores containing high amounts of sulfur/magnesium, highly toxic ions, high mud oxide, and sulfide. In addition, future studies should pay more attention to strain selection, leaching mechanism, leaching technology, and engineering control technology. For gold robbing ores, effective solutions for the inhibition or elimination of this effect during cyanidation must be determined. Furthermore, to achieve efficient high-temperature chalcopyrite heap bioleaching in low-temperature environments using low-sulfur content ore, research on leaching mechanism must be strengthened [20].

Since 2013, the price of nonferrous metals has continued to slump, causing a significant impact on the nonferrous metal industry and resulting in an urgent need for low-cost extraction technology. Owing to the limited and poor resources in China, more attention should be given to the application of biohydrometallurgy techniques to waste rocks and complex secondary resources. Based on the reality of increasing emphasis on low-carbon economy, the next-generation biohydrometallurgy techniques should be more cost-effective, more environment-friendly, involve gentler separation conditions (e.g., low leaching temperature, low leaching concentrations), and be suitable for the treatment of more complex resources.

Acknowledgements This study was financially supported by the National High Technology Research and Development Program (Nos. 2012AA061501, 2012AA061502, and 2012AA061504), and Youth Found Project of GRINM.

References

1. Fecht S (2011) Microbe miners: bacteria extract metals and clean up the mess afterward. Sci Am 305(6):46
2. Tyson GW, Lo I, Baker BJ, Allen EE, Hugenholtz P, Banfield JF (2005) Genome-directed isolation of the key nitrogen fixer *Leptospirillum ferrodiazotrophum sp.* nov. from an acidophilic microbial community. Appl Environ Microbiol 71(10):6319
3. Rawlings DE, Johnson DB (2007) The microbiology of biomining: development and optimization of mineral-oxidizing microbial consortia. Microbiology 153(2):315
4. Keeling SE, Palmer ML, Caracatsanis FC, Johnson JA, Watling HR (2005) Leaching of chalcopyrite and sphalerite using bacteria enriched from a spent chalcocite heap. Miner Eng 18(13–14):1289
5. Brierley CL (2016) Biological processing: biological processing of Sulfidic ores and concentrates—integrating innovations. In: Lakshmanan IV, Roy R, Ramachandran V (eds) Innovative process development in metallurgical industry. Springer International Publishing, Cham, p 109
6. Dew D, Rautenbach G, van Hille R, Davis-Belmar C, Harvey I, Truelove J (2011) High temperature heap leaching of chalcopyrite: method of evaluation and process model validation. In: Proceedings of percolation leaching: the status globally and in Southern Africa, Cape Town, p 201
7. Cameron RA, Lastra R, Mortazavi S, Bedard PL, Morin L, Gould WD, Kennedy KJ (2009) Bioleaching of a low-grade ultramafic nickel sulphide ore in stirred-tank reactors at elevated pH. Hydrometallurgy 97(3–4):213
8. Du Plessis CA, Slabbert W, Hallberg KB, Johnson DB (2011) Ferredox: a biohydrometallurgical processing concept for limonitic nickel laterites. Hydrometallurgy 109(3):221
9. Mo XL, Wen JK, Chen BW, Wu ML, Gao WC (2015) Breeding of bacteria for high concentration fluoride-tolerance. Chin J Rare Met 39(1):75
10. East China University of Technology, General Research Institute for Nonferrous Metals (2014) Report on key issues of uranium ore bioleaching technology, Beijing, p. 12
11. Chen JW, Gao CJ, Zhang QX, Xiao LS, Zhang GQ (2009) Leaching of nickel-molybdenum sulfide ore with *Sulfolobus metallicus.* Chin J Process Eng 9(2):257
12. Chen JW, Xiao LS, Gao CJ, Zhang GQ (2012) Bioleaching of nickel-molybdenum sulfide ore by reactor with membrane combined ion-exchange. J Cent South Univ Sci Technol 43(7):2473
13. Liao MX, Wang MH, Deng TL (2004) Stratagem on tellurium recovery from an unique independent tellurium ore deposit in the world by biohydrometallurgy. Sichuan Nonferrous Met 19(3):55
14. Zhu Y, Hu H, Guo SX (2003) Technology of leaching germanium from lignite by means of microorganisms aided. Chin J Rare Met 27(2):310
15. Xie K (2008) Associated indium zinc sulfide concentrate bioleaching-extractive separation process research. Kunming University of Science and Technology, Kunming, p 6
16. Xin B, Zhang D, Zhang X, Xia Y, Wu F, Chen S, Li L (2009) Bioleaching mechanism of Co and Li from spent lithium-ion battery by the mixed culture of acidophilic sulfur-oxidizing and iron-oxidizing bacteria. Bioresour Technol 100(24):6163
17. General Research Institute for Nonferrous Metals, Zijin Mining Group (2013) Report on investigation and industrial optimization of Zijinshan copper ore green circulation heap bioleaching process, Beijing, p. 7
18. General Research Institute for Nonferrous Metals (2014) Report on low-temperature heap bioleaching of complex nickel-cobalt-copper ore, Beijing, p. 12
19. Hao FL, Zhang GP, Su BC, Gao JC (2011) Industrial application of biological oxidation technology in cold area. Gold 32(11):46
20. Chen BW, Wen JK (2013) Feasibility study on heap bioleaching of chalcopyrite. Rare Met 32(5):524

Part IV
Process Development and Optimization for Rare Metals

A Highly Selective Metal–Organic Complex Collector for Efficient Mineral Flotation

Zhao Wei, Haisheng Han, and Wei Sun

Abstract The increasingly low-grade, fine particle size, and complex composition of mineral resources require higher efficiency and precision of flotation process. Flotation is one of the core technologies for the efficient utilization of complex low-grade resources, and the flotation reagents (especially collectors) are the scientific basis and the key of flotation technology. Metal ions have a good template effect and can regulate the structure of metal–organic complex to realize directional functional assembly. The metal–organic complex collector with metal ions as functional groups and organic ligands as hydrophobic groups has displayed excellent performance in collecting ability and selectivity, showing a new direction for the development of new flotation reagents. Metal ion-hydroxamic acid complexes have made important progress in the flotation process of cassiterite, rutile, ilmenite, and rare earth ore, providing a new approach for the efficient utilization of complex low-grade mineral resources.

Keywords Mineral resources · Flotation reagents · Collector · Metal–organic complex · Flotation technology

Introduction

The composition of mineral resources is increasingly complex, with lower grade and finer grain size. As the complex multicomponent mineral becomes the main body, the difficulty of mineral processing is further increased, which puts forward higher requirements for efficiency and precision in mineral processing. Flotation is one of the core technologies for the efficient development and utilization of complex low-grade resources [1, 2]. However, efficiently recovering certain mineral resources remains

Z. Wei · H. Han · W. Sun (✉)
School of Minerals Processing and Bioengineering, Central South University, Changsha 410083, China
e-mail: sunmenghu@126.com

Key Laboratory of Hunan Province for Clean and Efficient Utilization of Strategic Calcium-Containing Mineral Resources, Central South University, Changsha 410083, China

T. Ouchi et al. (eds), *Rare Metal Technology 2022*, The Minerals, Metals & Materials Series, https://doi.org/10.1007/978-3-030-92662-5_21

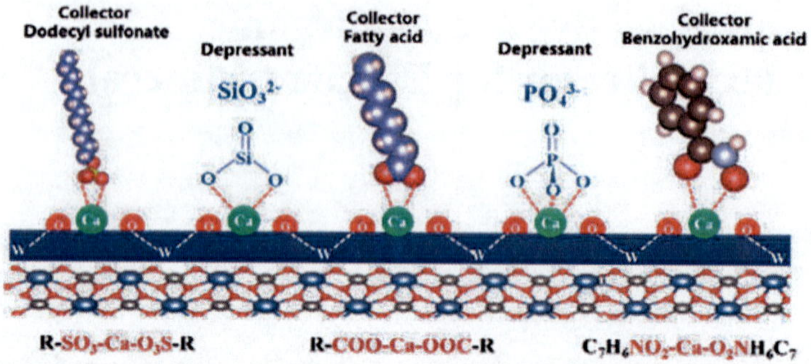

Fig. 1 The adsorption models of collectors and depressants on calcium minerals

challenging because of the lack of high-selectivity flotation reagents, resulting in a waste of resources and serious environmental problems. Scheelite is one of the most significant representatives of mineral resources, which is a major commercial tungsten mineral and associated with various calcium-containing minerals, including fluorite, dolomite, apatite, and calcite [3, 4]. However, the calcium-containing minerals are semi-soluble, exhibit similar structural and surface properties, and react similarly with various known collector families [5–7]. The collectors in scheelite flotation are categorized as anionic, cationic, and amphoteric collectors [8, 9]. The anionic collectors, including fatty acids, hydroxamic acids, dodecyl sulfonates, oleoyl sarcosine, and other anionic surfactants, are the most used collectors in scheelite flotation [10–13]. Furthermore, using these anionic surfactants to separate scheelite from calcium minerals is challenging because both of them chemically combine to the Ca^{2+} cations on the surfaces of the minerals (Fig. 1) [14, 15], resulting in poor flotation selectivity [16, 17]. Therefore, the unavailability of high-selectivity flotation reagents poses a challenge for the efficient separation of scheelite from other calcium minerals during mineral processing [4, 13].

These challenges can be overcome by developing novel collectors with high selectivity. Here, we focus on the challenge of separating scheelite from fluorite. Advances in mineral crystal and solution chemistry have revealed the microscopic differences between scheelite and fluorite in complex multiphase systems. The main differences between scheelite and fluorite surfaces are the diverse anionic groups and opposite surface charges in an alkaline solution (mostly pH = 7–10), indicating that scheelite possesses a negatively charged oxygen-rich surface, whereas fluorite possesses a positively charged calcium-rich surface [5, 7, 18, 19]. Therefore, we considered a design with a positively charged collector that can interact with the anionic groups on the mineral's surface rather than with the calcium atoms. Generally, a flotation collector comprises polar and nonpolar groups. If a collector exhibits absolute specificity for scheelite, its polar groups will combine with oxygen atoms rather than with calcium or fluorine. Therefore, we propose a metal-group collector, where the polar groups are metal ions. Hence, we selected Pb(II), which exhibits a large radius

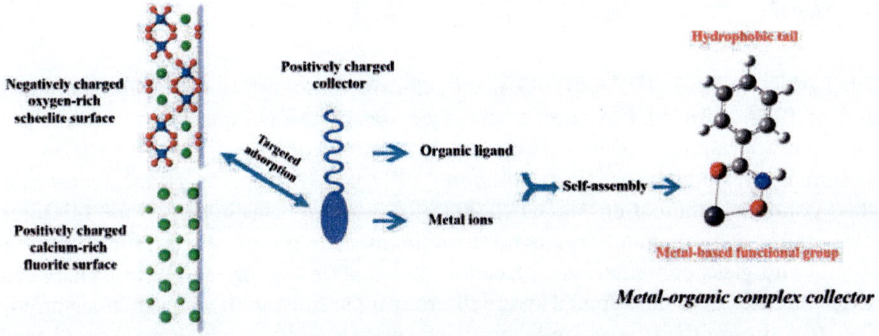

Scheme 1 Design of a metal–organic complex collector for separating scheelite from fluorite

and a flexible coordination environment, to facilitate the preparation of new classes of metal–organic complexes with novel structures. Benzohydroxamic acids (BHAs) were selected as the nonpolar group of the collector because of their highly selective chelating ability toward metal ions, similar to a hydrophobic tail behind the metal ions. We considered combining BHA with Pb(II) to develop a metal–organic complex in which organic linkers bind with metal centers via a metal–ligand coordination interaction. Here, we first introduce the metal–organic complex into the design of a flotation collector. Then, we propose a new design for the complex collector, which involves the self-assembly of BHA and Pb(II) (Scheme 1).

Materials and Methods

Materials and Reagents

Single mineral samples were acquired from the Shizhuyuan Mine in Hunan Province, China, and were ground to powder using a ball mill in a laboratory. The purities of single scheelite and fluorite mineral samples were >98.5%. The screened mineral samples with sizes of $-74+38$ μm were used for single mineral flotation, adsorption capacity measurements, TOF–SIMS, FTIR, and XPS. A fraction of the samples with a size of -2 μm was screened for measuring the zeta potential. The mineral sheet samples used for AFM and contact angle measurements were smoothed and polished repeatedly using a metallographic polisher to achieve a suitable roughness.

Analytical BHA (purity, 98.5%), lead nitrate (purity, 99.0%), and terpineol (purity, 98.0%) were obtained from Kemiou Co., Ltd. (Tianjin, China). A solution of NaOH and HCl was used as the pH regulator. All analytically pure starting materials were used as received. Deionized water was used in all experiments.

Synthesis of the Complex

$[Pb_6L_8(NO_3)_3]NO_3$ (HL = BHA) was synthesized through the coordination reaction of BHA with lead nitrate. A single crystal of the complex was cultured by solvent evaporation. A 10 mL of an aqueous solution containing $Pb(NO_3)_2$ (0.662 g, 2.0 mmol) was added to a BHA solution (0.137 g, 1.0 mmol) in 10 mL water in a glass container; thereafter, NaOH (0.008 g, 0.2 mmol) was added to the mixture. Air bubbles were manually removed from the mixture before stirring for 30 min at 50 °C. The glass container was placed at 25 °C after sealing. White crystals were collected after 22 d. The crystals were filtered and washed with water then air-dried. The structures of the synthesized compounds were identified, which were consistent with the results of the elemental analysis and infrared data.

X-ray Crystallography

A Rigaku 007 charge-coupled-device area detector using graphite-monochromated Mo Kα radiation (k = 0.71073 Å) was used to conduct the single-crystal XRD measurements. XRD pattern of single crystal of the complex was shown in Fig. 2. The structures were resolved using direct methods, employing the SHELXS-97 program [20–22]. The anisotropic refinement of all nonhydrogen atoms was performed using the full-matrix least-squares method of the SHELXL-97 crystallography package [23]. The molecular structure plots were prepared using Diamond [24]. Table 1 lists the crystallographic data and structure refinement for the complex.

Fig. 2 XRD pattern of single crystal of the complex

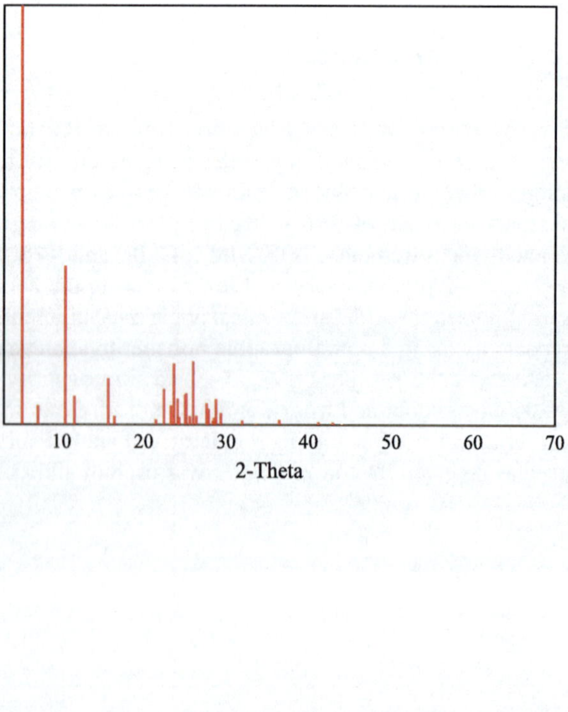

Table 1 Crystallographic data and structure refinement for the complex

	$[Pb_6L_8(NO_3)_3]NO_3$
Chemical formula	$C_{56}H_{48}N_{12}O_{28}Pb_6$
Formula weight	2580.20
Crystal system	Monoclinic
Space group	$P2_1/c$
Temperature (K)	150
a (Å), b (Å), c (Å)	17.9246(11), 18.1038(11), 21.4467(13)
α (°), β (°), γ (°)	90.000, 108.879(3), 90.000
V (Å^3)	6585.1(7)
Z	4
Goodness-of-fit on F2	1.009
θ range (°)	2.4–25.8
h/k/l indices	$-21 \rightarrow 21, -2 \rightarrow 20, -25 \rightarrow 25$
No. of measured, independent, and observed [$I > 2\sigma(I)$] reflections	51,061, 11,579, 8629
$R[F2 > 2\sigma(F2)]$	0.033
$\Delta\rho_{max}, \Delta\rho_{min}$ (eÅ^{-3})	1.09, −1.37

Flotation Tests

An XFG single mineral flotation machine was used to conduct single mineral flotation experiments using a 40 mL cell at 1,900 rpm. For each experiment, 2.0 g of the mineral samples were treated individually with the desired reagents (Fig. 3). After flotation, the concentrate and tailings were collected, dried, and weighed. We calculated the flotation recovery based on mass distribution. Each experiment was performed in triplicate and the results were averaged.

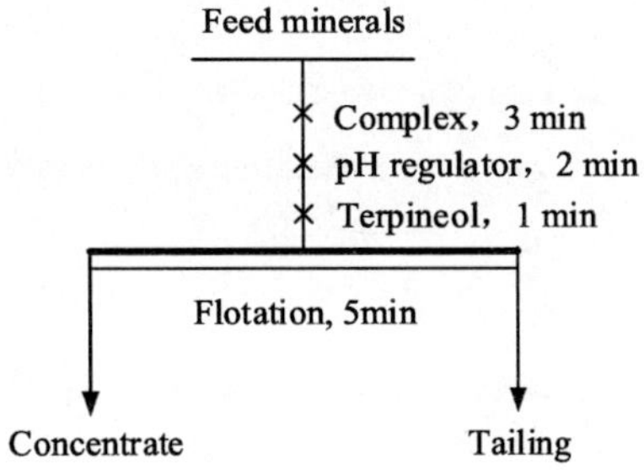

Fig. 3 Flowsheet of the flotation test

Characterization Methods

Coulter Delsa 440SX zeta analyzer (ZetaPlus, Bruker, Germany) was used for measuring the zeta potential. Each experiment was carried out for three times, and the mean value and standard deviation of the three experiments were used as the result and error. Alpha FTIR spectrometer (Bruker, Germany) was used to conduct the IR measurement at a range of 400–4,000 cm^{-1}. Thermo Fisher Scientific K-Alpha 1063 X system was used for XPS measurements, and 284.8 eV was used as the standard C1s binding energy to correct the data. PHI nanoTOFII TOF-SIMS instrument (JPN) was used for TOF–SIMS measurements. The ion source of the instrument was Bi_3^{++} (positive and negative modes). The adsorbed amount of the complex on the scheelite surface was investigated by the adsorption capacity measurements of Pb^{2+} and BHA, and the concentrations of BHA and Pb^{2+} in the supernatant were measured using a Shimadzu total organic carbon (TOC)-L CPH analyzer and a Perkin–Elmer ICP–AES instrument, respectively.

Results and Discussion

Crystal Structure Analysis and Description of the Complex

The $[Pb_6L_8(NO_3)_3]NO_3$ complex is observed to crystallize in the monoclinic crystal system with the space group $P2_1/c$ for which the lattice constants are $a = 17.9246(11)$ Å, $b = 18.1038(11)$ Å, and $c = 21.4467(13)$ Å. In the complex's molecular structure, six Pb(II) ions coordinate with eight L ligands and three NO_3^- anions; a free NO_3^- can also be observed in the structure (Fig. 4a). The Pb(II) ion coordinates with the oxygen atoms of L ligands to form nonplanar five-membered ring complexes. Figure 4b shows the coordination environment of Pb(II) ions in the complex. Six Pb(II) ion centers can be observed in the complex molecule, and all Pb(II) ions are

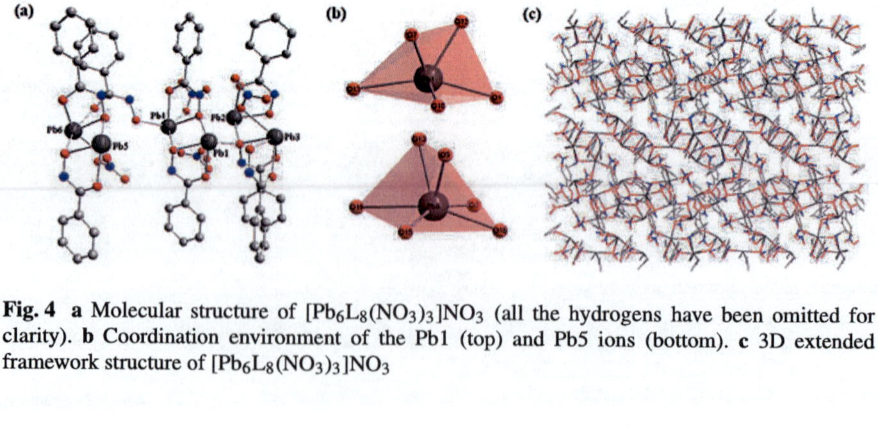

Fig. 4 **a** Molecular structure of $[Pb_6L_8(NO_3)_3]NO_3$ (all the hydrogens have been omitted for clarity). **b** Coordination environment of the Pb1 (top) and Pb5 ions (bottom). **c** 3D extended framework structure of $[Pb_6L_8(NO_3)_3]NO_3$

coordinated with the oxygen atoms with a Pb–O bond distance of 2.364(5)–2.751(5) Å. Among these ions, the Pb1, Pb2, Pb3, and Pb6 ions are five-coordinated, whereas the Pb4 and Pb5 ions are six-coordinated. The Pb1, Pb2, and Pb5 ions coordinate with the oxygen atoms of the L ligands, whereas the Pb3, Pb4, and Pb6 ions coordinate either with the oxygen atoms of the L ligands or NO_3^-. Figure 4b shows our selected ions (Pb1 and Pb5) that form polyhedrons centered on the Pb ions. The Pb(II) and oxygen atoms in the ligands join the adjacent Pb(II) complex molecules, generating a three-dimensional (3D) extended framework structure, which is a polymer with a repetitive structure (Fig. 4c).

Performance of $[Pb_6L_8(NO_3)_3]NO_3$ in Scheelite and Fluorite Flotation

After obtaining the structural information of $[Pb_6L_8(NO_3)_3]NO_3$ based on single-crystal structural analysis, we applied the obtained information to scheelite and fluorite flotation to verify our hypothesis. The performance of the complex in scheelite and fluorite flotation was investigated (Fig. 5). Figure 5a shows that an increase in the amount of collectors resulted in a rapid increase in scheelite recovery. When the amount of the complex was $>1.5 \times 10^{-4}$ mol/L, the flotation recovery of scheelite was >80%, whereas that of fluorite was <10%. Figure 5b shows that a scheelite recovery of >80% was obtained in the pH range 7.5–10, whereas the fluorite recovery was <10%. Thus, high-efficiency separation of scheelite from fluorite will be achieved by controlling the pH range of the flotation pulp. These results indicate that the complex exhibits good collecting ability and excellent selectivity in scheelite flotation.

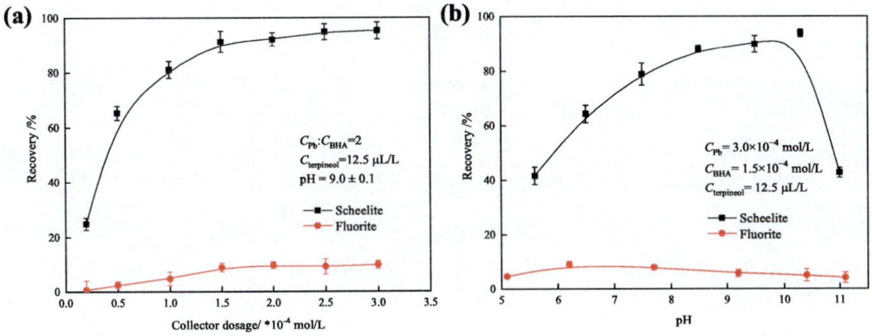

Fig. 5 **a** Effect of collector dosage on the scheelite and fluorite flotation, **b** Effect of pH on the flotation recovery of scheelite and fluorite

Adsorption Mechanism of [$Pb_6L_8(NO_3)_3$]NO_3 at the Solid–Liquid Interface

The flotation experiments prove that the complex collector exhibits absolute specificity for scheelite. Subsequently, the adsorption behavior of the complex on mineral/water interface was characterized to reveal its flotation mechanism. The mass spectroscopic analysis of alkaline liquid precipitates of the complex shows that the peaks at m/z of 822.07, 1022.04, 1165.09, 1372.07, 1508.10, and 1599.12 corresponded to $Pb_2L_3^+$, $Pb_2L_4(OH) + 2Na^+$, $Pb_3L_4^+$, $Pb_4L_4^+$, $Pb_4L_5^+$, and $Pb_4L_5(OH)_3 + K^+$, respectively (Fig. 6a), indicating that the complex existed in the form of a crystal structure fragment Pb_4L_5 complex with a hydroxyl group in alkaline liquid. Zeta potential measurements show that the complex colloids were positively charged and their adsorption caused the scheelite surface to be positively charged (Fig. 6b). This indicated that the complex was positively charged and electrostatic forces were

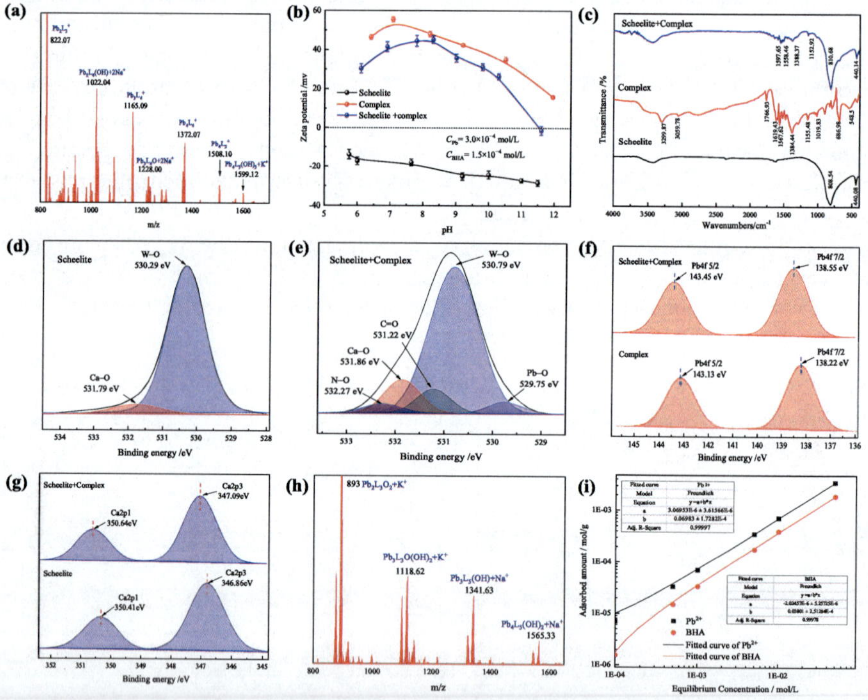

Fig. 6 **a** Mass spectrometry of alkaline liquid precipitates of complex, **b** Effect of complex on the zeta potentials of scheelite, **c** Infrared spectra of the scheelite surface treated with complex, **d–g** O_{1s}, Pb_{4f} and Ca_{2p} spectra of the scheelite treated with complex, **h** The TOF–SIMS of complex on the scheelite surface, **i** Adsorption isotherm of complex on the scheelite surface (equilibrium time = 60 min)

significant in the adsorption process, correlating with our assumption. The IR spectrum of the scheelite adsorbed by the collector showed absorption peaks of WO_4^{2-} at 808.54 and 440.08 cm^{-1}, and absorption peaks because of C = O and C–N could be observed at 1,597.65 and 1,152.92 cm^{-1}, respectively. Vibrations of the benzene rings could be observed at 1,558.46 and 1,388.37 cm^{-1} (Fig. 6c) [25], indicating chemical adsorption of the complex on the scheelite surface.

The binding-energy analysis was conducted to study the interaction sites between the collector and mineral surfaces. In O_{1s} spectra of scheelite, the peaks of the Ca–O and W–O groups were observed at 531.79 and 530.29 eV, respectively (Fig. 6d). After treatment with the complex, new peaks could be observed in the spectra, which can be attributed to the chemical states of the C = O, N–O, and Pb–O groups corresponding to the complex (Fig. 6e). The observed shifts of the Ca–O and W–O groups are 0.33 and 0.52 eV, respectively, indicating that the W–O bond shift was more obvious. Therefore, the complex is more inclined to the W–O groups. The split peaks of Pb_{4f} 5/2 and Pb_{4f} 7/2 were observed at 143.13 and 138.22 eV in the Pb_{4f} spectra of complex, respectively (Fig. 6f). After being treated with scheelite, both Pb_{4f} 5/2 and Pb_{4f} 7/2 peaks shifted by + 0.33 eV. The displacement was + 0.23 eV for Ca_{2p} 1/2 and Ca_{2p} 3/2 (Fig. 6g). Therefore, the more obvious shifts of the W–O bond, Pb_{4f}, and Pb–O bond revealed that the adsorption process was probably accomplished by the bonding of the lead ions in the complexes with the oxygen atoms on the scheelite surface. The Ca_{2p} shift can be attributed to the reaction of hydroxylated calcium atoms on the scheelite surface with the complex. In addition, the hydroxyl and N–H bonds in the complexes can form hydrogen bonds with the oxygen atoms on the mineral surface, which also contributes to the adsorption of the complexes on the mineral surface.

The TOF–SIMS spectra of the complex collector adsorbed on the scheelite surface showed that the peak with maximum molecular weight was 1,565.33, and its structure could be inferred to be $Pb_4L_5(OH)_2$. The structural fragments of $Pb_4L_5(OH)_2$ could be detected in the low molecular weight region (Fig. 6h). Thus, it can be inferred that the $Pb_4L_5(OH)_3$ was adsorbed on the surface of minerals to form a stable adsorbent. The amount of complex adsorbed on the scheelite surface shows that the adsorption amounts of Pb^{2+} and BHA increased synchronously as their concentrations increased (Fig. 6i). The fitting of the adsorption isotherms of Pb^{2+} and BHA demonstrated that the Freundlich model could fit the logarithm points well. The Freundlich equation shows that the adsorption process of the complex was multilayered [26]. Therefore, the mass spectroscopic, zeta potential, FTIR, XPS, TOF–SIMS spectra, and adsorption amount measurements indicate that the complex existed in the form of $Pb_4L_5(OH)_3$ in alkaline liquid, which was positively charged. The adsorption of the complex on the scheelite surface is a chemical adsorption process in which the lead atoms in the complexes coordinated with the oxygen atoms of scheelite, accompanied by electrostatic attraction. The expansion of the polymer structure of the complex mainly contributed to the multilayer adsorption of the complex on the mineral surface.

A Simplified Normal Temperature Flotation Process in the Industry for the Scheelite Using Complex Collector

In the industrial production of scheelite flotation, the flotation process includes roughing and cleaning flotation, which is long and complicated. Acquiring high-grade tungsten concentrate ($WO_3 \geq 40.0\%$) depends on a repeatedly cleaning process. A conventional method for cleaning flotation is Petrov's process [27], which is in an 80–90 °C agitation vat, with 2–4% sodium silicate solutions stirred for >30 min to desorb the collector on the surface of gangue minerals and then cooling and repulping for the repeatedly cleaning process. However, Petrov's process is costly, has high energy consumption, and is challenging to control; water containing many chemicals cannot be reused effectively, resulting in severe water pollution. Shizhuyuan Mine is a representative tungsten mineral deposit and characterized by low-grade and high-calcium gangue minerals. However, the WO_3 recovery of dressing plants via the fatty acid flotation process is only ~65% and faces a series of problems.

Based on the fundamental research on the complex collector, we applied it to the flotation of natural tungsten ores. A simplified normal temperature flotation process was designed to improve the flotation of natural scheelite ore from the Shizhuyuan Mine. In the new process, the complex was used as the only collector instead of fatty acids; the production process was significantly simplified. A high-grade concentrate was obtained using only minimum depressants in the cleaning flotation process, proving the complex possessed excellent selectivity between scheelite and fluorite flotation. It contributed to a 15% improvement in tungsten mineral recovery and substantial economic benefits for the Shizhuyuan Mine. Figure 7 compares the traditional process and new processes. The new process, which is simple, efficient, and

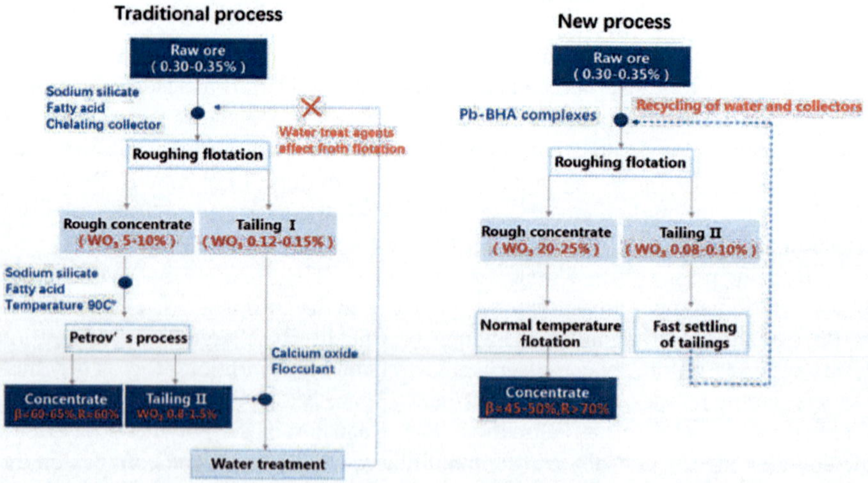

Fig. 7 The comparison of traditional process and new processes

low-energy consumption, replaced the high-energy and time-consuming, complex, and environmentally polluting Petrov's process. The recovery of tungsten concentrate was significantly improved, whereas the fluorite was almost not collected, which is beneficial for recovering fluorite from tungsten tailings. Currently, this process has been successfully applied in three large tungsten mines in China, significantly improving the utilization of tungsten resources and becoming a critical development direction in the tungsten flotation process.

High selectivity is the most significant advantage of Pb-BHA, which has shown strong applicability in the flotation of other oxidized ores, such as cassiterite [28], rutile [29], ilmenite [30], spodumene [31], and other minerals. In addition, some scholars have studied other metal ion or ligand metal–organic complex collectors, such as Fe^{3+} and BHA complex collectors [32]. Owing to the variety of metal ions and ligands, the structures of metal–organic collectors are diverse. Therefore, the development and application of complex collectors has great potential and will become an important future research direction.

Conclusion

The X-ray crystal structural analyses indicate that the novel Pb(II) complex ($[Pb_6L_8(NO_3)_3]NO_3$) is an eight-ligand structure containing six Pb(II) ions as the center of a polymer structure formed by connecting the complex molecules through Pb–O bonds. The flotation experiments confirmed our hypothesis that the complex exhibited strong collecting ability and outstanding selectivity in scheelite flotation. The experimental results indicate that the complex existed in the form of positively charged $Pb_4L_5(OH)_3$ in alkaline liquid. The adsorption of the complex onto the scheelite surface is a chemical adsorption process aided by electrostatic attraction and the functional group of the complex is a metal group, which binds with the oxygen atoms of scheelite. The Freundlich adsorption isotherm indicates the occurrence of multilayered adsorption on the original monolayer by expanding the polymer structure and hydrogen bond, and the adsorption of the complex renders the scheelite surface strongly hydrophobic. Thus, the complex collector exhibited the targeted adsorption ability for oxygen-containing minerals, and their successful industrial applications indicate that they are conducive to the accurate and efficient recovery of mineral resources, saving costs, and contributing to the recycling of water resources. The successful industrial applications of the complex collector indicate that they can be a new class of flotation reagents with metal-based functional groups, which are different from the traditional ion collectors.

References

1. Hubbard A (2004) Colloidal science of flotation. J Colloid Interface Sci 273(1):1–850
2. Rubio J, Souza ML, Smith RW (2002) Overview of flotation as a wastewater treatment technique. Miner Eng 15(3):139–155
3. Kupka N, Rudolph M (2018) Froth flotation of scheelite–a review. Int J Min Sci Technol 28(3):373–384
4. Yang X (2018) Beneficiation studies of tungsten ores–a review. Miner Eng 125:111–119
5. Gao Y et al (2018) Adsorption of a novel reagent scheme on scheelite and calcite causing an effective flotation separation. J Colloid Interface Sci 512:39–46
6. Gao Z, Fan R, Ralston J, Sun W, Hu Y (2019) Surface broken bonds: an efficient way to assess the surface behaviour of fluorite. Miner Eng 130:15–23
7. Gao Z et al (2018) Probing anisotropic surface properties and surface forces of fluorite crystals. Langmuir 34(7):2511–2521
8. Hu Y, Yang F, Sun W (2011) The flotation separation of scheelite from calcite using a quaternary ammonium salt as collector. Miner Eng 24(1):82–84
9. Wills Ba N-MTJ (2006) Wills' mineral processing technology: an introduction to the practical aspects of ore treatment and mineral recovery. 7th ed. Oxford: Butterworth-Heinemann Elsevier: 444.
10. Arnold R, Warren LJ (1974) Electrokinetic properties of scheelite. J Colloid Interface Sci 47(1):134–144
11. Deng L, Zhao G, Zhong H, Wang S, Liu G (2016) Investigation on the selectivity of N-((hydroxyamino)-alkyl) alkylamide surfactants for scheelite/calcite flotation separation. J Ind Eng Chem 33:131–141
12. Meng Q, Feng Q, Shi Q, Ou L (2015) Studies on interaction mechanism of fine wolframite with octyl hydroxamic acid. Miner Eng 79:133–138
13. Filippov LO, Foucaud Y, Filippova IV, Badawi M (2018) New reagent formulations for selective flotation of scheelite from a skarn ore with complex calcium minerals gangue. Miner Eng 123:85–94
14. Gallios GP, Matis KA (1992) Flotation of salt-type minerals. In: Mavros P, Matis KA (eds) Innovations in flotation technology. Springer Netherlands, Dordrecht, pp 357–382
15. Hu Y, Xu J, Qiu G, Wang D (1994) Effects of dissolved mineral species on the surface chemical characteristic, electrokinetic property and flotation behavior of fluorite and scheelite. J Cent South Univ Technol 1(1):63–67
16. Han H et al (2017) Fatty acid flotation versus BHA flotation of tungsten minerals and their performance in flotation practice. Int J Miner Process 159:22–29
17. Changgen L, Yongxin L (1983) Selective flotation of scheelite from calcium minerals with sodium oleate as a collector and phosphates as modifiers. II. The mechanism of the interaction between phosphate modifiers and minerals. Int J Miner Process 10(3):219–235
18. Gao Z, Sun W, Hu Y, Liu X (2013) Surface energies and appearances of commonly exposed surfaces of scheelite crystal. Trans Nonferrous Metal Soc China 23(7):2147–2152
19. Hiçyilmaz C, Özbayoglu G (1992) The effects of amine and electrolytes on the zeta-potential of scheelite from Uludag, Turkey. Miner Eng 5(8):945–951
20. Dolomanov OV, Bourhis LJ, Gildea RJ, Howard JAK, Puschmann H (2010) OLEX2: a complete structure solution, refinement and analysis program. J Appl Crystallogr 42(2):339–341
21. Millán C et al (2015) Combining phase information in reciprocal space for molecular replacement with partial models. Acta Crystallogr 71(9):1931–1945
22. Sheldrick GM (2008) A short history of SHELX. Acta Crystallogr 64(1):112–122
23. Sheldrick GM (2015) SHELXT-integrated space-group and crystal-structure determination. Acta Crystallogr A Found Adv 71(Pt 1):3–8
24. Nejati K, Bakhtiari A, Bikas R, Rahimpour J (2019) Molecular and electronic structure, spectroscopic and electrochemical properties of Copper(II) complexes: experimental and DFT studies. J Mol Struct 1192:217–229

25. Yongxin L, Changgen L (1983) Selective flotation of scheelite from calcium minerals with sodium oleate as a collector and phosphates as modifiers. I. Selective flotation of scheelite. Int J Miner Process 10(3):205–218
26. Hu Y, Wang D, Xu Z (1997) A study of interactions and flotation of wolframite with octyl hydroxamate. Miner Eng 10(6):623–633
27. Gallios GP, Matis KA (1992) Flotation of salt-type minerals. Springer, Netherlands
28. Tian M et al (2018) Novel insights into adsorption mechanism of benzohydroxamic acid on lead (II)-activated cassiterite surface: an integrated experimental and computational study. Miner Eng 122:327–338
29. Cao M, Gao Y, Bu H, Qiu X (2019) Study on the mechanism and application of rutile flotation with benzohydroxamic acid. Miner Eng 134:275–280
30. Fang S et al (2018) Adsorption of Pb(II)/benzohydroxamic acid collector complexes for ilmenite flotation. Miner Eng 126:16–23
31. Tian M, Gao Z, Khoso SA, Sun W, Hu Y (2019) Understanding the activation mechanism of Pb2+ ion in benzohydroxamic acid flotation of spodumene: Experimental findings and DFT simulations. Miner Eng 143
32. Tian M et al (2018) Effects of the preassembly of benzohydroxamic acid with Fe (III) ions on its adsorption on cassiterite surface. Miner Eng 127:32–41

Tellurium Recovery—Development of a Novel Hydrometallurgical Process

Shijie Wang

Abstract Cadmium telluride (CdTe) photovoltaic modules have become among the most inexpensive technologies to produce solar electricity, and therefore tellurium recovery is a hot minor metals topic in today's bull market. This paper summarizes the recovery of tellurium and its processing technology from electrolytic copper refinery slimes, which almost invariably contain tellurium, selenium, copper, silver, gold, and traces of platinum group metals. A novel hydrometallurgical process (Wang's process) is presented that was developed in bench scale experiments and pilot tests and led to commercial applications. The process chemistry is also published for the first time in this paper.

Keywords Anode slimes · Cementation · Hydrometallurgy · Leaching · Selective precipitation · Tellurium

Introduction

Similar in appearance to tin, tellurium (Te, Z = 52) is conventionally used in a variety of industries, primarily as an additive to several compounds and alloys. Since the innovation of using tellurium in solar cell manufacture was commercialized in 2002, the growth area for tellurium demand in solar cells has swiftly changed its traditional application. As solar cell technology has improved, cadmium telluride (CdTe) photovoltaic modules have become among the most inexpensive producers of solar electricity. Consequently, tellurium recovery is a hot minor metals topic in today's bull market.

In the Chinese character for tellurium (碲), the left stone (石) implies that tellurium is a non-metal, but in fact, along with boron (B), silicon (Si), and arsenic (As), it is a "metalloid." It has some properties regarded as characteristic of metals and other properties regarded as characteristic of non-metals. Tellurium is widely distributed in the earth's crust in many types of deposits, especially those associated with

S. Wang (✉)
Coeur Mining Inc., 104 S. Michigan Ave., Chicago, IL, USA
e-mail: swang@coeur.com

T. Ouchi et al. (eds), *Rare Metal Technology 2022*, The Minerals, Metals & Materials Series, https://doi.org/10.1007/978-3-030-92662-5_22

epithermal gold and silver deposits. Small concentrations but large quantities of tellurium are present in copper porphyries, massive pyritic copper sulfides, nickel sulfide deposits, and, frequently, lead sulfide deposits. Interestingly, tellurium and selenium are chemically closely related, but they behave somewhat differently in ore formation. Although widely disseminated like selenium, tellurium minerals do not form an ore. Hence, there are no deposits that can be mined for tellurium alone.

Given the limited availability of tellurium, a novel hydrometallurgical process was developed that not only enables us to efficiently recover tellurium, but also processes it in an environmentally benign way. Tellurium as a byproduct of electrolytic copper refining—where it preferentially concentrates into anode slimes to a concentration of up to 5%—accounts for over 90% of the global tellurium supply [1]. The process for tellurium recovery from anode slimes begins with leaching the raw slimes and precipitating tellurium as insoluble copper telluride from the leaching solution. Then, the copper telluride is further treated to obtain the final product, tellurium dioxide (TeO_2). Typical process parameters, such as dosage of copper powder (or chips and cut rod), concentrations of chemical reagents, and kinetic critical temperature need to be defined and controlled such that tellurium reduction will occur.

Tellurium Recovery

Tellurium usually occurs in the form of compounds and is considered a byproduct in the mining industry. Despite efforts to enrich tellurium content in new gold and bismuth mines, the main tellurium resource in metallurgy today remains copper sulfide ores. During flotation of copper ores, the bulk of tellurium is practically proportioned to the copper concentrates. Tellurium recovery is normally related to the recoveries of copper, gold, and bismuth. Although it is also recovered from flue dusts and gases generated during copper, bismuth, and lead ore smelting and from lead refinery skimming, nearly all commercial tellurium is produced from electrolytic copper refinery slimes.

Slime composition varies considerably among refineries, thus slime treatments also differ. Metallurgists traditionally focus only on precious metals in slimes processing: non-precious metal constituents are removed using various pyrometallurgical refining steps to obtain an impure gold-silver alloy known as doré. The doré metal is then subjected to electrolytic refining operations to recover silver and gold. The soda slag collected from slimes smelting is further treated via grinding, leaching, and filtering to produce tellurium dioxide. The filtrate is carried forward to granular selenium production.

Hydrometallurgical Process Development

Leaching is conducted in an autoclave with commercial oxygen at an elevated temperature to permit the copper concentration in acid-leached slimes to fall well below 5% and facilitate compliance with increasingly stringent environmental regulations. The leach solution is then collected and treated with efforts to control the process for tellurium recovery (Fig. 1).

Process Flowsheet

The new hydrometallurgical process comprises the following steps, which are associated with the block flow diagrams (Fig. 2a) and the alternative acidic leaching process (Fig. 2b).

1. Decopperization and detellurization of the raw anode slimes via sulfate oxidative pressure leaching.
2. Tellurium removal from the leachate via heating and copper cementation.
3. Copper telluride production via precipitation in a reaction tank and collection of sellable copper telluride in a filter press.
4. Further treatment of copper telluride to produce tellurium dioxide via (i) caustic leaching, selective precipitation, and filtration; and (ii) acidic leaching and precipitation, followed by caustic leaching, selective precipitation, and filtration.

Process Chemistry

Because copper is particularly difficult to remove by atmospheric decopperizing techniques, pressure leaching techniques were developed. Pressure leaching not only readily reduced the copper content of the slimes to approximately 1%, but it also removed most of the tellurium from the slimes, a metal that is nearly as difficult to remove by fire refining as copper.

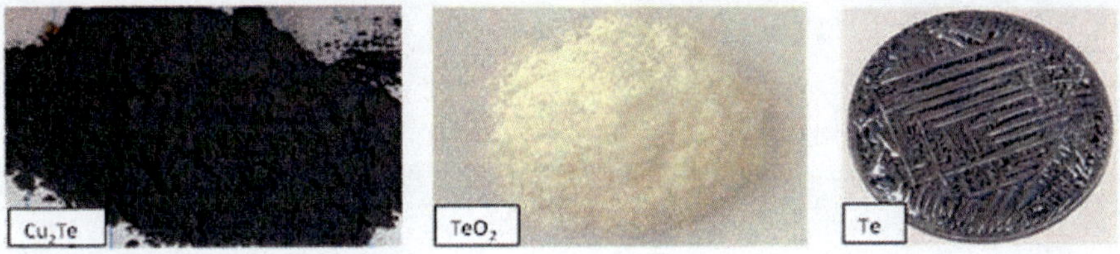

Fig. 1 Photographs showing copper telluride, tellurium dioxide, and tellurium metal (*source* https://en.wikipedia.org/wiki/Tellurium)

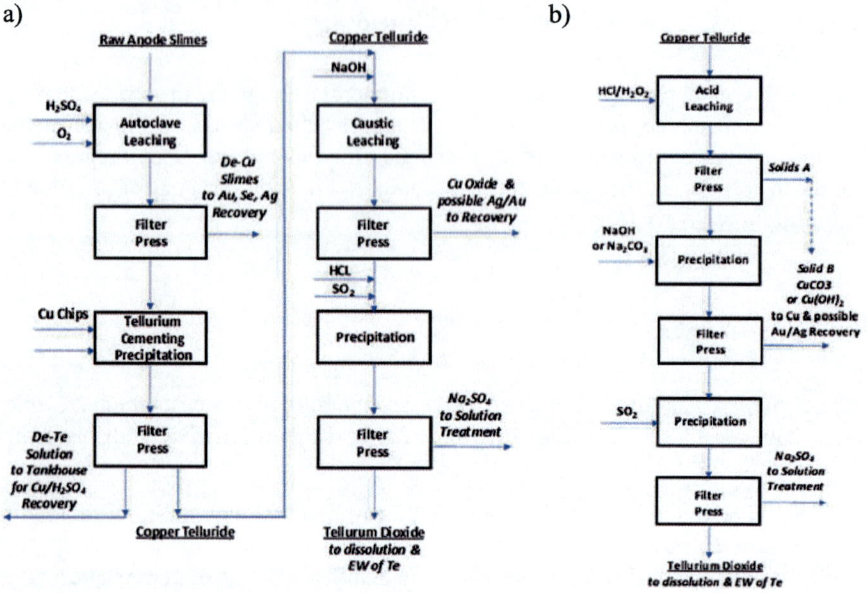

Fig. 2 Block flow diagram of the novel process to convert copper telluride to tellurium dioxide via **a** cementation and caustic leaching and **b** cementation and acid leaching

Copper and Tellurium Removal

The goal of the pressure leach is to extract the maximum amount of copper without attacking precious metals such as silver and gold. The principle chemical reactions in pressure leaching of anode slimes containing copper, copper-silver selenides, and tellurides may be summarized as by reactions (1)–(4):

$$Cu_2O + H_2SO_4 \rightarrow Cu + CuSO_4 + H_2O \quad (1)$$

$$Cu + H_2SO_4 + O_2 \rightarrow CuSO_4 + H_2O \quad (2)$$

$$2CuAgSe + 2H_2SO_4 + O_2 \rightarrow 2CuSO_4 + Ag_2Se + 2Se + H_2O \quad (3)$$

$$Cu_2Te + 2H_2SO_4 + 1/2O_2 \rightarrow 2CuSO_4 + H_2TeO_3 + H_2O \quad (4)$$

Making Copper Telluride

After the autoclave leach solution is transferred to the tellurium circuit, the chemical reaction involved in the conversion of the tetravalent tellurous acid to copper telluride is

$$4Cu + H_2TeO_3 + H_2SO_4 \rightarrow Cu_2Te + 2CuSO_4 + H_2O \quad (5)$$

If the oxidative pressure leaching is elevated by oxygen or is more prolonged than required, hexavalent tellurium (telluric acid) can form according to following reaction:

$$5Cu + H_6TeO_6 + 3H_2SO_4 \rightarrow Cu_2Te + 3CuSO_4 + 6H_2O \quad (6)$$

Reactions (5) and (6) will occur rapidly only when the cementation is conducted at elevated temperature with sufficient acid concentration, and the reaction temperature reaches close to the boiling point.

Making Final Tellurium Product

To make the final tellurium product, copper telluride is leached in aerated sodium hydroxide solution to dissolve tellurium as sodium telluride, leaving cuprous oxide as a residue according to reactions (7) and (8):

$$2Cu_2Te + 4NaOH + 3O_2 \rightarrow 2Na_2TeO3 + 2Cu_2O + 2H_2O \quad (7)$$

$$Te + NaOH + O_2 \rightarrow Na_2TeO_3 + H_2O \quad (8)$$

The leachate is then neutralized with sulfuric acid to precipitate tellurium as tellurium dioxide (reaction (9)):

$$Na_2TeO_3 + H_2SO_4 \rightarrow TeO_2 + Na_2SO_4 + H_2O \quad (9)$$

In the study, copper telluride is also leached via oxidation with hydrogen peroxide in a hydrochloric acid medium according to reactions (10) and (11):

$$Cu_2Te + 4HCl + 4H_2O_2 \rightarrow H_2TeO_3 + 2CuCl_2 + 5H_2O \quad (10)$$

$$Te + 2H_2O_2 \rightarrow H_2TeO_3 + H_2O \quad (11)$$

The copper tellurium-bearing solution is then treated with sodium hydroxide or sodium carbonate to precipitate copper as cupric hydroxide and retain tellurium as soluble sodium tellurite in solution, as illustrated in reaction (12):

$$CuCl_2 + 2NaOH \rightarrow 2NaCl + Cu(OH)_2 \quad (12)$$

The filtrate and wash water from the copper salts separation are then treated with sulfuric acid to precipitate tellurium as tellurium dioxide (reaction (13)):

$$Na_2TeO_3 + H_2SO_4 \rightarrow TeO_2 + Na_2SO_4 + H_2O \quad (13)$$

Using hydrochloric acid to adjust pH, tellurium dioxide can be dissolved in sodium hydroxide solution with sulfur dioxide to precipitate tellurium (reaction (14)).

$$TeO_2 + 2SO_2 + 2H_2O \rightarrow Te + 2H_2SO_4 \quad (14)$$

Tellurium is then recovered as pure cathode metal by electrowinning (reactions (15) and (16)).

$$TeO_2 + 2NaOH \rightarrow Na_2TeO_3 + H_2O \quad (15)$$

$$Na_2TeO_3 + H_2O \rightarrow Te + 2NaOH + O_2 \quad (16)$$

Some oxidation of tellurium to the hexavalent state such as sodium tellurate (Na_2TeO_4) may occur during electrowinning. This insoluble compound can be charged to the decopperizing autoclave to recover the tellurium.

Experimental Methods

Making Copper Telluride

Given that the reaction is strongly temperature dependent (apparent activation energy 25 kcal/mol), operation at the boiling point should serve to provide both the highest rate of precipitation and an inert atmosphere steam. All cementation tests were conducted at 95 °C with a heating supply. Stock solutions were collected from the surge tank of autoclaving at the study refinery. Sulfuric acid was added if the initial acid concentration was less than 100 g/L. Copper powder is an efficient cementing agent, but due to impurity and moisture concerns with the mine copper powder samples, copper anode milling chips (1.25 cm anode shavings) were used in the tests (Fig. 3). The required mass of chips was calculated based on the stoichiometric ratio of copper to cement of all tellurium and selenium in the solution.

Fig. 3 Cementation test: Cu chips, cementing, reacted chips, and copper telluride filter cake

Cu:Te mass ratios of 5, 10, and 30 were tested. After the chips were weighed, stock solution of autoclave leachate was added to the reactor with a heat source. Heating and stirring were then initiated at a mixing speed of 450 rotations per minute. Zero reaction time was taken when the set point temperature was reached.

For the kinetic studies, solution samples were taken at set intervals throughout the cementation period (10–14 h). At each sampling point, the evaporated volume was replaced with deionized water while cementation continued. When analytical data showed that tellurium concentration was less than 10 mg/L, the slurry was rapidly cooled to 70 °C. The reactor was disassembled, and the slurry was decanted and filtered. The filtrate volume was measured, and the solution was saved for chemical analysis. The sludge (filter cake) was dried, weighed, and chemically analyzed.

Making Tellurium Dioxide

Since all copper telluride is to be treated internally at the operations rather than sold, a more direct and efficient process for producing tellurium dioxide was proposed during development of the novel hydrometallurgical process. Both caustic and acid leaching were separately tested to make tellurium dioxide from copper telluride filter cake samples via dissolution of the cake and selective precipitation of tellurium dioxide from the solution. The reactions are also temperature dependent; therefore, all tests were temperature controlled with a heating supply and sufficient mixing with an overhead mixer.

1. Caustic leaching: tellurium was dissolved with sodium hydroxide to form sodium telluride, leaving a residue of cuprous or cupric oxide. Tellurium and copper were separated in leaching step.
2. Acid leaching: copper telluride is leached in an acidic solution with hydrogen peroxide. In theory, most of the tellurium and copper will be dissolved as sodium tellurite and cuprous chloride. Tellurium and copper can then be separated from the leach solution by two precipitation steps.
3. Both leaches require the production of tellurium dioxide as an intermediate material, which must be processed to obtain tellurium.

Tellurium Recovery Test Results and Discussion

Copper Telluride Cementation Tests

The tellurium concentration in the cementation solution must be reduced because the final solution concentration of tellurium must be controlled in a scheduled time before it is recycled to the tankhouse. Since copper chips would be a variable cost if the copper telluride was sold as a byproduct, benchscale tests on variable copper chip dosage were conducted. Initial tellurium concentrations were reduced from approximately 0.8 g/L to less than 0.005 g/L in 5 h at Cu:Te mass ratios of 5, 10, and 30 (Fig. 4a). Tellurium reduction from tellurous acid to insoluble copper telluride was very effective at the highest Cu:Te ratio at boiling point. In a pilot test, the initial tellurium concentration was reduced from 5.1 g/L (initial) to less than 0.005 g/L in the final solution in 6 h (Fig. 4b).

At Cu:Te mass ratios of 5, 10, and 30, 48.5, 64.2, and 92.6% of the copper, respectively, were consumed in a 12-h cementation period (Fig. 5). The consumption was reduced to 13.9, 47.8, and 69.1%, respectively, when the retention time was shortened to 4 h.

In theory, the copper telluride produced should match its composition of 50:50 per the chemical formula of Cu_2Te, where the Cu weight is 127 (63.5×2) and the Te weight is 127.6. In reality, the tellurium content in the copper telluride byproduct can be as low as 35% due to the presence of co-cemented impurities. In benchscale tests, when the Cu:Te ratio was increased from 5 to 30, the tellurium content increased from 34.2 to 47.4%, and the copper content increased from 47.4 to 48.1% (Table 1). It seems that sufficient copper chips dosage is essential to produce the desired quality of copper telluride byproduct.

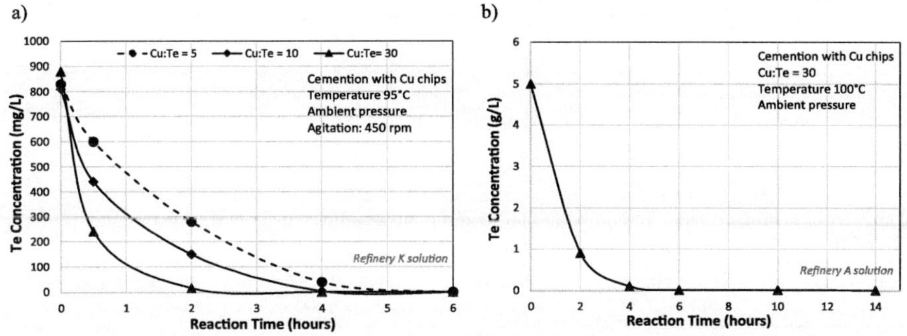

Fig. 4 Cementation **a** Te recovery at three Cu:Te ratios; **b** Te recovery at high initial concentration (note change in scale)

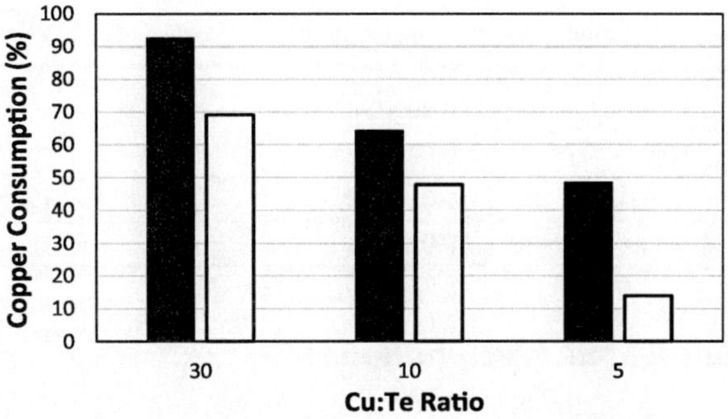

Fig. 5 Copper consumption versus Cu:Te ratio during a 12-h (black) and 4-h (white) Cementation period

Table 1 Copper telluride assay results

Cu:Te mass ratio	Ag	Au	As	Cu	Se	Te
	(%)					
5	0.38	0.01	0.84	47.4	0.37	34.2
10	0.30	0.01	0.46	46.7	0.48	44.5
30	0.21	0.01	0.20	48.1	0.42	47.4

Tellurium Dioxide Leaching and Selective Precipitation Tests

The copper and tellurium contents of copper telluride cake samples from a copper refinery in the US were 36.4 and 48.8%, respectively (Table 2). X-ray diffraction analysis of the as-received refinery copper telluride cake samples indicated that copper telluride predominated the sample, with a high level of crystallinity compared to elemental tellurium (data not shown).

Table 2 Content of the copper telluride sample used for TeO_2 tests

Element	Bi	Cu	Se	Te
	(wt.%)			
	2.18	36.4	1.31	48.8

Table 3 Bismuth, copper, selenium, and tellurium content after caustic leaching and precipitation after 4 h

Leaching	Leachate (g/L)	Wash (g/L)	Residue (wt.%)	Extraction (%)
Bi	0.054	0.0002	2.77	1.9
Cu	0.035	0.0015	46.5	0.08
Se	0.012	0.0006	1.72	0.7
Te	27.4	14.6	35.5	56.3
Precipitation	Precipitate (wt.%)	Final solution (g/L)		
Bi	0.05	0.00005		
Cu	0.07	0.0029		
Se	0.01	0.014		
Te	65.0	0.018		

Caustic Leaching and Precipitation

At ambient pressure and 85 °C, ~56% of the tellurium was dissolved by caustic leaching for 4 h (Table 3). Copper was present in the residue as copper oxide (~47%). The leachate solution was neutralized with sulfuric acid with a final pH of 5.6.

Acid Leaching and Precipitation

Ambient pressure acid leaching at 70 °C followed by two stages of precipitation resulted in only ~12% of tellurium extraction after 4 h (Table 4). This is due to the reason hydrochloric acid was not amenable to copper telluride treatment.

Conclusions

1. Since the cementation reaction is strongly temperature dependent, the temperature must be at the solution boiling point to initiate the cementation reaction and make the reaction swift.

Table 4 Bismuth, copper, selenium, and tellurium content after acid leaching and precipitation

Leaching	Leachate (g/L)	Wash (g/L)	Residue (wt.%)	Extraction (%)
Bi	2.06	0.0025	0.19	94.7
Cu	35.8	3.55	0.55	99.1
Se	0.026	0.0025	2.16	2.1
Te	5.6	0.47	74.9	11.5
Precipitation	Precipitate 1 (wt.%)	Precipitate 2 (wt.%)	Final solution (g/L)	
Bi	32.6	0.01	0.00005	
Cu	10.3	47.3	0.0029	
Se	0.35	0.01	0.014	
Te	20.0	6.54	0.018	

2. Increasing the Cu:Te ratio rapidly lowers the tellurium concentration independent of the initial concentration.
3. When trying to maximize copper and tellurium removal during autoclaving, some impurities such as selenium are co-extracted in the solution and are 100% precipitated in the copper telluride.
4. Manufacturing tellurium dioxide from copper telluride would be a more economical approach to maximize net present value via recoveries of copper, silver, gold, and residual selenium in tellurium dioxide circuits.
5. Acid leaching was not effective for tellurium dioxide production. Although hydrochloric acid was not amenable to copper telluride treatment, it could be used to selectively precipitate tellurium dioxide.
6. Preliminary benchscale test results indicated that copper telluride was readily amenable to caustic leaching and precipitation: approximately 85% of the tellurium was recovered. A commercial application demonstrated that tellurium dioxide was obtained and met with specifications after optimizing the operational parameters.
7. Given the limited availability of tellurium and the need to achieve reduction of CO_2 emissions, all copper refineries should consider and pursue the recovery of tellurium from anode slimes.
8. This novel hydrometallurgical process not only enables us to recover tellurium efficiently and economically, but also facilitates processing it in an environmentally benign way.

Reference

1. Wang S (2011) Tellurium, its resourcefulness and recovery. JOM 63(8):90–93

CO_2-Optimized Recovery of Special Metals from Precipitation Residue by Selective Chlorination

S. Steinlechner and L. Höber

Abstract Millions of tons of iron precipitation residues, predominantly jarosite, are accumulating in the primary zinc or primary precious metals industry every year. Regardless of environmental concerns the material is land filled in almost any case, although valuables such as indium, silver, gold, nickel, or zinc are present in considerable amounts. Within the presented research, CO_2-optimized multi-metal recovery from the residue jarosite by means of a selective chlorination extraction has been evaluated not only by executed experiments but also by a multidimensional simulation of possible process parameters, utilizing a Python algorithm in combination with automated FactSage process step simulation. This allows a simultaneous iteration of relevant reaction parameters such as temperature, pressure, stoichiometry, or variation of additives and with this offers a high degree of freedom in the choice of evaluated reactants. The paper will outline a selection of possibly recovered special metals and the best choice of additives and process parameters.

Keywords Selective chlorination · Jarosite · Iron precipitate · Nickel industry · Zinc industry · Silver · Zinc · Lead · Indium · Nickel

Introduction

Throughout the nonferrous industry several examples are known where the element iron must be removed in course of the production of the main metal. In most cases, it forms or is a component of a generated residue, such as red mud from aluminium industry, fayalitic slag from copper industry, or precipitation residues from zinc or also nickel industry. While slags can find their utilization in various fields, such as building material, as sand blasting agent, or substitutes in cement clinker, the sludges obtained during hydrometallurgical precipitation processes are subject to be are often sent to landfill.

S. Steinlechner (✉) · L. Höber
Christian Doppler Laboratory for Selective Recovery of Minor Metals Using Innovative Process Concepts, Montanuniversität, Franz Josef-St. 18, 8700 Leoben, Austria
e-mail: stefan.steinlechner@unileoben.ac.at

T. Ouchi et al. (eds), *Rare Metal Technology 2022*, The Minerals, Metals & Materials Series, https://doi.org/10.1007/978-3-030-92662-5_23

A common method to precipitate iron from sulphatic solution is the jarosite process, named after its formed mineral during the removal step [11]. This method is applied, next to other possible forms of precipitated mineral structure such as goethite or hematite, in the hydrometallurgical zinc production or also after the leaching of the nickel matte in course of the nickel and precious metals production [1]. Aside of the separation of iron as a basic iron sulphate also other present metals can be co-precipitated or adsorbed on the iron product and consequently removed from the remaining sulphatic solution. It is documented that an incorporation of valuable elements, such as indium, silver, nickel, or also arsenic and cobalt into the jarosite structure, can occur in the precipitation product [3, 4, 6, 8] in the same way as in naturally occurring jarosite minerals [2]. As a consequence, valuable and technologically important metals are lost, due to the dumping of such process residues.

Several strategies were investigated in the past to face increasing needs regarding ecological and economical improvements of the current situation, including concepts for immobilizing hazardous elements, recovering valuable elements, and minimizing the residual amount by utilizing the precipitate, for instance, as a substitute material in the building industry [7]. Till today only the jarofix process gained industrial scale but represents only an immobilizing strategy without any recovery of valuable elements.

Based on a previous pyrometallurgical investigated concept ([13], Wegscheider und Steinlechner) for the recovery of valuable metals from iron precipitation residues by mixing with electric arc furnace dust as a chlorine source and carbon as reducing agent, the present paper describes a concept, which waived the use of carbon as a reducing agent and therefore represents a CO_2-optimized recovery strategy for the recovery of selected valuable elements. The approach followed is a selective chlorination of specific metals, while leaving behind the main constituent of jarosite, the iron compounds. Therefore, the thermochemical behavior that iron chlorides are only moderately stable compared to silver or indium chloride is fundamental.

Materials and Methods

The investigated materials from both zinc and nickel industries are iron precipitates in the form of a jarosite mineral, generated in course of an iron extraction from sulphatic process solution. While the adherent or co-precipitated elements are differing, the general mineral formular for both is the same, which is $XFe_3(SO_4)_2(OH)_6 \cdot mH_2O$.

The basic iron sulphate has a monovalent cation on the X-position, which can be represented by K^+, Na^+, NH_4^+, Ag^+, ½ Pb^{2+}, and H_3O^+, for instance. Similar than in nature also the precipitate forms different kinds of jarosite depending on what cation is added or what impurities are present in the process solution.

The investigated jarosite from zinc industry represents mainly a sodium and potassium jarosite with a small rest amount of plumbo- and hydronium-jarosite. The one from nickel industry was precipitated as ammonium jarosite indicated by the

Table 1 Range of elemental analysis of investigated jarosites from zinc- and nickel industry

Element [wt-%]									
Jasrosite from	Fe^{3+}	Fe^{2+}	S	Na + K	Zn	Ag	In	Ni	Pb
Zinc industry	21–26	0–1	11–12.5	4–4.5	3.7	0.022	0.023	–	1–1.5
Nickel industry	42–46	0–2	6.5–8	0.1	–	–	–	3	–

absence of sodium and potassium, which can be seen in Table 1 summarizing the main elements from elemental analysis.

The related XRD plots summarizing the phases in the material are shown in Fig. 1, where (a) is the jarosite from zinc industry and (b) the one from nickel industry.

It is known from literature [6], Wegscheider und Steinlechner [5], that a thermal treatment leads to a stepwise decomposition of the jarosite structure, which was proved by own investigations carried out at a Netzsch STA 409 Pc under inert as well as oxidizing atmosphere for both materials. Figure 2 shows the result for the jarosite from nickel industry under inert (Argon) atmosphere, which was heated up to 1000 °C. Typically, three main steps can be observed, which partly vary in their temperature range because of different occurring jarosite types [7].

$$NaFe_3(SO_4)_2(OH)_6 \cdot mH_2O \rightarrow NaFe_3(SO_4)_2(OH)_6 + m\{H_2O\} \tag{1}$$

$$NaFe_3(SO_4)_2(OH)_6 \rightarrow NaFe(SO_4)_2 + Fe_2O_3 + 3\{H_2O\} \tag{2}$$

$$NaFe(SO_4)_2 \rightarrow Na_2SO_4 + Fe_2O_3 + 3\{SO_3\} \tag{3}$$

However, the evaporation of moisture can be observed below 200 °C, the following split and removal of OH-group follows up to 450 °C and between 600 and 850 °C the SO_3 is removed [10], illustrated by Eqs. 1, 2 and 3 exemplarily for a sodium jarosite. This in parallel leads to a release of incorporated elements in the jarosite structure, such as indium sitting on the iron position or silver as well as lead incorporated at the cation position. Also, other reactions of decomposition can occur in parallel, which then overlay the jarosite decomposition.

The decomposition of the jarosite structure forms the base for a number of accessible valuable side elements for the investigated selective chlorination process. The chlorination reactions take place without any addition of carbon as reducing agent which entails a significant reduction of the process greenhouse gas emissions compared to other research concepts.

The target of the present investigation is the evaporation of valuable elements from a jarosite material as a result of the formation of a volatile chloride compound. Furthermore, the iron should be left behind to form a high Fe-containing side product next to the collected valuable metal chloride dust.

Due to the number of possible process parameter combinations and the aim to determine the optimum reaction mixtures, parameters, and influencing factors, the

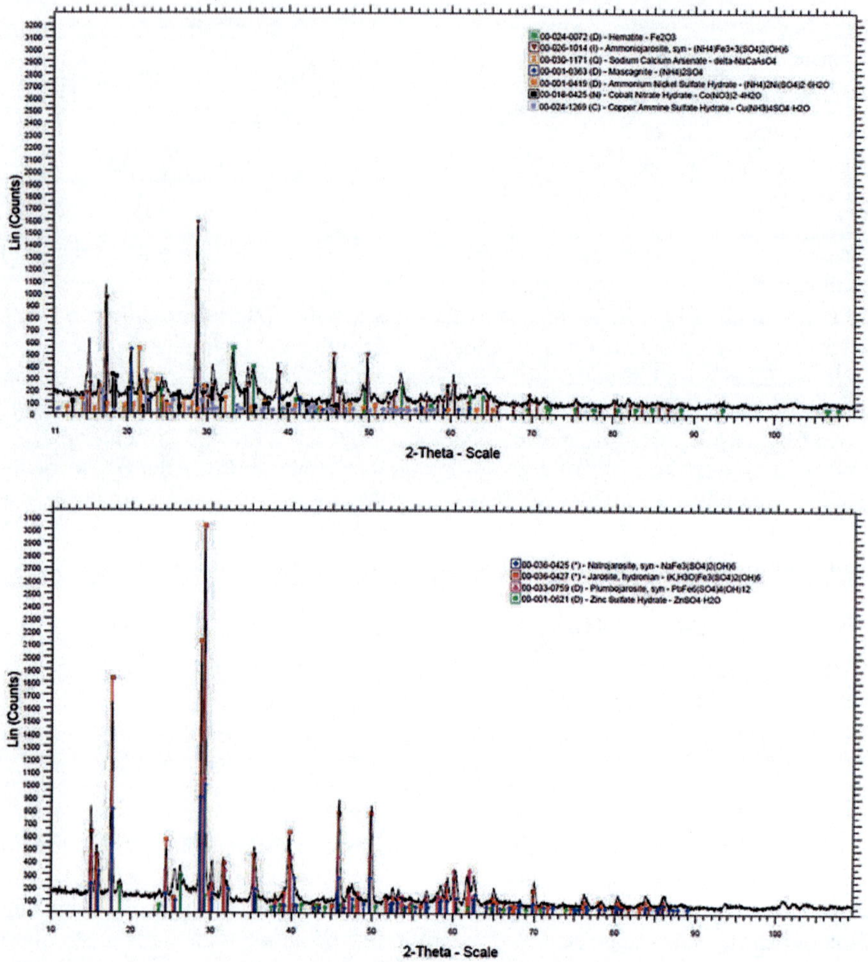

Fig. 1 **a** Left: XRD jarosite material from zinc industry **b** Right: XRD jarosite material from nickel industry

thermochemical software package FactSage with its integrated Equilibrium module was used together with a Python algorithm for data evaluation [9]. Non-chloride-containing additives, atmospheres, chlorine carriers, treatment temperatures, and stoichiometry factors were evaluated. The calculation assumes a thermal decomposed jarosite as starting condition. For the easier evaluation of the best conditions for a high extraction of valuables, a so-called Valex-factor was introduced, representing a monetarily weighted extraction rate of all available valuable metals, illustrated in Fig. 3 by the solid green circles. Additionally, the loss of iron to the gas phase was calculated, represented by the solid red circle. The black line represents a theoretical extraction of 100%.

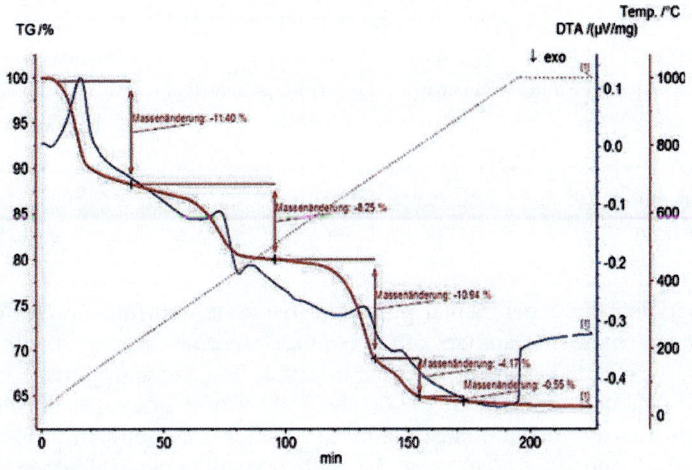

Fig. 2 Thermogravimetric analysis in a Netzsch STA 409 Pc of the jarosite from nickel industry under argon atmosphere

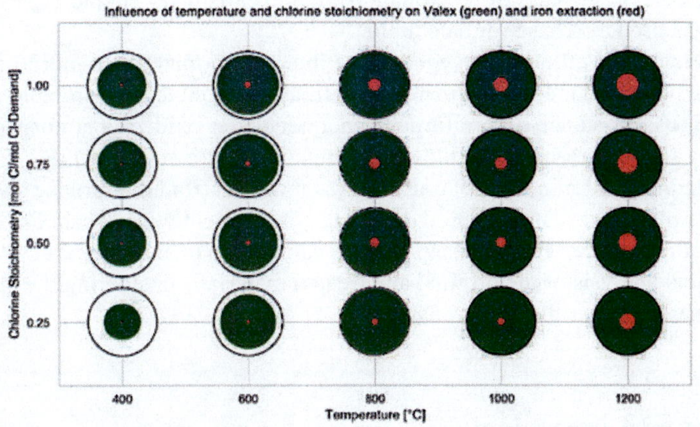

Fig. 3 Influence of temperature and chlorine stoichiometry on valex (green) and iron extraction (red) for a jarosite from zinc industry, the black circle represents a value of 1.00 [9] (Color figure online)

The parameter study in FactSage [9] showed that a temperature above 800 °C is required to obtain a satisfying valuable metal extraction yield by chlorination. It further was shown that trivalent chlorides have a better performance than monovalent or bivalent chlorides.

To verify the theoretical results, lab-scale experiments were performed. $AlCl_3$ and $MgCl_2$ were selected for further investigations. The design of experiment is shown in Table 2.

Table 2 DoE for the selective chlorination and evaporation of valuables from Jarosite

Temperature [°C]	Investigated jarosite material		Chlorine carrier	
	From zinc industry	From nickel industry	$AlCl_3$	$MgCl_2$
			Stoichiometry of addition	
900	X	–	2, 4, 6	2, 4, 6
1100	X	X		

Summarizing the experimental procedure twofolds, fourfolds, or sixfolds, the calculated stoichiometric amount of the required chlorine compound, either $AlCl_3$ or $MgCl_2$, were added to the dry jarosite material. In the case of jarosite from zinc industry, a treatment temperature of 900 and 1100 °C was investigated, while in the case of jarosite from nickel industry only 1100 °C was chosen. All used chlorides were technical laboratory quality and due to their stability in hexahydrate form.

For the jarosite from zinc industry, the required amount is based on a total chlorination of Zn, Pb, Cu, Ag, and In and additionally assumed 5% of the iron amount, as this chlorination reaction cannot be avoided completely although it is the least stable chloride. In the case of precipitate from nickel industry Ni and 5% of the iron content was used as a base for calculation.

The jarosite and chlorination agent were intensively mixed in a mortar and 80 g of mixture per experiment filled in a MgO refractory crucible. The crucible was then charged into a resistance heated furnace operated under oxidizing atmosphere (air), which was already pre-heated to the treatment temperature of 900 or 1100 °C. After a treatment time of 30 min, the hot crucible was removed from the furnace and cooled down. The mass loss was measured during the experiment with a scale linked to the bottom of the furnace. The final step was the calculation of extraction rates based on the chemical analysis of the original and treated material, considering the decreased mass of material after the experiment.

Results and Conclusion

Figure 4a–c shows the obtained extraction rates of the lab-scale experiments of Pb and Zn for the treated jarosite from zinc industry, for the addition of magnesium chloride and aluminium chloride, at 900 as well as 1100 °C. Deviating from the FactSage calculation, a temperature of 900 °C does not lead to a complete extraction of valuables, which might be a subject of not perfectly mixed reaction material or also due to imperfect conditions leading to kinetic effects. As can be seen in the left segments in Fig. 4a, b, the extraction of iron in trials with $AlCl_3*6H_2O$ is not given as it could not be qualitatively confirmed by analytical fluctuations. The values fluctuate around the value 0, it can therefore be stated that the extraction of iron was very low, compared to the experiments with $MgCl_2*6H_2O$. The increase of temperature in both cases of additives lead to a significant increase in the extraction of Zn and Pb. The iron loss increases with higher temperature and tendentially is higher in case of $MgCl_2*6H_2O$, at the same time the extraction rates of Zn and Pb in the case of magnesium chloride addition are significantly higher.

Figure 4d shows the result of the experiment with nickel jarosite at 1100 °C. Higher extraction of both iron and nickel was achieved when using $MgCl_2*6H_2O$ compared to the addition of $AlCl_3*6H_2O$, similar to the results with material from zinc industry. The highest nickel extraction (45%) was achieved using the sixfold of the stoichiometric ratio.

Comparing the results of zinc and lead with the nickel extraction rate it is obvious that the extraction via selective chlorination is more efficient at given conditions for zinc and lead.

Considering the theoretical achievable iron content of the remaining iron compound (treated jarosite) also an iron chloride addition is meaningful from the

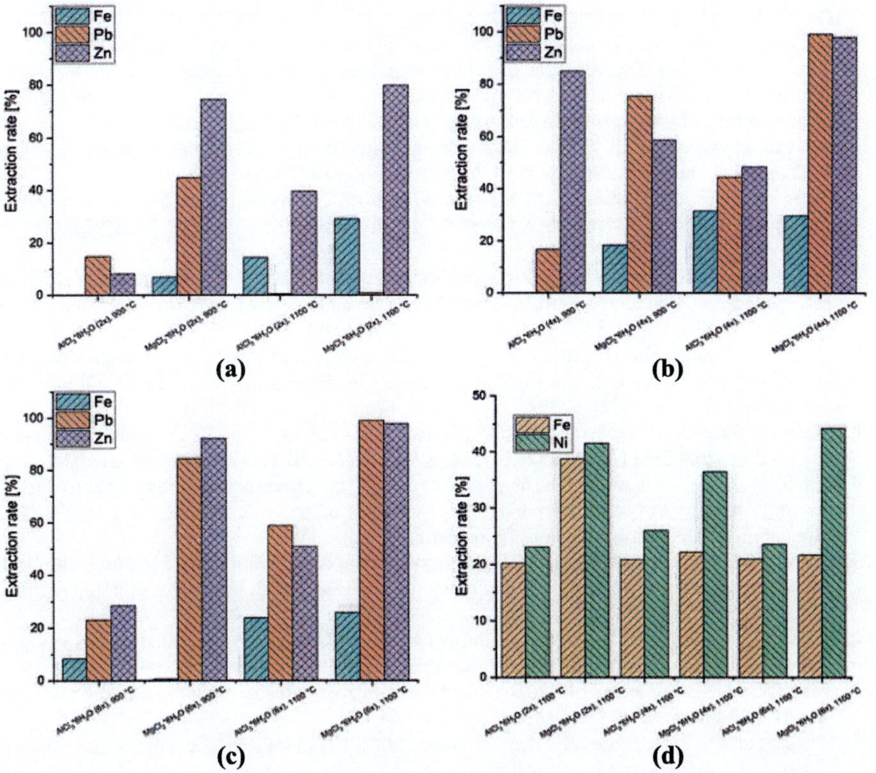

Fig. 4 **a-c** Extraction rates for Zn, Pb, and Fe for the experiments with jarosite from zinc industry at 900 and 1100 °C for **a** two times, **b** four times, and **c** six times stoichiometric addition of chlorine carrier; **d** Extraction rates for Ni and Fe for the experiments with jarosite from nickel industry at 1100 °C for two to six times stoichiometric addition of chlorine carrier

point of view of no new introduced impurity, which limits a possible utilization. Also, the low stability of iron chloride compared to the other targeted formed valuable metal chlorides and its trivalent state promise good extraction yields but must be proofed by an additional campaign.

Funding This work was funded by the Christian Doppler Research Association; the Austrian Federal Ministry for Digital and Economic Affairs; and the National Foundation for Research, Technology and Development.

References

1. Davenport WG (2011) Extractive metallurgy of nickel, cobalt, and platinum group materials. Elsevier, Amsterdam
2. Desborough GA, Smith KS, Lowers HA, Swayze GA, Hammarstrom JM, Diehl SF et al (2010) Mineralogical and chemical characteristics of some natural jarosites. In: Geochimica et Cosmochimica Acta 74 (3):1041–1056. https://doi.org/10.1016/j.gca.2009.11.006
3. Dutrizac JE (1984) The behavior of impurities during Jarosite Precipitation. In: Renato G. Bautista (Hg.): Hydrometallurgical process fundamentals. Tre behavior of impurities during jarosite precipitation. Boston, MA: Springer US, S. 125–169
4. Dutrizac JE, Chen TT (2013) The behaviour of gallium during jarosite precipitation. Canadian Metallurgical Quarterly 39(1):S. 1–14. https://doi.org/10.1179/cmq.2000.39.1.1
5. Frost RL, Wills RA, Kloprogge JT, Martens WN (2006) Thermal decomposition of hydronium jarosite (H3O)Fe3(SO4)2(OH)6. J Therm Anal Calorim 83 (1):S. 213–218. https://doi.org/10.1007/s10973-005-6908-0
6. Frost RL, Wills RA, Weier ML, Musumeci AW, Martens W (2005) Thermal decomposition of natural and synthetic plumbojarosites. Importance in 'archeochemistry'. Thermochimica Acta 432 (1):S. 30–35. https://doi.org/10.1016/j.tca.2005.04.001
7. Hoeber L, Steinlechner S (2021) A comprehensive review of processing strategies for iron precipitation residues from zinc hydrometallurgy. Cleaner Eng Technol 4:S. 100214. https://doi.org/10.1016/j.clet.2021.100214
8. Pappu A, Saxena M, Asolekar SR (2006) Jarosite characteristics and its utilisation potentials. Sci Total Environ 359(1–3):S. 232–243. https://doi.org/10.1016/j.scitotenv.2005.04.024
9. Lerche R (2021) Evaluation of the mechanisms of CO2-optimised chlorination reactions during the multi-metal recovery from precipitation residues of the zinc industry. Master's thesis. Montanuniversität, Leoben. Chair of Nonferrous Metallurgy
10. Steinlechner S, Antrekowitsch J (2018) Thermodynamic considerations for a pyrometallurgical extraction of indium and silver from a jarosite residue. Metals 8. https://doi.org/10.3390/met8030xxx
11. Sinclair RJ (2005) The extractive metallurgy of zinc. 1. ed. Carlton, Vic.: AusIMM (Spectrum series/Australasian Institute of Mining and Metallurgy, 13)
12. Wegscheider S, Steinlechner S (2016) Residues from zinc industry-a potential future resource for silver and indium. In: Proceedings of COM 2016
13. Wegscheider S, Steinlechner S, Leuchtenmüller M (2017) Innovative concept for the recovery of silver and indium by a combined treatment of Jarosite and electric arc furnace dust. JOM 69(2):S. 388–394. https://doi.org/10.1007/s11837-016-2192-7

Comprehensive Processing of Mica Concentrate from Yaroslavsky GRK for Lithium Carbonate, Potassium Sulfate, Aluminum Sulfate, Rubidium, and Cesium

Aleksandr Suss, Nataliya Kuznetsova, Sergey Ostashin, and Evgeniy Isakov

Abstract UC RUSAL's Yaroslavsky Mining Company (Yaroslavsky GRK) operates two of the world's largest fluorite deposits in the East Coast of Russia. The fluorite extraction process leaves rare alkali metals (Li, Rb, Cs, and K) in the tailings and becomes a potential raw material source for these elements. In 2015–2018, Irgiredmet (Irkutsk, Russia) jointly with Central South University (Changsha, China) developed and pilot tested the technology of beneficiation of the tailings from Yaroslavsky GRK to extract fluorite and produce mica concentrate with rare alkali metals content $\Sigma Li_2O + Rb_2O + Cs_2O = 1.5 \div 1.7$ mass %. To process mica concentrate the energy-efficient technology of sulfuric acid–fluoride low-temperature atmospheric leaching was developed enabling to exclude the preliminary roasting commonly used for processing such ores. The technology was proven in pilot scale resulting in a range of products meeting customer requirements: battery-grade lithium carbonate ($Li_2CO_3 \geq 99.9\%$), first-grade potassium sulfate, rubidium and cesium formates, and aluminum sulfate.

Keywords Mica concentrate · Sulfuric acid–fluoride leaching · Lithium carbonate · Rubidium · Cesium · Potassium sulfate · Aluminum sulfate

Introduction

At present, over 70% of global lithium reserves exist in ground waters and salt lake brine in the arid climate, such as the Salar de Atacama salt flat (Chile) containing ~4 mln. tonnes of lithium metal reserves, Bolivia's Salar de Uyuni comprising up to

A. Suss (✉) · N. Kuznetsova
RUSAL Engineering and Technology Center, Sredny Pr., 86, 199106 Saint Petersburg, Russia
e-mail: Aleksandr.Suss@rusal.com

S. Ostashin
JSC RUSAL Management, Vasilisy Kozhinoy St., 1, 121096 Moscow, Russia

E. Isakov
Yaroslavsky GRK LLC, Yubileynaya st., 66 Primorky Territory, 692271 Yaroslavsky, Russia

T. Ouchi et al. (eds), *Rare Metal Technology 2022*, The Minerals, Metals & Materials Series, https://doi.org/10.1007/978-3-030-92662-5_24

100 mln. tonnes of lithium metal, lithium salt lakes in Argentina, the USA (Nevada), etc.

Russia's climate does not allow natural lithium concentrating by sun evaporation. Other processes, such as, for example, developed by GK Rostekh and CJSC Ecostar-Nautekh lithium production by sorption from chloride solutions of Kovytkinsk and other oil deposits have poor prospects due to environmental and economic considerations. The solutions contain very little lithium chloride, i.e. <1 g/dm^3 (at total dissolved content of ~500 g/dm^3), while disposing a large amount of soluble salts negatively affects the environment [1–3]. The regions in Latin America where lithium is obtained from the brine are usually referred to as "white death" as 1 tonne of lithium carbonate generates several hundred tonnes of salt wastes.

Main lithium reserves in bedding granitoid rocks are located in the USA, China, Canada, and Australia. Russia's lithium reserves amount to ~900,000 tonnes and are mainly located in the ground waters. However, Russia should focus on obtaining lithium from the rocks as they contain ~25% of all explored reserves.

Lithium industry in Russia enjoys rich history. In the USSR, the first lithium mine was put into operation in 1941 in the Eastern Trans-Baikal Region on the Zavitinsk spodumene deposit (Trans-Baikal Mining and Processing Plant). The plant had been operating for 56 years and then was closed in 1997 because of reserves depletion and economic changes in Russia. In the 1980s, this mine produced over 30% of all global lithium.

Granitoids contain lithium in several minerals, including lepidolite, spodumene, amblygonite, and petalite. Lithium production from mineral raw material always comprises preliminary roasting of the concentrate at a high temperature (~1,100 °C for spodumene, ~800 °C for lepidolite). Roasting activates Li-containing minerals to provide for mild conditions of sulfuric acid leaching.

Li-bearing micas in the wastes of the Yaroslavsky Mining Company differ in their mineral composition from traditional lithium raw material in the granitoid rocks, pegmatite, etc. Due to this peculiar composition, RUSAL Engineering and Technology Center (RUSAL ETC) developed an energy-saving roasting-free process. This process will be implemented upon retrofitting a cryolite shop of JSC Cryolite (the Polevskoy Cryolite Plant, RUSAL's Group).

Raw Material Base

UC RUSAL's Yaroslavsky Mining Company (Yaroslavsky GRK) is located in the south of the Primorsky Territory, 150 km from Vladivostok. Yaroslavsky GRK was commissioned in 1949 and starting from 1964 it has been producing fluorite concentrate, fluorite briquettes, zinc concentrate, etc. Earlier Yaroslavsky GRK supplied 92% of Russia's demand for fluorides raw material. Yaroslavsky GRK operates two world's largest fluorite deposits, i.e. Voznesenskoe and Pogranichnoe deposits with ~19 mln. tonnes reserves (Fig. 1). Rear alkalis (Li, Rb, Cs, and K) in the fluorites were not extracted but concentrated in the residue of three tailing storage facilities of

Fig. 1 Fluorite mines, Yaroslavsky GRK

Fig. 2 Tailing storage facilities, Yaroslavsky GRK

the beneficiation plant. These tailings represent a man-made deposit of rear metals and fluorites with the reserves of ~30 mln. tonnes (Fig. 2).

Growing competition became a serious challenge for Yaroslavsky GRK. Among other reasons, competition growth was attributed to market entry of low-cost fluorides from Chine and Mongolia, as well as emerging of a cost-effective dry process of production of aluminum fluoride from high-purity fluoride concentrate. To remain profitable, Yaroslavsky GRK had to improve the quality of the main product, i.e. fluorite concentrate with ≥95% CaF_2 (FF-95) and launch deep comprehensive raw material processing to produce commercially successful value-added products (including lithium carbonate, potassium sulfate, rubidium and cesium formates, and aluminum sulfate). For these reasons, Yaroslavsky GRK was closed for upgrade.

It was resolved to relaunch Yaroslavsky GRK for comprehensive processing of the tailings. Use of fine tailings allows reducing operating expenses for mining, fine milling, and other processes and adopting a new concentration technology using the existing equipment to produce high-quality fluorite concentrate FF-95 and a number of other products, including mica concentrate, which contains rear alkalis.

In 2015–2018, JSC Irgiredmet (Irkutsk, Russia) jointly with minerals processing and bioengineering school, Central South University (Changsha, Hunan, China) developed and pilot tested the technology of beneficiation of the tailings from

Yaroslavsky GRK to produce fluorite concentrate FF 93 ÷ 95 and mica concentrate containing the following rare alkali metals ΣLi2O + Rb2O + Cs2O = 1.5 ~ 1.7 mass. %.

Tables 1 and 2 present chemical and mineral compositions of the mica concentrate (MC) obtained during the pilot tests at JSC Irgiredmet' pilot plant from a representative large-tonnage sample from Yaroslavsky GRK.

Table 2 demonstrates a quantitative mineral composition of the mica concentrate.

Mica in the mica concentrate is represented by the mixture comprising the following:

- predominantly lepidolite $KLi_2(Al, Li)$ $[(Al, Si)_3O_{10}](F, OH)_2$;
- 8–10% muscovite KAl_2 $[AlSi_3O_{10}]$ $(OH)_2$; and
- some percentage of biotite K $(Mg, Fe)_3[AlSi_3O_{10}]$ $(OH, F)_2$, ephesite $NaLiAl_2$ $[Al_2Si_2O_2]$ $(OH)_2$, and zinwaldite $K(Al, Fe, Li)_3$ $[(SiAl)_4O_{10}]$ (OH, F).

Particle size distribution of the mica concentrate after beneficiation is 100% of minus 150 μm with average particle size of 27–29 μm and the moisture content of 10%.

The obtained mica concentrate is clay-like so it can be transported over long distances in gondola cars without any dusting or congelation issues (even at a temperature of up to –15 °C) and is practically ready for processing without any additional expenditures for milling.

Lithium content in the mica concentrate is relatively low, i.e. ~0.9 mass.% but mica concentrate is a complex raw material. Comprehensive processing of the concentrate yields a number of products, including rubidium and cesium formates?, which possess unique properties, i.e. freezing temperature of below 45 °C and high density of $\rho \geq 1.7$ g/cm^3. Moreover, they can be used as non-corrosive inhibitors. They are highly demanded for well drilling of the seams under high pressure and their cost exceeds the cost of lithium carbonate.

Treatment Process

The developed technology for producing lithium carbonate, rubidium and cesium compounds, and other products from the mica concentrate comprises several processes (see Fig. 3):

- preparation of thick slurry by mixing the mica concentrate with 72% sulfuric acid at a L: S ratio of ≈ 1: 1 (with account of MC moisture) and sulfidation to remove carbon dioxide and convert fluorine into the hydrofluoric acid;
- sulfuric acid—fluoride leaching of the mica concentrate;
- filtration, blowing, washing, and "squeezing" the Si-residue on the filter-press followed by disposal thereof to the residue storage area or shipment to prospective customers;

Table 1 Chemical composition of mica concentrate, mass.%

Li_2O	Rb_2O	Cs_2O	K_2O	Na_2O	SiO_2	Al_2O_3	Fe_2O_3 total	incl FeO	TiO_2	P_2O_5	CaO	MgO	MnO	BeO	ZnO	SO_3	F	LOI	incl CO_2
0.87	0.49	0.026	6.2	0.61	43.3	22.5	4.0	1.6	0.15	0.08	4.65	4.85	0.02	0.08	0.1	0.70	2.78	6.6	1.34

Table 2 Phase and mineral composition of a mica concentrate sample (data from JSC Irgiredmet)

Mineral	Content, mass.%
Micas	70–73
Plagioclase, mainly albite $Na[AlSi_3O_8]$	7.0
Quartz SiO_2	6.5
Fluorite CaF_2	2–4
Chlorite	3.5
Carbonates (calcite, dolomite, ankerite)	1.5–3.5
Amphibole	Sparse grains
Pyrite FeS_2	0.5
Arsenic pyrite FeAsS	0.2
Sphalerite ZnS	0.1
Pyrrhotine Fe_nS_{n+1}, chalcopyrite $CuFeS_2$, galenite PbS, magnetite Fe_3O_4, hematite Fe_2O_3, goethite FeOOH, titanium oxides (anatase, rutile), gypsum $CaSO_4 \times 2H_2O$, carbonaceous matter	Sparse and single grains

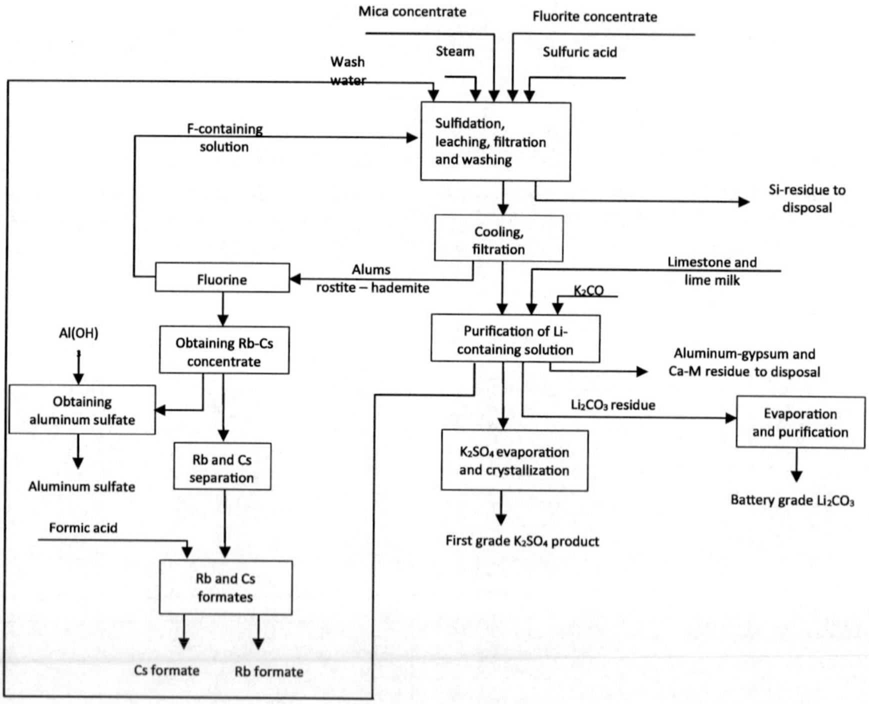

Fig. 3 Process flow diagram of processing mica concentrate from Yaroslavsky GRK to obtain lithium carbonate and other products

- cooling of Li-containing concentrate, crystallization, and extraction of aluminum-potassium-rubidium-cesium alums, separation of fluorine as rostite–hademite. Selective dissolution of rostite and hademite and return of lithium- and fluorine-containing solution to the head of the process, i.e. mica concentrate leaching;
- three-stage purification of Li-containing solution with limestone and lime to remove detrimental and harmful impurities with steady increase of pH value from ~1 to ~11. Removal of some gypsum, aluminum–limestone–gypsum, and magnesium residue from the process;
- concentrating evaporation of purified Li-containing solution and removal of calcium impurity;
- crystallization of crude lithium carbonate with potash;
- purification of crude lithium carbonate by re-precipitation to obtain Li_2CO_3 product (battery grade) and drying thereof;
- cooling and crystallization of the mother liquor from the lithium side of the process to obtain potassium sulfate product;
- drying of a combined stream from lithium and alum processes;
- dissolution and decomposition of washed Al(K,Rb,Cs) alums and removal of impurities from the liquor;
- two-stage evaporation and crystallization of potassium sulfate in the alum side of the process;
- precipitation of sulfates and other low-soluble hydroxides with lime;
- conversion of Rb and Cs hydroxides into carbonates with carbon dioxide;
- evaporation of the liquor and extraction of the concentrate containing rubidium and cesium carbonates, separation of rubidium and cesium carbonates to obtain rubidium and cesium formates; and
- obtaining of aluminum sulfate crystalline hydrate $Al_2(SO_4)_3 \times 18\ H_2O$.

Majority of process operations and stages are common for production of lithium carbonate, rubidium and cesium compounds, and other products [4]. Therefore, this paper discusses only some distinctive features of the processes for the specific raw materials.

Sulfuric Acid–Fluoride Leaching of the Mica Concentrate

Mica concentrate (MC) comprises 2–4 mass.% of residual fluorite. Moreover, alums contain fluorine as an additional anionic group. When mica concentrate is leached with sulfuric acid fluorine dissolves from the fluorite and alums into the solution as hydrofluoric acid by the following reaction:

$$CaF_2 + H_2SO_4 = 2HF + CaSO_4$$

Thus, sulfuric acid leaching becomes sulfuric acid–fluoride leaching at the stage of sulfatisation of the thick slurry.

Leaching is carried out at L: S ≈ 2.5: 1. After sulfidation, the thick slurry with L: S ≈ 1: 1 is diluted with the recycled fluorine-containing solution from the rubidium–cesium side after dissolution of rostite–hademite, as well as wash water from Si-residue washing on the filter-press. The concentration of fluorine ions during the sulfuric acid–fluoride leaching does not exceed 28÷30 g/dm^3 and shall be limited due to possible formation of lithium fluoride (LiF) that will result in the loss of some lithium at the first purification stage of Li-containing solution.

When mica concentrate is leached in the agitated tank at atmospheric pressure and at a temperature of 108÷110 °C, lepidolite, muscovite, and other alums react with a sulfate-fluorine solution. Alum structure changes to form illite (hydromuscovite), and ≥ 90% of lithium, rubidium, cesium, and potassium dissolve into the solution. Illite is generated as a pseudomorph of lepidolite and muscovite so the obtained precipitate preserves the structure of the initial mica and can be easily filtered using the filter-press. The precipitate is nominally called Si-residue but it comprises mainly illite (hydromuscovite) instead of silica gel. Leaching of 1 tonne of the mica concentrate (as dry matter) under the specified parameters yields ~0.75 t of Si residue.

Increase of leaching temperature of up to ≥120 °C adversely affects the process, i.e.

- silicon tetrafluoride and hydrogen fluoride start to generate;
- Si-residue filtration deteriorates due to breakage of layered silicate structure.

Tables 3 and 4 present chemical and phase composition of Si-residue obtained from the pilot testing of this process in the pilot shop of SC Rosatom at JSC PIMCU, PJSC (Krasnokamensk, Trans-Baikal Territory). Figure 4 shows XRD characterization data of the samples.

Crystallization and Extraction of Aluminum–Potassium–Rubidium–Cesium Sulfates, Rostite–Hademite from Li-Containing Solution. Selective Dissolution of Rostite and Hademite and Return of Lithium- and Fluorine-Containing Solution to the Head of the Process

Lithium can be separated from rubidium and cesium, as lithium does not form the alums [4]. The filtrate from MC leaching has a density of $\rho = 1.39–1.43$ g/cm^3 so when the temperature decreases to ~25 °C it yields a crystalline precipitate consisting of the following mixture:

- coarse-crystalline alums $(K, Rb, Cs)_1Al(SO_4)_2 \times 12H_2O$;
- crystalline hydrate of basic aluminum sulfate of rostite–hademite type $Al(SO_4)(OH) \times 5H_2O - Al(SO_4)F \times 5H_2O$.

Alum precipitate removes ~95% rubidium and cesium from the Li-containing solution.

Table 3 Chemical composition of Si-residue and other products obtained from purification of Li-containing solution, mass %

Sample	Chemical composition, mass.%														
	SiO_2	Al_2O_3	Fe_2O_3	TiO_2	P_2O_5	CaO	MgO	Na_2O	K_2O	SO_3	F	Li_2O	Rb_2O	Cs_2O	LOI
Si-residue (W = 20÷35%) pH minus 1	33.8	14.6	2.6	0.12	0.02	5.0	2.4	0.36	4.4	17.1	0.5	0.169	0.181	0.006	29.2
Gypsum residue W = 35% pH ≈ 3	<0.06	2.3	0.87	0.06	0.09	35.5	1.5	<0.15	0.66	36.3	0.7	0.163	0.007	0.001	13.8
Aluminum–limestone–gypsum residue W = 35% pH ≈ 6÷8	<0.06	4.9	2.7	0.06	0.06	39.8	1.6	<0.15	0.5	13.6	1.7	0.131	0.018	0.001	31.2
Magnesium residue W = 38% pH ≈ 11	<0.06	0.21	0.12	<0.05	<0.02	49.1	10.3	<0.15	<0.15	16.9	0.3	0.1097	0.0063	–	20.4

Table 4 Semi-quantitative phase composition of Si-residue and other products obtained from purification of Li-containing solution, mass.%

Sample	Phases, in descending order
Si-residue at pH minus 1	Illite, montmorillonite, traces of zinnwaldite, lepidolite, anhydrite, quartz
Gypsum residue at pH ≈ 3	Bassanite $CaSO_4*0.5H_2O$, gypsum $CaSO_4*2H_2O$, traces of quartz
Aluminum-limestone-gypsum residue at pH ≈ 6~8	Calcite, bassanite, some gypsum
Magnesium residue at pH ≈ 11	$Ca(OH)_2$, bassanite, brucite $Mg(OH)_2$

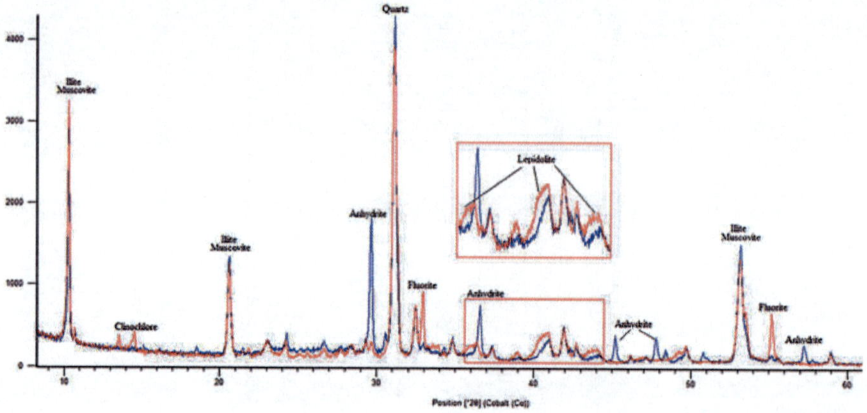

Fig. 4 XRD diagrams of initial concentrate (red) and silica residue (blue). Amorphisation of sample as a result of micas structure decomposition and lepidolite breakage is shown. After filtration, squeezing, and washing on a filter-press, the obtained Si-residue (moisture content of 25÷35%) is disposed to the residue storage area for dry stacking. Said technology of sulfuric acid–fluoride leaching is protected by the patent granted to UC RUSAL [5] (Color figure online)

Hademite and rostite are more soluble as compared with the alums. Therefore, when the precipitate is treated with some water at a temperature of ~25–30 °C rostite–hademite dissolves quickly. The obtained Li- and F-containing solution is returned to the head of the process for MC leaching to provide for the specified fluorine concentration during the leaching process thus reducing the specific consumption of sulfuric acid. The remaining alum precipitate enriched in Cs, Rb, and K is treated with water and filtered. The second filtrate is used for production of aluminum sulfate.

Filtered alums are directed to produce rubidium formate, cesium formate, and first-grade potassium sulfate as per the proposed technology.

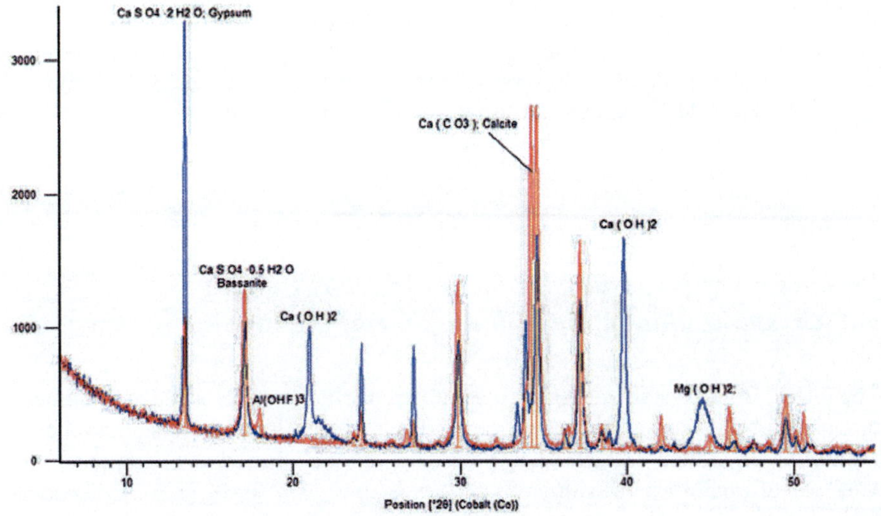

Fig. 5 XRD diagrams of aluminum-limestone-gypsum residue at pH ≈ 6~8 (red) and magnesium residue at pH ≈ 11 (blue) (Color figure online)

Purification of Li-Containing Solution to Remove Impurities

The process is based on the bonding of impurities to form solids at a gradual increase of pH value from minus 1 to 10~11. The proposed process differs from the others as it uses low-cost materials, i.e. limestone and lime, as well as it establishes the recycling of residue wash water from the high pH area to the low pH area. This counter-current purification allows reducing the consumption of limestone and lime, reducing the residue amount, and improving lithium total recovery ≥ 85% by decreasing its secondary losses.

Tables 3, 4, and Fig. 5 present chemical and phase compositions of the residues from purification of Li-containing solution.

Concentrating Evaporation of Purified Li-Containing Solution, Crystallization of Crude Lithium Carbonate with Potash, Purification of Crude Lithium Carbonate to Obtain Li_2CO_3 Product (Battery Grade)

Lithium concentration by evaporation is a typical process of lithium production [4]. The proposed process differs from the conventional one by further extraction of the first-grade potassium sulfate from the mother liquor of lithium carbonate crystallization [6]. For this purpose, potash (K_2CO_3) is used to obtain primary crude lithium carbonate instead of sodium carbonate (Na_2CO_3), which is typically used to crystallize Li_2CO_3. The reasons for using the potash instead of sodium carbonate are as follows: abundance of K versus Na in the mica concentrate raw material and greater market value of potassium sulfate versus sodium sulfate.

The proposed process for lithium carbonate re-precipitation was pilot tested to produce a sample of battery-grade Li_2CO_3 ($Li_2CO_3 \geq 99.9$ mass. %).

Production of Potassium Sulfate Product by Cooling and Crystallization in the Lithium Side of the Process

The filtrate from crystallization of crude lithium carbonate with potash at a temperature of ~95 °C is fed to the thickener and cooled to ~25–35 °C to crystallize coarse-crystalline potassium sulfate. Then potassium sulfate is dried at a temperature of <200 °C to obtain a potassium sulfate that is used as a fertilizer for greenhouse facilities.

Dissolution and Decomposition of Washed Al(K,Rb,Cs) Alums and Removal of Impurities from the Liquor. Conversion of Rb and Cs Hydroxides into Carbonates with Carbon Dioxide. Evaporation of the Liquor and Extraction of the Concentrate Containing Rubidium and Cesium Carbonates, Separation of Rubidium and Cesium Carbonates to Obtain Rubidium and Cesium Formates

Al(K,Rb,Cs) alum precipitate is dissolved in the water at a temperature of 90–95 °C, then treated with milk of lime to separate aluminum in the form of aluminum-gypsum residue at a pH of $\approx 9 \div 10$, evaporated and cooled to 25 °C. Potassium sulfate precipitates from the cool solution, and the resulting precipitate is filtered and mixed with potassium sulfate from the lithium production side.

The Rb- and Cs-rich solution is treated with barium hydroxide to convert rubidium and cesium into hydroxides $Rb(OH)_2$ and $Cs(OH)_2$. Furthermore, they are converted into carbonates by carbonation with CO_2, separated and converted into formates by well-known processes.

Conclusions

The proposed sulfuric acid–fluoride leaching of lithium and other rear alkalis from the mica concentrate allows processing the raw materials without preliminary roasting and pressure leaching.

Co-crystallization of aluminum–potassium–rubidium–cesium alums and rostite and hademite enables to recycle fluorine and prevents the secondary losses of lithium with lithium fluoride during the purification of Li-containing solution.

Counter-current purification of Li-containing solution with limestone and lime improves lithium total recovery and minimizes the amount of residue generated from the purification.

The financial and economic model developed for comprehensive processing of the tailings from Yaroslavsky GRK to obtain fluorite concentrate that can be used for producing fluorites by dry methods and other commercially successful products (lithium carbonate, potassium sulfate, rubidium and cesium formates, and aluminum sulfate) proved the commercial viability of the process.

Russia relaunches the production of lithium carbonate from domestic raw material and fluorites to improve lithium self-sufficiency of the country. LiO_2 reserves at Yaroslavsky GRK amount to >1120 thousand tonnes thus fulfilling Russia's demand in lithium for 40 years to come.

This project will provide for producing ~2.5 thousand tonnes of lithium carbonate compliant with the requirements of technical specifications TU 95.1951-89 starting from 2025. The availability of domestic low-cost production of lithium salts will allow RUSAL to promote manufacture of Al-Li and Li-Mg alloys for aviation. Moreover, allied industries will be able to increase production of Li-ion batteries for renewable energy sources, lithium ceramics and glass, lithium greases, etc.

Comprehensive processing of the raw materials at Yaroslavsky GRK will enable the company to reduce the amount of disposed tailings by ~40%.

Acknowledgements The authors acknowledge Tatyana Golovanova, Tatyana Mukina, Yulia Chernyshova, Irina Paromova, and Aleksandr Damaskin (Engineering and Technology Department of LLC RUSAL ETC in St. Petersburg); Marina Garmash (Yaroslavsky GRK); Mikhail Tiunov (JSC Irgiredmet); and Anatoly Bugaev (JSC Criolite Plant), who made a great contribution to this work.

References

1. Ryabtsev A et al (2012) Method of obtaining lithium concentrate from lithium-bearing natural brines and processing thereof. RU. Patent 2516538. Accessed 17 Feb 2012
2. Kotsupalo NP, Ryabtsev AD (2008) Chemistry and process for production of lithium compounds from lithium-bearing hydromineral raw materials. GEO, Novosibirsk
3. Ryabtsev AD, Kotsupalo NP et al (2014) Comprehensive processing of polycomponent lithium-bearing brines with preliminary lithium concentration therein. GEO, Novosibirsk

4. Plyushchev VE, Stepin BD (1970) Chemistry and technology of lithium, rubidium, and cesium compounds. Khimiya, Moscow
5. Suss A, Kuznetsova N (2020) Method of processing mica concentrate. RU. Patent 27495998. Accessed 30 Nov 2020
6. National Standard GOST 4145–74 Reagents. Potassium sulphate. Specifications

Selective Sulfidation for Rare Earth Element Separation

Caspar Stinn and Antoine Allanore

Abstract Rare earth metals and compounds are critical components of advanced materials for energy, structural alloys, and transportation. These low-tonnage elements are sourced together as by- and co-products, and presently require complete hydrometallurgical dissolution followed by liquid–liquid separation for their isolation and production. There is great interest in developing alternatives to those hydrometallurgical processes in order to limit the environmental impact of rare earth element supply. Herein, we present selective sulfidation as a novel, high-temperature alternative to facilitate physical separation of rare earth by- and co-product elements. We explore the thermodynamics of rare earth oxide sulfidation with elemental sulfur, and discuss the role of carbon in controlling sulfidation selectivity. We apply these findings to the demonstration of selective sulfidation for iron-rare earth and lanthanide-lanthanide separations.

Keywords Sulfidation · Rare earth separation · Lanthanides · Scandium · Pyrometallurgy · Physical separation

Introduction

Rare earth elements are required for a broad range of technologies essential to modern society, ranging from neodymium and dysprosium for high-strength magnets to cerium for automotive catalytic converters to scandium for aerospace aluminum alloys [1]. These elements, consisting of the lanthanides, scandium, and yttrium, are found together in nature and are produced as co- and by-products of one another [2]. Despite being classified as “rare earth” elements, they are relatively common within

C. Stinn
Department of Materials Science and Engineering, Massachusetts Institute of Technology, 77 Massachusetts Avenue, Office 13-5095, Cambridge, MA 02139-4307, USA
e-mail: stinn@mit.edu

A. Allanore (✉)
Department of Materials Science & Engineering, Massachusetts Institute of Technology, 77 Massachusetts Avenue, Office 13-5066, Cambridge, MA 02139-4307, USA
e-mail: allanore@mit.edu

T. Ouchi et al. (eds), *Rare Metal Technology 2022*, The Minerals, Metals & Materials Series, https://doi.org/10.1007/978-3-030-92662-5_25

the earth's crust—on par with many base metals. For example, the crustal abundance of neodymium is on the same order of magnitude of nickel, zinc, and copper, twice that of cobalt, three times that of lead, and twenty times that of tin [3]. Nevertheless, commercially viable deposits of rare earth elements are geographically concentrated, with China accounting for over 60% of recorded global production in 2020, followed by the United States at about 15% [4]. Concerns over supply stability, and the lack of suitable substitutes for rare earth elements in many applications, have led rare earth elements to be classified as critical and strategically significant materials by numerous governments [5]. Challenges surrounding rare earth element supply are further compounded by the "balance problem" [6, 7]. The "balance problem" refers to the reality that the distribution of demand between individual rare earth elements does not match up with their relative abundances in commercially relevant mineral bodies. This disconnect between supply and demand results in production being driven by a few, high-demand elements such as neodymium, with oversaturated supplies of lower demand rare earth elements.

Rare earth elements may be grouped into two main categories: light (La, Ce, Pr, Nd, Pm, Sm) and heavy (Eu, Gd, Tb, Dy, Ho, Er, Tm, Yb, Lu, Sc, Y). Light rare earth elements are most commonly produced from the fluorocarbonate mineral bastnaesite and phosphate mineral monazite, whereas heavy rare earth elements are more concentrated in the phosphate mineral xenotime and in lateritic ion absorption clays [2]. No distinct mineral phases of individual lanthanide elements exist; they are all present, substituted for one another within the mineral. Actinides including thorium and uranium are also present as normally occurring radioactive materials (NORMs) in rare earth minerals through atomic substitution with the lanthanides [8]. Scandium meanwhile may also be found and produced as a by-product of nickel mining from lateritic ore deposits [9–11]. Furthermore, significant attention has been directed toward recovery of scandium from the red mud tailings of the Bayer process for alumina production from bauxite [12–14], yet separating scandium from iron remains a difficult task [10].

Conventional rare earth element production from fluorocarbonate and phosphate mineral sources begins with mining, comminution, and physical separation to produce a rare earth mineral concentrate [15, 16]. The rare earth concentrate is chemically digested, or "cracked", via acid or alkaline treatment, with the rare earth elements leached into an aqueous solution for individual separation via aqueous-organic liquid–liquid solvent extraction. During the cracking and leaching process, the majority of NORMs and entrained transition and alkaline earth metal impurities remain with the leach residue or are selectively precipitated [2]. Due to the chemical similarity of rare earth elements, solvent extraction is a very energy- and reagent-intensive process, with individual rare earth separations each generally requiring on the order of 50–150 extraction and stripping stages for separation [2]. The separated rare earth element is then precipitated as a pure compound. The pure compound may then be calcined to form an oxide, halogenated, and/or reduced to metal via molten salt electrolysis or metallothermic reduction [17].

The environmental impact of rare earth metal reduction is greatly dependent on the process employed [18]. In the rare earth element separation pathway upstream of reduction, the environment impact is largely inherited from the solvent extraction steps [19]. For rare earth element separation from bastnaesite, solvent extraction contributes approximately 77% to the terrestrial acidification; 63% to the global warming potential; and 70% to the water resource demand of the entire rare earth element mining, beneficiation, and separation pathway [8]. Solvent extraction in monazite processing exhibits similar environmental burdens. When processing steps in the rare earth element mining, beneficiation, and separation pathway are normalized by material throughput, solvent extraction is likewise the largest contributor to process capital cost by an order of magnitude [20–23]. The economic and environmental impact of liquid–liquid processing motivates efforts to develop new methods of rare earth element separation with reduced environmental and capital footprints.

Numerous hydrometallurgical methods have been explored to improve the thermodynamics and sustainability of solvent extraction, ranging from new organic collector phases [24] to introducing magnetic fields during the hydrometallurgical process [25]. However, even highly optimized solvent extraction systems for readily separated base metals, such as iron-copper or nickel-cobalt, still exhibit large energy and environmental impacts [26]. This suggests that incremental improvements to hydrometallurgy ultimately have limited promise in greatly reducing the overall environmental burden of rare earth element separation. In general, physical processes to separate elements isolated in distinct mineral phases exhibit lower environmental impacts than chemical processes to separate elements mixed in multi-element compounds or solid solutions [27, 28]. Therefore, a low-cost, low-energy, and sustainable pretreatment to partition ("crack") mixed rare earth element compounds to distinct, physically separable phases of individual rare earth elements is an attractive method to reduce the energy and economic burden of rare earth element separation.

A variety of halide-, sulfate-, and borate-based pretreatments to partition rare earth elements from a multi-element compound feed into distinct phases for physical separation have been explored. Chemical vapor transportation has been studied to separate rare earth elements through differences in volatility and stoichiometry of their respective rare earth aluminum chloride vapor complexes [29]. Meanwhile, reactive vacuum distillation has been explored for separation of divalent and trivalent rare earth chlorides and iodides, also based on differences in volatility [30]. However, halogenation chemistry for metal compounds exhibits significant environmental impacts and costs associated with toxic reagents such as chlorine, hydrogen fluoride, and phosgene gasses [31, 32]. Selective acid roasting or sulfation with sulfur dioxide has been considered to promote selective leaching of rare earth elements [12, 13, 33–35], yet acid roasting processes too exhibit significant environmental impacts when applied to rare earth element processing [8]. Finally, while borate crystallization or slag refining shows good selectivity for rare earth elements [36, 37], burdensome chemical treatments remain necessary to separate, downstream, the rare earth elements from boron.

A promising pretreatment to support physical separation of individual rare earth elements is selective sulfidation [38], due to the well-known differences in physical properties between oxide and sulfide phases. A generalized pseudo-binary phase diagram is depicted in Fig. 1 between a rare earth oxide (Ln–O) and sulfide (Ln-S),

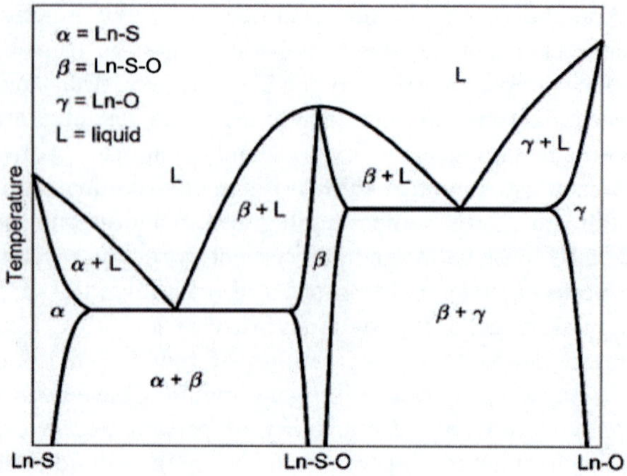

Fig. 1 Pseudo-binary phase diagram between a rare earth oxide (Ln–O) and sulfide (Ln-S), with an oxysulfide (Ln–O-S) intermediate. Rare earth oxides, oxysulfides, and sulfides are immiscible, thereby supporting liberation from one another via comminution and subsequent physical separation

with a rare earth oxysulfide (Ln–O-S) intermediate. Rare earth elements are reported to form eutectic mixtures of immiscible oxide, oxysulfide, and sulfide phases [39, 40]. Numerous rare earth oxysulfide stoichiometries are known [39–41]. If individual rare earth elements can be selectively partitioned between oxide, oxysulfide, and sulfide phases, they could be selectively liberated from one another via comminution. Following comminution, oxides, oxysulfides, and sulfides are readily separated using economical and environmentally benign physical methods, such as magnetic separation and froth flotation. Indeed, selective sulfidation may promote separation of individual rare earth elements using physical methods in place of complete hydrometallurgical dissolution and solvent extraction.

Sulfidation has previously been applied for numerous transition metal separations from mixed compounds or solid solutions [38]. Examples encompass iron-nickel from laterite [42], nickel-manganese-cobalt-lithium from battery cathodes [43], and zinc-iron-silicon from lead smelter slag [44, 45]. Furthermore, sulfidation chemistry utilizes proven reactor technologies [46], and is conducted in industrial settings for nickel extraction [47]. Nevertheless, despite encouraging results for other system chemistries, selective sulfidation has until recently [38] never been attempted for rare earth element processing. Herein, we present a thermodynamic framework for the selective sulfidation of individual rare earth oxides, oxysulfides, and sulfides to enable physical separation of individual rare earth elements. We consider the presence of transition metal impurities such as iron and explore the role of carbon as a process lever to control sulfidation selectivity. For the first time, we then experimentally demonstrate selective sulfidation and desulfidation in the iron-scandium oxide, iron-lanthanum sulfide, and praseodymium-neodymium-dysprosium oxide systems as a potential pretreatment for isolation of individual rare earth elements via physical separation methods.

Methods

Thermodynamic Framework

While the most commercially relevant rare earth element (RE) minerals are the fluorocarbonate bastnaesite and the phosphates monazite and xenotime, the mixed anion thermodynamics of RE-C-O-F-S and RE-P-O-S systems are thus far unexplored, but presumably different from one another. The rare earth sesquioxide is chosen as a mineral agnostic system for thermodynamic and experimental study of selective sulfidation for rare earth elements. Numerous technologies exist to convert fluorocarbonates and phosphates to oxides without complete hydrometallurgical dissolution [2, 15, 16, 48]. Pyrometallurgical sulfidation of a rare earth sesquioxide (Ln_2O_3) with gaseous, diatomic elemental sulfur (S_2) to form sulfur dioxide and a rare earth sesquisulfide (Ln_2S_3) via a rare earth oxysulfide intermediate (Ln_2O_2S) may be described by the following reactions:

$$S_2(g) + \frac{4}{3}Ln_2O_3 = \frac{2}{3}SO_2(g) + \frac{4}{3}Ln_2O_2S \tag{1}$$

$$S_2(g) + \frac{2}{3}Ln_2O_2S = \frac{2}{3}SO_2(g) + \frac{2}{3}Ln_2S_3 \tag{2}$$

Equation 1 may be rebalanced for the oxides of cerium and praseodymium, which are more stable with metal valencies of +4 and +3, +4, respectively. While hydrogen sulfide and carbon disulfide have classically been employed for sulfidation of rare earth oxides [49], elemental sulfur is employed in this study due to fact that those reagents have previously been shown to be non-selective in sulfidation[50]. The predominant allotrope of sulfur above 800 °C is a diatomic gas [51]. Sulfidation of a transition metal impurity, here iron oxide (Fe_2O_3), is described by similar reactions to Eqs. 1–2. Sulfidation of iron oxide is known to progress through an iron(II,III) oxide intermediate (Fe_3O_4) [52]:

$$S_2(g) + 12Fe_2O_3 = 2SO_2(g) + 8Fe_3O_4 \tag{3}$$

$$S_2(g) + \frac{2}{5}Fe_3O_4 = \frac{4}{5}SO_2(g) + \frac{6}{5}FeS \tag{4}$$

At equilibrium, the diatomic sulfur and sulfur dioxide partial pressures (P_{S_2} and P_{SO_2}, respectively) required for the reactions in Eqs. 1–4 to be thermodynamically spontaneous may be related to the standard Gibbs energy of reaction ($\Delta_r G°$) gas constant (R), absolute temperature (T), and activities of condensed metal compound reactants and products (a).

$$\frac{P_{S_2}}{P_{SO_2}^{\frac{2}{3}}} = \frac{e^{\frac{\Delta_{r1} G^{o}}{RT}} a_{Ln_2O_2S}^{\frac{4}{3}}}{a_{Ln_2O_3}^{\frac{4}{3}}} \tag{5}$$

$$\frac{P_{S_2}}{P_{SO_2}^{\frac{2}{3}}} = \frac{e^{\frac{\Delta_{r2} G^{o}}{RT}} a_{Ln_2S_3}^{\frac{2}{3}}}{a_{Ln_2O_2S}^{\frac{2}{3}}} \tag{6}$$

$$\frac{P_{S_2}}{P_{SO_2}^{2}} = \frac{e^{\frac{\Delta_{r3} G^{o}}{RT}} a_{Fe_3O_4}^{8}}{a_{Fe_2O_3}^{12}} \tag{7}$$

$$\frac{P_{S_2}}{P_{SO_2}^{\frac{4}{5}}} = \frac{e^{\frac{\Delta_{r4} G^{o}}{RT}} a_{FeS}^{\frac{6}{5}}}{a_{Fe_3O_4}^{\frac{2}{5}}} \tag{8}$$

Assuming the metal compound reactants and products are non-volatile, and that the gas phase only contains sulfur-oxygen species at a given temperature and pressure, the Gibbs phase rule indicates the equilibrium sulfur dioxide partial pressure is a direct function of diatomic sulfur partial pressure. Isobars depicting equilibrium sulfur dioxide partial pressure as a function of diatomic sulfur partial pressure at a total pressure of 1 atmosphere for temperatures of 800, 1000, 1200, and 1400 °C can be calculated, here using FactSage 8.0 with the FactPS database. From these isobars, a stoichiometric-independent ratio of the partial pressure of diatomic sulfur to sulfur dioxide may be defined and related to the stoichiometrically dependent sulfur-to-sulfur dioxide ratios in Eqs. 5–8 [38]. The sulfur-to-sulfur dioxide ratio required to sulfidize an oxide or oxysulfide, or similarly "desulfidize" a sulfide or oxysulfide, may be defined based on the stoichiometrically dependent ratios of the activities. Selective sulfidation of rare earth or transition metal elements from a mixed metal compound feedstock promotes the formation of sulfide phases that are naturally immiscible with oxide and oxysulfide feedstock phases. Similarly, oxide and oxysulfide phases may be precipitated from mixed sulfide feeds via selective desulfidation of target elements. Elements partitioned into different phases may thereby be liberated and physically separated from one another using technologies such as magnetic separation or froth flotation, providing an opportunity to replace expensive, energy-intensive hydrometallurgical separation with cleaner, cheaper, and physical separations. In Fig. 2, we compare the diatomic sulfur-to-sulfur dioxide partial pressure ratios to sulfidize various mineral phases of interest; iron oxide (Fe_3O_4), scandium oxysulfide, and lanthanum oxysulfide. Temperatures of 800, 1000, and 1200 °C are considered, and the partial pressure ratio is presented as a function of metal compound activity ratios, calculated using FactSage 8.0 with the FactPS database supplemented with literature data [53]. Due to the differences in sulfidation reaction stoichiometry between iron oxide and rare earth oxysulfides, a stoichiometric-independent activity ratio of product (A_{M-S}) to reactant (A_{M-O-S}) is employed, illustrated here for Eq. 6:

$$\frac{A_{M-S}}{A_{M-O-S}} = \frac{a_{Ln_2S_3}^{\frac{2}{3}}}{a_{Ln_2O_2S}^{\frac{2}{3}}} \tag{9}$$

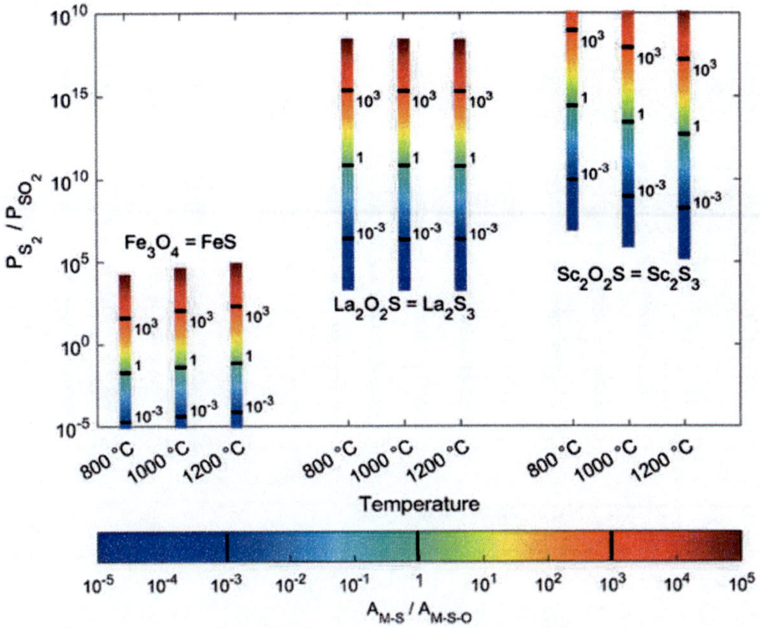

Fig. 2 Sulfidation series for iron and rare earth oxides. The sulfur to sulfide dioxide partial pressure ratio at a product and reactant activity ratio of unity is defined as the *pure-phase sulfur-to-sulfur dioxide ratio*, corresponding to the gas ratio required for sulfidation, as reported in a Kellogg [54] or Pourbaix [55] diagram

A similar relation may be defined for Eq. 8. At an activity ratio of 1, the diatomic sulfur-to-sulfur dioxide partial pressure ratio for sulfidation or desulfidation is equivalent to that determined for pure, immiscible reactants and products as found in a Kellogg [54] or Pourbaix [55] diagram. This sulfur-to-sulfur dioxide partial pressure ratio is defined as the *pure-phase sulfur-to-sulfur dioxide ratio.* As shown in Fig. 2, formation of a sulfide from iron oxide necessitates lower diatomic sulfur-to-sulfur dioxide partial pressure ratios than from a rare earth oxysulfides at a given ratio of reactant and product activities. This is consistent with Goldschmidt geochemical classification [56] of elements, where iron is observed to be more chalcophilic than the highly lithophilic rare earth elements. In Fig. 3, we compare the diatomic sulfur-to-sulfur dioxide partial pressure ratios to sulfidize rare earth oxysulfides at 1400 °C as a function of metal compound activity ratios, calculated using FactSage 8.0 with the FactPS database supplemented with literature data [53, 57]. The series in which individual rare earth oxysulfides form sulfides with increasing diatomic sulfur-to-sulfur dioxide partial pressure ratios is highly dependent on the oxysulfide and sulfide activities in the system. Presently, solution thermodynamics of mixed RE-O-S systems are not well-established, yet documented mixed metal oxide and

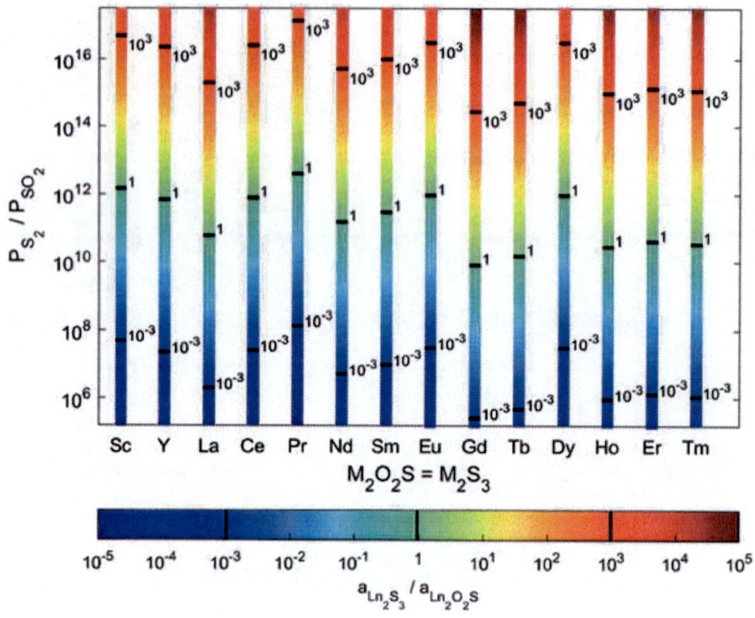

Fig. 3 Sulfidation series for rare earth oxysulfides. The sulfur-to-sulfide dioxide partial pressure ratio at a product and reactant activity ratio of unity is defined as the *pure-phase sulfur-to-sulfur dioxide ratio*, corresponding to the gas ratio required for sulfidation, as reported in a Kellogg [53] or Pourbaix [54] diagram

sulfide compound formation [39, 40, 58, 59] suggests mixing behavior with large deviations from ideality that are difficult to accurately capture through predictive computational methods [60].

Thermodynamically spontaneous sulfidation of rare earth oxysulfides at 1400 °C with a reactant and product activity ratio of unity requires a diatomic sulfur-to-sulfur dioxide partial pressure ratio on the order of 10^{10} –10^{12}. This suggests that conversion of rare earth oxysulfide to sulfide will be thermodynamically limited by accumulation of sulfur dioxide within the reactor. One option to minimize sulfur dioxide accumulation in a sulfidation reactor is to carbothermically reduce sulfur dioxide in situ back into diatomic sulfur gas, forming a *carbothermically driven sulfur reflux* within a sulfidation reactor [38]:

$$SO_2(g) + (x + y)C = \frac{1}{2}S_2(g) + xCO(g) + yCO_2(g) \tag{10}$$

The reducing power of carbon for sulfur dioxide, effectivity the carbon monoxide to carbon dioxide ratio (x/y in Eq. 10), is in turn described by the Boudouard equilibrium. Equations 5–8 show that the selectivity of sulfidation is therefore determined from the thermodynamic activities of the reactant and product compounds, as well as possibly by the diatomic sulfur-to-sulfur dioxide partial pressure ratio. The later can be modulated via the extent of the carbothermically driven sulfur reflux. Since the mixing thermodynamics of mixed RE-O-S systems are not well understood and will likely vary based on fluctuations in composition between mineral feedstocks and ore bodies, we herein explore the role of the diatomic sulfur-to-sulfur dioxide partial on sulfidation and desulfidation selectivity independent of mineral solution thermodynamics. In the following sections, we experimentally demonstrate selective sulfidation and desulfidation in the absence of carbon for mixed iron and rare earth compounds. We then explore the role of carbon content on sulfide product formation and selectivity in mixed rare earth oxides.

Experimental Framework

Sulfidation and desulfidation experiments were conducted in an aluminum oxide packed bed reactor at temperatures ranging from 800 to 1400 °C, following previously described designs and methodology [38]. The sulfidizing agent was vaporized elemental sulfur (S_X, 99.5%, purity, Acros Organics). Argon (Ar, 99.95%, UHP300, Airgas) was used as a carrier gas for vaporized sulfur, with sulfur partial pressure monitored by calibrating evaporation rate to an H_2S tracer. For desulfidation experiments, volatilized sulfur was combusted upstream of the reactor with oxygen (O_2, 99.95%, UHP300, Airgas) to produce sulfur dioxide to be used in conjunction with argon as carrier gas.

For demonstration of selective sulfidation with mixed pure oxides, iron(III) oxide (Fe_2O_3, 99.85% metals basis, Alfa Aesar), scandium (III) oxide (Sc_2O_3), neodymium(III) oxide (Nd_2O_3, 99.9%-Nd REO basis, Strem Chemicals), praseodymium (III,IV) oxide (Pr_6O_{11}, 99.9%-Pr REO basis, Strem Chemicals), and dysprosium(III) oxide (Dy_2O_3, 99.9% Dy, REO basis, Strem Chemicals) were utilized as oxide precursors. When selective sulfidation was conducted in the presence of carbon, activated carbon (C, Alfa Aesar, 20 μm) was utilized. For demonstration of selective desulfidation of mixed pure sulfides, iron sulfide (FeS, 99.95%, Alfa Aesar) and lanthanum sulfide (La_2S_3, 99.5% metals basis purity, Strem Chemical) were utilized as precursors.

Selective sulfidation/desulfidation campaigns at temperatures ranging from 800 to 1400 °C were conducted at scales ranging from 2 to 3 g of feed material. While not strictly representative of natural abundances, equal molar or mass ratios of oxides are utilized so that trends in product purity and impurity solubility are readily apparent for different sulfidation conditions. Heating and cooling rates of 5 °C/min were utilized. Reaction times were 50 min, with a 5-min hold at temperature before and after sulfidation/desulfidation. Argon carrier gas flow rates were determined so that the gas residence time through the packed bed was on the order of 0.1 s. The compositions of product gasses were analyzed for oxygen, carbon monoxide, carbon dioxide, sulfur dioxide, and hydrogen sulfide content via an infrared/electrochemical gas analyzer (IR208, Infrared Industries). The compositions of sulfidation/desulfidation products were analyzed via SEM/EDS (SEM: JEOL JSM-6610LV, JEOL Ltd., EDS: Sirius SD detector, SGX Sensortech Ltd.) and QXRD (Panalytical X'Pert MPD diffractometer).

Selective sulfidation of iron from mixed iron and scandium oxides at a 1:1 mass ratio was conducted at temperatures of 800 and 1000 °C. Oxide precursors were ground to a particle size of sub-44 microns and loaded into the reactor bed at a porosity of approximately 80%. Sulfur gas was supplied at a partial pressure on the order of 0.1 atm. The sulfidation reaction occurred at such a rate that the sulfur-to-sulfur dioxide molar ratio within the reactor was approximately 10:1. From Fig. 2, under such conditions iron was thermodynamically predicted to form a sulfide, whereas scandium was predicted to be stable as an oxysulfide.

Selective sulfidation was also conducted for a multi-element lanthanide system of equimolar mixed praseodymium oxide, neodymium oxide, and dysprosium oxide. Sulfidation was conducted at a temperature of 1400 °C with and without the addition of carbon, with oxide particle sizes on the order of 10 microns and bed porosities on the order of 80%. Sulfur gas was supplied at a partial pressure of approximately 0.05 atm to 0.2 atm. In the absence of carbon, the sulfidation reaction occurred at such a rate that the sulfur-to-sulfur dioxide molar ratio within the reactor was approximately 10:1. For sulfidation in the presence of carbon, activated carbon was mixed in with the oxide charge at carbon-to-oxide molar ratios of 1.25:1, 2.5:1, 3.75:1, and 6:1. From Fig. 3, by varying the carbon addition and therefore the sulfur-to-sulfur dioxide molar ratio, oxysulfide, and sulfides of rare earth elements are predicted to form.

Selective desulfidation for mixed iron and lanthanum sulfides at a 1:1 mass ratio was demonstrated at a temperature of 1000 °C. Sulfide powders were loaded into the reactor bed at a porosity of approximately 80%. Sulfur and sulfur dioxide gases were introduced to the reactor at mole ratios varying from 10:1 to 1:10 over the course of the experiment, constituting a total partial pressure of between 0.05 atm and 0.2 atm in the argon carrier gas. From Fig. 2, under such conditions iron was thermodynamically predicted to remain stable as a sulfide, whereas lanthanum sulfide was predicted to desulfidize to form an oxysulfide. The results of selective sulfidation and desulfidation trials are presented in the following section.

Results and Discussion

Selective Sulfidation of Iron-Scandium Oxides

An equimolar mixture of scandium and iron oxides (46 and 54 wt%, respectively) was selectively sulfidized as a potential pretreatment to facilitate physical separation of iron and scandium, following the methodology described above. Sulfidation products were analyzed via QXRD. Following sulfidation, the only remaining crystalline compound of iron was iron sulfide, making up 28 and 29 wt% after selective sulfidation at 800 and 1000 °C, respectively. Meanwhile, the only crystalline phase of scandium observed after sulfidation was scandium oxide, decreasing in composition from 38 to 30 wt% as sulfidation temperature rose from 800 to 1000 °C. The preferential partitioning of scandium and iron to separate oxide and sulfide phases, respectively, suggests that selective sulfidation is a promising pretreatment to facilitate physical separation of scandium and iron. At the higher sulfidation temperature, the amorphous content correspondingly increased from 33 to 42 wt%. As shown in Fig. 2, thermodynamically scandium oxide sulfidation to oxysulfide is expected to accompany iron sulfide formation. The presence of crystalline scandium oxide following the sulfidation treatment suggests that sulfidation of scandium oxide to oxysulfide is kinetically sluggish compared to sulfidation of iron oxide.

Figure 4 illustrates the distribution of oxygen, sulfur, iron, and scandium following sulfidation at 1000 °C, as measured via EDS element mapping. Scandium is revealed to partition into oxygen-rich phases, whereas iron is observed to partition into sulfur-rich phases, confirming the results of QXRD analysis. Under the sulfidation conditions of the experiment, scandium oxide phases are observed to exhibit sulfur and iron solubilities of approximately 5 and 15 wt%, respectively. Meanwhile, iron sulfide phases exhibit solubilities for oxygen and scandium of approximately 5–10 wt% each. Solubilities between scandium oxide and iron sulfide phases are a function of their mixing thermodynamics, defined by the activities of iron and scandium oxide

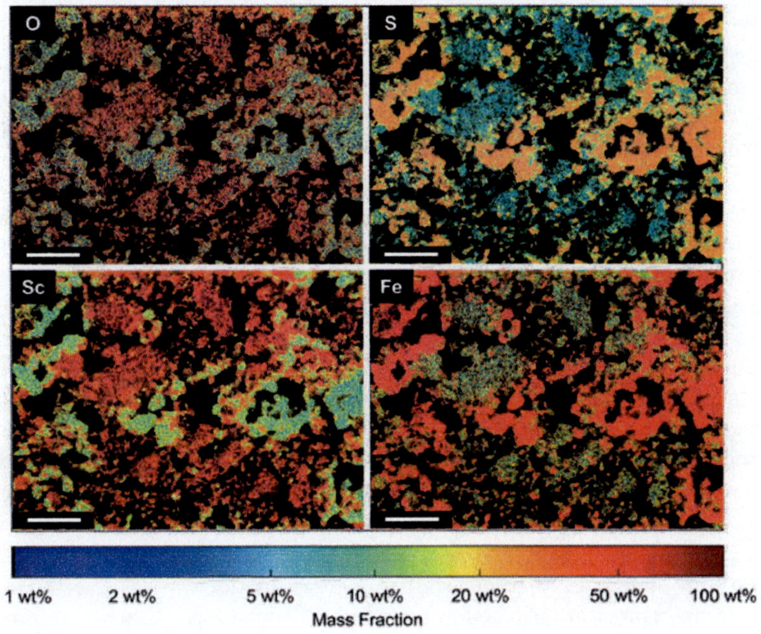

Fig. 4 EDS map detailing distribution of O, S, Sc, and Fe following selective sulfidation of mixed iron-scandium oxides. Following selective sulfidation, iron was observed to partition into a sulfide phase, while scandium remained as an oxide. Scale bars correspond to 50 μm

and sulfide compounds in the system. As shown in Eqs. 5–8, at thermodynamic equilibrium, ratios of reactant and product activities are set by the sulfur-to-sulfur dioxide partial pressure ratio in the gas phase. Therefore, the solubility between scandium oxide and iron sulfide products is controlled by the composition of the sulfidation atmosphere. As shown in Fig. 2, at increasing temperatures, the difference in the *pure-phase sulfur-to-sulfur dioxide ratio* between sulfidation of iron oxide and scandium oxysulfide falls, corresponding to a decrease in sulfidation selectivity and increase in solubility between oxide and sulfide products. This predicted drop in selectivity likely corresponds to the decrease in crystalline scandium oxide content and increase in multi-metal amorphous compounds at increasing sulfidation temperature. Further thermodynamic analysis of sulfidation selectivity is hindered by the lack of relevant mixing thermodynamic data for the Sc-Fe-O-S system. Similarly, limited knowledge of the thermal activation of scandium oxide sulfidation kinetics leads to difficulties in distinguishing the roles of thermodynamics and kinetics in selective sulfidation for iron-scandium separation.

Selective Sulfidation of Mixed Lanthanide Oxides

For selective sulfidation of mixed lanthanide (Ln) oxides as a pretreatment to facilitate their physical separation, an equimolar mixture of praseodymium, neodymium, and dysprosium oxides was used as a feedstock. Carbon content was varied as a means to control the sulfur-to-sulfur dioxide partial pressure ratio in the gas atmosphere. The conversions at 1400 °C of equimolar mixed praseodymium, neodymium, and dysprosium oxides to oxysulfide and sulfide products, and amorphous carbon to carbon monoxide are depicted in Fig. 5 as functions of carbon content and supplied sulfur gas partial pressure. While challenges remain in consistently controlling the sulfur partial pressure at the scale of this experiment, the fractional molar conversion of oxide sulfidation is shown to generally increase with increasing carbon feed content. This confirms the role of the *carbothermically driven sulfur reflux* (Eq. 10) in increasing the sulfur-to-sulfur dioxide molar gas ratio. Surprisingly, the conversion of carbon-free sulfidation of the mixed rare earth oxides is observed to be 50% higher than with an oxide to carbon molar feed ratio of 1.25. Further research into Ln–O-S-C thermodynamics are likely needed to understand the origin of this behavior. The fractional molar conversion of carbon to carbon monoxide meanwhile showed a slight increase with carbon content, generally remaining on the order of 30–35%. Whether this conversion was limited due to the formation of by-product carbon–sulfur gases [61], or intrinsic to the thermodynamics and kinetics of the *carbothermically driven sulfur reflux* [38], remains to be determined. Carbon dioxide was not detected to form at these temperatures in measurable quantities due to the Boudouard equilibrium. Under the reaction and mass transport conditions of these experiments, the sulfidation rate of the mixed rare earth oxide was observed to vary to the first order with sulfur partial pressure, independent of carbon content, until the maximum conversion

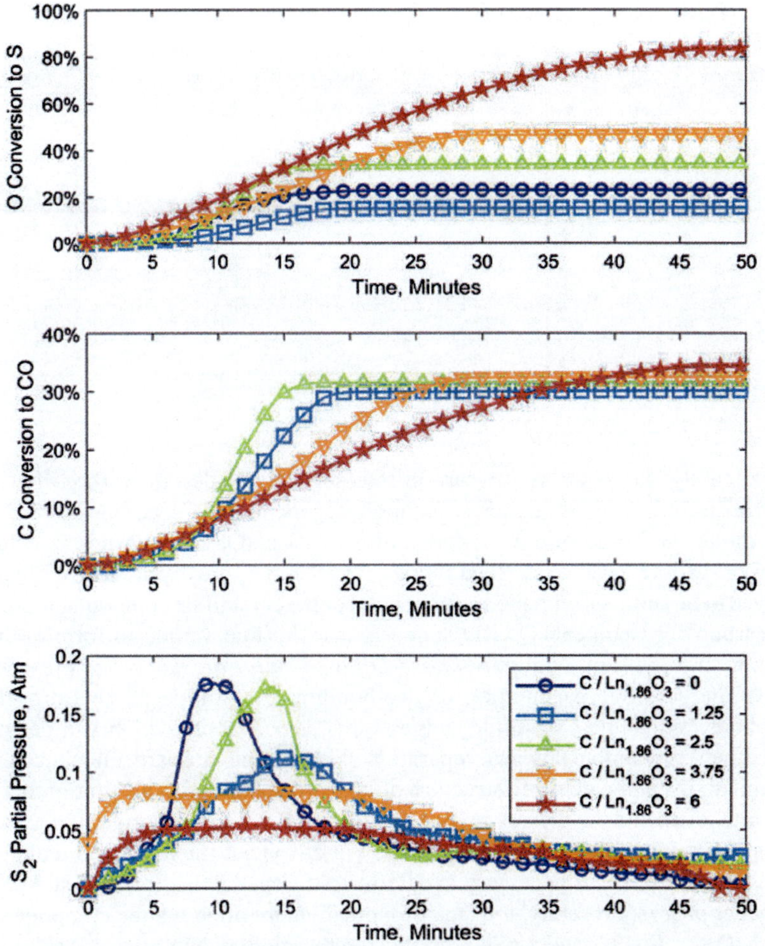

Fig. 5 Conversion of oxide to sulfide and carbon to carbon monoxide as functions of carbon content and sulfur gas partial pressure during selective sulfidation of mixed praseodymium-neodymium-dysprosium oxides. Increasing carbon content in the feed leads to higher conversion of oxides to sulfides by increasing the sulfur-to-sulfur dioxide partial pressure ratio, demonstrating the *carbothermically driven sulfur reflux* (Eq. 8)

was reached. However, detailed analysis of internal and external mass transport limitations[62–64] would be required to rigorously relate sulfidation kinetics observed in this experiment to the intrinsic chemical kinetics of rare earth oxide sulfidation reactions [38].

Sulfidized product phases as a function of carbon content are identified via QXRD and depicted in Table 1. With increasing carbon feed content, the total amorphous oxysulfide content of the sulfidized product increases. At a carbon-to-rare earth oxide molar feed ratio of 1.25, the crystalline content of Ln_2O_2S oxysulfide phases is

Table 1 QXRD analysis of praseodymium-neodymium-dysprosium oxide (REO) sulfidation products

QXRD wt %	C/REO = 0	C/REO = 1.25	C/REO = 2.5	C/REO = 3.75	C/REO = 6
Pr_2O_3	–	1%	–	–	–
Nd_2O_2S	17%	12%	–	–	–
$Nd_{10}OS_{14}$	–	-	25%	–	–
Dy_2O_2S	27%	28%	12%	–	–
Pr_2O_2S	3%	1%	–	–	–
Nd_3S_4	–	–	–	9%	–
Nd_2S_3	–	–	–	16%	16%
NdS_2	–	–	–	–	1%
Amorphous	53%	58%	63%	75%	83%

approximately the same as for carbon-free sulfidation, despite carbon-free sulfidation exhibiting a 50% higher sulfidation conversion. At a carbon-to-rare earth oxide molar feed ratio of 2.5, a more sulfur-rich $Ln_{10}OS_{14}$ oxysulfide is detected. At carbon-to-rare earth oxide feed ratios of 3.75 and 6, the crystalline products are observed to be sulfides, with the sulfur content of the crystalline compound increasing with carbon feed content. QXRD suggests that the first sulfide to form is that of neodymium, consistent with *pure-phase sulfur-to-sulfur dioxide ratios* presented in Fig. 3. The selective partitioning of neodymium to a sulfide phase from mixed, amorphous lanthanide oxysulfide suggests that selective sulfidation is a promising pretreatment to promote physical separation of individual rare earth elements. Due to the similar size and electronic structure of lanthanides, atomic substitution between rare earth elements in the crystalline sulfide product and amorphous oxysulfide phases is likely to occur. Further analysis via advanced micrographic techniques, such as WDS/EPMA, is necessary to determine the spatial distributions and purities of product phases. Presently, the lack of lanthanide thermodynamic compound data for $Ln_{10}OS_{14}$, Ln_3S_4, and LnS_2 complicates thermodynamic predictions of sulfidation selectivity, as these compounds are presently excluded from the sulfidation series in Figs. 2 and 3. Furthermore, the absence of mixing thermodynamic data for Ln–O-S systems presently negates the ability to inform solubility and product purity through activity models. Nevertheless, *carbothermically driven sulfur reflux* is an attractive avenue to control the sulfur-to-sulfur dioxide partial pressure ratio to maximize sulfidation selectivity, subject to the solution thermodynamics of rare earth feedstock compounds and sulfidation products.

Selective Desulfidation of Iron-Lanthanum Sulfide

Due to differences in oxide and sulfide solution behavior, some separation challenges may be easier to overcome starting from a mixed sulfide instead of a mixed oxide. Under such a scenario, the thermodynamic framework presented herein can simultaneously be employed to describe a selective desulfidation process. Reversing Eqs. 1 and 2, the oxygen source for desulfidation is sulfur dioxide, with desulfidation potential still controlled by a sulfur-to-sulfur dioxide molar gas ratio. Selective desulfidation of lanthanum from a 1:1 mass ratio mixture of iron and lanthanum sulfides (FeS and La_2S_3) was conducted herein at a temperature of 1000 °C, with sulfur-to-sulfur dioxide partial pressure ratios ranging from 10:1 to 1:10. In Fig. 6, SEM/EDS element maps depict the spatial distribution of oxygen, sulfur, lanthanum, and iron in the selectively desulfidized product. Iron was found to remain stable in sulfide phases, as predicted from Fig. 2 for the conditions of the experiment. Meanwhile, lanthanum was found to selectively desulfidize under these conditions as predicted from Fig. 2, partitioning preferentially into oxysulfide phases with stoichiometries of approximately La_2O_2S and $La_{10}OS_{14}$. The observed solubility of iron in the La_2O_2S and $La_{10}OS_{14}$ product phases was approximately 2 and 10 at%, respectively, whereas the solubilities of lanthanum and oxygen in the iron sulfide phases were approximately 1–2 at% each. As shown in Fig. 2 and Eqs. 5–8, decreasing the sulfur-to-sulfur dioxide partial pressure ratio would work to decrease the activity of iron sulfide and increase the activity of iron oxide. An optimum sulfur-to-sulfur

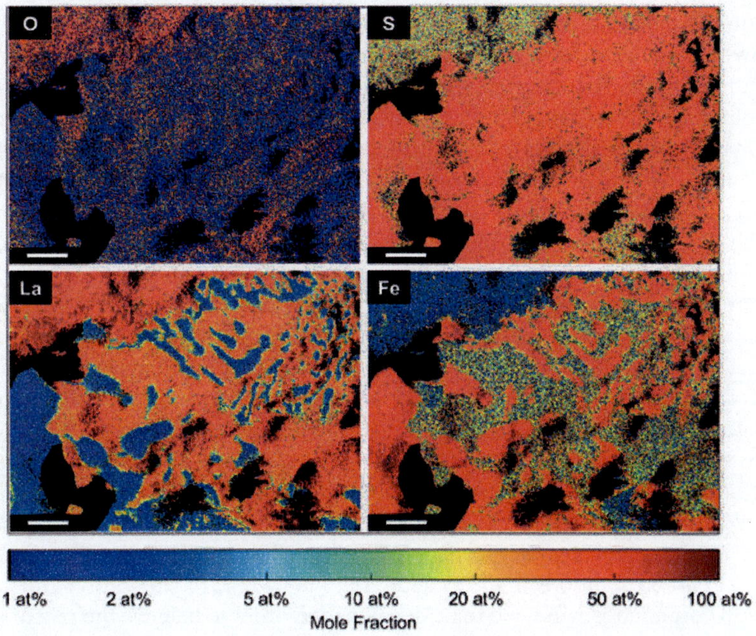

Fig. 6 EDS map detailing distribution of O, S, La, and Fe following selective desulfidation of mixed iron-lanthanum sulfides. Following selective desulfidation, lanthanum was observed to partition into oxysulfide phases, while iron remained as a sulfide. Scale bars correspond to 15 μm

dioxide partial pressure ratio therefore exists to minimize the solubility of iron in lanthanum-rich phases and lanthanum in iron-rich phases. While determination of the optimal sulfur-to-sulfur dioxide partial pressure ratio is mired by a lack of La-Fe-O-S solution data, these results suggest that selective desulfidation is a feasible pretreatment to accomplish challenging material separations via physical methods.

Summary

In this study, mixed oxides were used as feedstocks for selective sulfidation/desulfidation in order to provide minerals-agnostic test cases. In practice, rare earth minerals most commonly occur as fluorocarbonates or phosphates, with rare earth elements contained at grades that may differ by orders of magnitude between one element and the next. While a variety of methods exist to convert fluorocarbonates and phosphates to oxides, understanding the sulfidation behavior of oxyhalides and phosphates will be necessary to design a rare earth physical separation process from mineral feedstocks [38]. Equal molar or mass ratios of oxides were employed in this study so that metallic elements were at sufficient concentrations to readily observe trends in product purity and impurity solubility with different sulfidation conditions. Solution thermodynamics, if available, may be used to predict sulfidation behavior as shown herein even for feedstocks of very different grades. However, knowledge of solution thermodynamics will have to be developed for the mineral chemistries relevant to a given separation process. For design of physical separation processes for elements following selective sulfidation or desulfidation, an understanding of product phase nucleation, growth, and liberation phenomenon will be necessary. Fortunately, liberation and physical separation of sulfides from oxides remains a vibrant area of research due to the growing need to adapt existing mineral beneficiation processes to declining ore grades [65, 66].

Overall, our preliminary results suggest that selective sulfidation and selective desulfidation are promising new avenues to promote physical separation of individual rare earth elements (RE) and their impurities. We present a thermodynamic framework for modelling selectivity in sulfidation and desulfidation processes that proves viable to predict overall trends in sulfidation behavior and selectivity. However, predicting product phase purities and impurity solubilities is hindered by an absence of mixing thermodynamics and activity models for RE-O-S solutions and their impurities. We identify the ratio of sulfur-to-sulfur dioxide partial pressures as a key process lever for selective sulfidation, which we show may be modulated through the use of carbon additions in the system. Yet, knowledge of sulfidation kinetics for rare earth compounds will need to be developed in order to balance the rate of sulfur dioxide production and its rate of carbothermic reduction in achieving a target sulfur-to-sulfur dioxide partial pressure ratio in the reactor. We successfully demonstrate proof-of-concept selective sulfidation and desulfidation for iron-scandium, iron-lanthanum, and lanthanide-lanthanide separations. These results illuminate a path forward for competitive rare earth separation, where physical separations become a new possibility, alleviating the need for complete hydrometallurgical dissolution and multi-stage liquid–liquid solvent extraction.

Acknowledgements The authors thank Dr. Hirotaka Higuchi for providing samples of scandium oxide. Part of this work was funded from the DOE-AMO-EERE Office under project DE-EE0008316.

References

1. Cheisson T, Schelter EJ (2019) Rare earth elements: Mendeleev's bane, modern marvels. Science 363(6426):489–493. https://doi.org/10.1126/science.aau7628
2. Zhao B, Zhang J, Schreiner B (2016) Separation hydrometallurgy of rare earth elements. Springer International Publishing AG Switzerland, Cham
3. Haynes WM (2016) Abundance of elements in the earth's crust and in the sea. In CRC Handb Chem Phys, 97th edn (2016), pp 14–17
4. U.S. Geological Survey (2021) Mineral commodity summaries 2021
5. U.S. Department of Energy (2020) Critical minerals and materials
6. Binnemans K, Jones PT, Müller T, Yurramendi L (2018) Rare earths and the balance problem: How to deal with changing markets? J Sustain Metall 4(1):126–146. https://doi.org/10.1007/s40831-018-0162-8
7. Jowitt SM, Werner TT, Weng Z, Mudd GM (2018) Recycling of the rare earth elements. Curr Opin Green Sustain Chem 13:1–7. https://doi.org/10.1016/J.COGSC.2018.02.008
8. Bailey G, Joyce PJ, Schrijvers D, Schulze R, Sylvestre AM, Sprecher B, Vahidi E, Dewulf W, Van Acker K (2020) Review and new life cycle assessment for rare earth production from bastnäsite, ion adsorption clays and lateritic monazite. Resour Conserv Recycl 155:104675. https://doi.org/10.1016/j.resconrec.2019.104675
9. Kim J, Azimi G (2020) An innovative process for extracting scandium from nickeliferous laterite ore: Carbothermic reduction followed by NaOH cracking. Hydrometallurgy 191:105194. https://doi.org/10.1016/j.hydromet.2019.105194
10. Wang W, Pranolo Y, Cheng CY (2011) Metallurgical processes for scandium recovery from various resources: A review. Hydrometallurgy 108(1–2):100–108. https://doi.org/10.1016/j.hydromet.2011.03.001
11. Anawati J, Yuan R, Kim J, Azimi G (2020) Selective recovery of scandium from nickel laterite ore by acid roasting–water leaching. Miner Met Mater Ser 77–90 (Springer)
12. Meng F, Li X, Shi L, Li Y, Gao F, Wei Y (2020) Miner Eng 157:106561. https://doi.org/10.1016/j.mineng.2020.106561
13. Borra CR, Mermans J, Blanpain B, Pontikes Y, Binnemans K, Van Gerven T (2016) Selective recovery of rare earths from bauxite residue by combination of sulfation, roasting and leaching. Miner Eng 92:151–159. https://doi.org/10.1016/j.mineng.2016.03.002
14. Narayanan RP, Kazantzis NK, Emmert MH (2018) Selective process steps for the recovery of scandium from jamaican bauxite residue (Red Mud). ACS Sustain Chem Eng 6:1478–1488. https://doi.org/10.1021/acssuschemeng.7b03968
15. Jordens A, Cheng YP, Waters KE (2013) A review of the beneficiation of rare earth element bearing minerals. Miner Eng 41:97–114. https://doi.org/10.1016/j.mineng.2012.10.017
16. Kumari A, Panda R, Jha MK, Kumar JR, Lee JY (2015) Process development to recover rare earth metals from monazite mineral: A review. Miner Eng 79(August):102–115. https://doi.org/10.1016/j.mineng.2015.05.003
17. Vogel H, Friedrich B (2018) An estimation of PFC emission by rare earth electrolysis. Light Met. Springer, Berlin, Heidelberg, pp 1507–1517

18. Lee JCK, Wen Z (2018) Pathways for greening the supply of rare earth elements in China. Nat Sustain 1:598–605. https://doi.org/10.1038/s41893-018-0154-5
19. Vahidi E, Zhao F (2017) Environmental life cycle assessment on the separation of rare earth oxides through solvent extraction. J Environ Manage 203:255–263. https://doi.org/10.1016/j.jenvman.2017.07.076
20. Hall LCON, Kelley S (2015) Yangibana Project
21. Ciuculescu T, Foo B, Gowans R, Hawton K, Jacobs C, Spooner J (2013) Nechalacho rare earth elements project
22. Limited NM (2014) Browns range project
23. Limited PR (2012) Nguallar rare earth project. Q Rep
24. Li XZ, Zhou LP, Yan LL, Dong YM, Bai ZL, Sun XQ, Diwu J, Wang S, Bünzli JC, Sun QF (2018) A supramolecular lanthanide separation approach based on multivalent cooperative enhancement of metal ion selectivity. Nat Commun 9:547. https://doi.org/10.1038/s41467-018-02940-7
25. Higgins RF, Cheisson T, Cole BE, Manor BC, Carroll PJ, Schelter EJ (2020) Magnetic field directed rare-earth separations. Angew Chemie Int Ed 59(5):1851–1856. https://doi.org/10.1002/anie.201911606
26. Norgate T, Jahanshahi S (2010) Low grade ores–Smelt, leach or concentrate? Miner Eng 23(2):65–73. https://doi.org/10.1016/j.mineng.2009.10.002
27. Skinner BJ (1976) A second iron age ahead? The distribution of chemical elements in the earth's crust sets natural limits to man's supply of metals that are much more important to the future of society than limits on energy. Sci Am 64(3):258–269
28. Skinner BJ (1979) Earth resources (minerals/metals/ores/geochemistry/mining). Proc Natl Acad Sci USA 76(9):4212–4217
29. Murase K, Shinozaki K, Hirashima Y, ichi Machida K, ya Adachi G (1993) Rare earth separation using a chemical vapour transport process mediated by vapour complexes of the $LnCl_3AlCl_3$ system. J Alloys Compd 198(1–2):31–38. https://doi.org/10.1016/0925-8388(93)90139-E
30. Uda T, Jacob KT, Hirasawa M (2000) Technique for enhanced rare earth separation. Science 289(5488):2326–2329. https://doi.org/10.1126/science.289.5488.2326
31. Elgin JC (2019) Scandium solvent extraction. Light Met 2019:1395–1401. https://doi.org/10.1021/ie50481a018
32. Andrade-Gamboa J, Andrade-Gamboa J, Pasquevich DM (2000) A model for the role of carbon on carbochlorination of TiO_2. Metall Mater Trans B Process Metall Mater Process Sci 31(6):1439–1446. https://doi.org/10.1007/s11663-000-0028-9
33. Onghena B, Borra CR, Van Gerven T, Binnemans K (2017) Recovery of scandium from sulfation-roasted leachates of bauxite residue by solvent extraction with the ionic liquid betainium bis (trifluoromethylsulfonyl) imid. Sep Purif Technol 176:208–219. https://doi.org/10.1016/j.seppur.2016.12.009
34. Wang J, Hu H (2021) Selective extraction of rare earths and lithium from rare earth fluoride molten-salt electrolytic slag by sulfation. Miner Eng 160 (November 2020):106711. https://doi.org/10.1016/j.mineng.2020.106711
35. Anawati J, Yuan R, Kim J, Azimi G (2020) Selective recovery of scandium from Nickel Laterite ore by acid roasting–water leaching. Springer International Publishing
36. Yin X, Wang Y, Bai X, Wang Y, Chen L, Xiao C, Diwu J, Du S, Chai Z, Albrecht-Schmitt TE, Wang S (2017) Rare earth separations by selective borate crystallization. Nat Commun 8. https://doi.org/10.1038/ncomms14438
37. Firdaus M, Rhamdhani MA, Durandet Y, Rankin WJ, McGregor K (2016) Review of high-temperature recovery of rare earth (Nd/Dy) from magnet waste. J Sustain Metall 2(4):276–295. https://doi.org/10.1007/s40831-016-0045-9
38. Stinn C, Allanore A (2021) Selective sulfidation of metal compounds. Nature 2021:1–9. https://doi.org/10.1038/s41586-021-04321-5
39. Bakhtiyarov IB (1987) Preparation of Eu_2S_3 and investigation of the system Eu_2S_3-Eu_2O_3. Azerbaidzhanskii Khimicheskii Zhurnal 4(156–160)

40. Asadov MM, Abbasov MM (1996) Phase diagrams of the Ho_2O_3-In_2S_3, Ho_2O_3-HO_2S_3 and Ho_2O_2S-In_2S_3 systems. Inorg Mater 32:280–283
41. Stewart VJ, Chamorro JR, McQueen TM (2021) Integer vs. half-integer spin on an approximate honeycomb lattice. Phys Rev Mater 5(094401):1–7. https://doi.org/10.1103/PhysRevMaterials.5.094401
42. Harris CT, Peacey JG, Pickles CA (2013) Selective sulphidation and flotation of nickel from a nickeliferous laterite ore. Miner Eng 54:21–31
43. Stinn C, Allanore A (2021) Selective sulfidation and electrowinning of nickel and cobalt for lithium ion battery recycling. In: C Andersoneditor (ed) Ni-Co 2021: The 5th international symposium on nickel and cobalt. Springer Nature Switzerland AG, Cham, pp 99–110
44. Han J, Liu W, Wang D, Jiao F, Qin W (2016) Metall Mater Trans B Process Metall Mater Process Sci 47:344
45. Han J, Liu W, Wang D, Jiao F, Qin W (2015) Selective sulfidation of lead smelter slag with sulfur. Metall Mater Trans B Process Metall Mater Process Sci 47(1):344–354. https://doi.org/10.1007/s11663-015-0526-4
46. Adham K, Lee C (2012) Fluid bed roasting of metal ores and concentrates for arsenic removal. Towar Clean Metall Process Procfit Soc Environ Steward, pp 109–116
47. Crundwell FK, Moats MS, Ramachandran V, Robinson TG, Davenport WG (2011) Extractive metallurgy of nickel, cobalt, and platinum-group metals. Elsevier, Oxford
48. Merritt RR (1990) High temperature methods for processing monazite: II. Reaction with sodium carbonate. J Less Common Met 166:211–219
49. Kaneko T, Yashima Y, Ahmadi E, Natsui S, Suzuki RO (2020) Synthesis of Sc sulfides by CS_2 sulfurization. J Solid State Chem 285:121268. https://doi.org/10.1016/j.jssc.2020.121268
50. Afanasiev P, Fischer L, Beauchesne F, Danot M, Gaborit V, Breysse M (2000) Preparation of the mixed sulfide $Nb_2Mo_3S_{10}$ catalyst from the mixed oxide precursor. Catal Lett 64(1):59–63. https://doi.org/10.1023/A:1019058113931
51. Meyer B (1976) Elemental sulfur. Chem Rev 76:367–388. https://doi.org/10.1021/cr60301a003
52. Chakraborti N (1983) Modified predominance area diagrams for the Fe-S-O system. Can J Chem Eng 61(5):763–765. https://doi.org/10.1002/cjce.5450610520
53. Dwivedi RK, Kay DAR (1984) Thermodynamics of the oxidation of rare earth oxysulfides at high temperatures. Metall Trans B 15(3):523–528. https://doi.org/10.1007/BF02657383
54. Kellogg HH (1964) A critical review of sulfation equilibria. Trans Metall Soc AIME 230:1622–1634
55. Pourbaix MJN, Rorive-Bouté MCM (1948) Graphical study of metallurgical equilibria. Discuss Faraday Soc 4(0):139–154. https://doi.org/10.1039/DF9480400139
56. Goldschmidt VM (1937) The principles of distribution of chemical elements in minerals and rocks. The seventh Hugo Müller Lecture, delivered before the Chemical Society on March 17th, 1937. J Chem Soc (0):655–673. https://doi.org/10.1039/JR9370000655
57. Kriklya AI, Bolgar AS, Pribyl'skii NY (1992) Heat capacity and enthalpy of y-Dy2S3 over a wide range of temperature. Sov Powder Metall Met Ceram 31:697–700
58. Kuznetsov AK, Tikhonov PA, Kravchinskaya MV, Iuneev VM, Ivanov LM (1976) Phase relations in the systems DyO1.5-PrO1.5 and DyO1.5-PrOx. Dokl Akad Nauk SSSR 231:882–884
59. Tichonov PA, Proks I, Kosa L (1982) Determination of the mixing enthalpy of solid solutions in the system Y_2O_3–Nd_2O_3. Silikaty 26:165–168
60. Chen M, Hallstedt B, Gauckler LJ (2005) CALPHAD modeling of the La_2O_3-Y_2O_3 system. Calphad Comput Coupling Phase Diagrams Thermochem 29(2):103–113. https://doi.org/10.1016/j.calphad.2005.06.006
61. Ahmadi E, Suzuki RO (2021) Tantalum metal production through high-efficiency electrochemical reduction of TaS_2 in molten $CaCl_2$. J Sustain Metall 72(7):437–447. https://doi.org/10.1007/S40831-021-00347-1
62. Sohn HY, Szekely J (1974) The effect of intragrain diffusion on the reaction between a porous solid and a gas. Chem Eng Sci 29(2):630–634. https://doi.org/10.1016/0009-2509(74)80076-X
63. Sohn HY, Szekely J (1972) A structural model for gas-solid reactions with a moving boundary—III. Chem Eng Sci 27(4):763–778. https://doi.org/10.1016/0009-2509(72)85011-5

64. Sohn HY, Fan D-Q (2017) On the initial rate of fluid-solid reactions. Met Mater Trans B 48B (June 2017):1827–1832. https://doi.org/10.1007/s11663-017-0940-x
65. Rötzer N, Schmidt M (2018) Decreasing metal ore grades—is the fear of resource depletion justified? Resources 7(4):88. https://doi.org/10.3390/RESOURCES7040088
66. Calvo G, Mudd G, Valero A, Valero A (2016) Decreasing ore grades in global metallic mining: A theoretical issue or a global reality? Resources 5(4):36. https://doi.org/10.3390/RESOURCES5040036

Upgrading of Riruwai Cassiterite (Tin Ore) from Doguwa Local Government, North West Nigeria, Using Gravity Beneficiation Technique

Furqan Abdulfattah, Daniel Markus Bwala, Ibrahim Abdullah Rafukka, and Suleiman Bolaji Hassan

Abstract Beneficiation of Cassiterite found in Riruwai, Doguwa Local Government, Kano State, Nigeria, was done using the gravity separation technique. Characterization of the crude ore was carried out using X-ray fluorescence (XRF) analysis which shows that the ore contains 9.352% Sn, and other related elements such as 42.802% Si, 28.786% Fe, and 6.232% Al. The presence of the associated minerals implies the need for upgrading the cassiterite to an appreciable grade for potential smelting. Fractional and grain sieve size results revealed that the liberation size of the ore is at $-210 + 180$ μm containing the highest percentage assay of tin (18.59% Sn). The ore was ground to the liberation size and then fed to the air-float machine. The resulting products (concentrate and tailings) were analysed using XRF, which shows an upgrade in tin to 26.534% Sn in the concentrate with a percentage recovery of 63.575%.

Keywords Gravity separation · Cassiterite · Mineral · Beneficiation

Introduction

Cassiterite (SnO_2) is the most common tin ore mineral, and it is often found alongside other minerals. Magmatic, hydrothermal, and/or replacement processes produce primary tin deposits. Weathering of primary deposits, followed by movement and deposition of the resultant sediments in a new environment, results in secondary deposits. China, Malaysia, Indonesia, Peru, Thailand, Brazil, Bolivia, DRC Congo,

F. Abdulfattah (✉) · D. M. Bwala · S. B. Hassan
Nigerian Institute of Mining and Geosciences, Jos, Nigeria
e-mail: fa1700001.mme@buk.edu.ng

F. Abdulfattah · I. A. Rafukka
Bayero University Kano, Kano, Nigeria

S. B. Hassan
University of Lagos, Lagos, Nigeria

T. Ouchi et al. (eds), *Rare Metal Technology 2022*, The Minerals, Metals & Materials Series, https://doi.org/10.1007/978-3-030-92662-5_26

Australia, Nigeria, Myanmar (Burma), Russia, Zimbabwe, Rwanda, South Africa, the United Kingdom, Mongolia, Colombia, Germany, Canada, Mexico, and Morocco all have significant tin reserves [1–5].

Nigeria is rich in solid mineral deposits, especially Cassiterites, which are virtually unexplored. In Nigeria, tin deposits may be found in two different provinces. The older granite is related to the Pan-African pegmatite belt. This Sn–Ta–Nb mineralized pegmatite belt stretches for over 400 km from southwestern (Ijebu area) to northern Nigeria (through Wamba-Jema'a to Zuru-Gusau areas) [6]. Nigeria was a large exporter of cassiterite concentrates prior to the oil boom, with a peak of approximately 11,000 tonnes. However, due to neglect, the value has plummeted to around 2000 tonnes. The inaccessibility of placer deposits and the current prohibitive cost of mining the ores beneath the basalt flows are among the causes that have contributed to the tin industry's collapse in Nigeria [7].

Due to its intrinsic traits such as great malleability and ductility, low melting point, softness, corrosion resistance, non-toxicity, anti-friction qualities, and silvery-white appearance, tin is one of the oldest metals known to man, due to its wide variety of uses [6]. Tin is used in the manufacturing of corrosion-resistant alloys (bronze, brass, solder), tin plates, chemicals, tin foil, ceramics, and glass, among others. Even though tin is usually a minor component in alloys, it is an important one because of its unique qualities and benefits to the matrix metal [5]. Other properties of tin include silvery-white or shiny black colour, brilliant luster, high ductility and malleability at 100 °C (212 °F), low tenacity, specific gravity of 6.89–7.3, low melting point of 232 °C (449 °F), high boiling point of 2,625 °C (4,757 °F), and allotropy. Some of the chemical properties of tin are non-toxicity, resistance to oxidation and corrosion, and existence in two oxidation states (+4 and +2) [5, 6].

The cost of tin increased for the past 6 years (2016–2021). The price of tin per metric ton being about 35,760USD/ton, with a significant increase in price from about 15,000USD/ton in January 2020 to more than 30,000USD/ton in July 2021 [8].

Because of the differences in density of the minerals in the gangue (typically silicate), gravity separation methods and froth flotation techniques are commonly used to concentrate cassiterite [9–12]. China and Indonesia produce around 70% of the world's tin, and most of the world's tin smelters are located in these two nations. Cassiterite's high relative density in comparison to its gangue elements makes it an attractive mineral for gravity-separation procedures, as seen by its physical features. Unfortunately, its relative hardness is about the same as steel, but it has the bad property of being extremely fragile. This aspect must be considered throughout the crushing and grinding processes prior to concentration, and it leads to the general guideline that cassiterite grains should be recovered as soon as possible and in their greatest size possible. Gravity concentration methods become significantly less efficient if particle sizes are decreased to less than 43 μm [13].

There are two types of gravity techniques used to recover cassiterite: those used on soft-rock deposits and those used on hard-rock deposits. Soft-rock deposits, which mostly consist of alluvial deposits and represent 70–80% of global reserves, have comparatively simple beneficiation processes due to the high degree of particle liberation. Panning and other hand-washing equipment were the simplest and perhaps the first method of recovery and concentration for alluvial deposits. For hard-rock ores, traditional crushing and grinding are used to separate the cassiterite from the gangue minerals. Due to cassiterite's exceptional brittleness, large numbers of very small particles can be created at this stage, resulting in tin losses in subsequent processing stages. It is worth noting that froth flotation, which is used to upgrade particles smaller than 43 microns, cannot handle particles smaller than 6 microns. The unrecoverable minus 6 microns cassiterite can account for up to 6% of the metallic tin entering the plant in some situations. To produce liberation, most hard-rock tin concentrators employ various combinations of the same crushing and milling equipment, which are subsequently followed by a range of gravity-concentration systems for additional beneficiation [13]. In half a century of practice around the world, gravity separation (air-floating technique) has been developed as an efficient method to remove impurities from tin ore [6].

The aim of this study is to upgrade the Riruwai cassiterite to a concentrate grade that meets the physical and chemical characteristics of Nigerian tin for smelting. The study involved physical and chemical characterization of Riruwai cassiterite in Doguwa Local Government Area and to upgrade the ore to a suitable grade concentrate by gravity separation (air-floating) technique.

Materials and Methods

Experimental Materials/Equipment

The following are the experimental materials and equipment used in this research.

i. Tin Ore (Cassiterite)
ii. Reference Ore sample (silica)
iii. Set of Sieves/Laboratory Sieve Shaking Machine—RO-TAP TYLER Model: RX-30–10 in Nigerian Institute of Mining and Geosciences (NIMG), Jos.
iv. Denver Pulverizing Machine
v. Weighing Balance
vi. X-ray Florescence (XRF) Machine—Hand-held XRF Thermo-scientific Model 3065 in Materials and Metallurgical Laboratory, Bayero University Kano
vii. Global Positioning System (GPS)
viii. SEPOR Laboratory Ball Mill—at Nigerian Institute of Mining and Geosciences (NIMG), Jos
ix. Laboratory Air-Float Machine—Kip Kelly Air floating Machine model My-1151 at National Metallurgical Development Centre (NMDC), Jos

Sample Collection

A sample of 25 kg of the Cassiterite sample was collected from various points of the deposit located at Riruwai village in Doguwa Local Government Area, Kano State. Grab method of sampling was adopted to collect the sample at 5 m depth. The sample was obtained from Artisanal and small-scale miners that are presently at the deposit site.

Sample Preparation

Sample preparation involves crushing and grinding processes (comminution process). The lump sizes of the ore sample were reduced to 1500 μm size which is acceptable by the crusher using sledgehammer. The samples were crushed using a jaw crusher to sieve size of 1000 μm. The discharge from the crusher was then ground to fine particle size of 600 μm using the ball milling process.

Ore Sampling

Coning and quartering sampling method was used to divide the pulverized sample to smaller portions, for us to have a true representation of the ore in the analysis. This sampling method is a traditional way for manually extracting a sample increment, done by dividing the sample into four equal sub-lots each time, and repeating the process 8 times to obtain the final mass needed for analysis [14]. The tests carried out are as follows.

Chemical Composition Analysis of the Crude Riruwai Cassiterite Sample

The X-ray Fluorescence a Hand-held Model 3065 Thermo-scientific Analyzer, from the Materials Laboratory of Mechanical Engineering Department in Bayero University Kano.

Samples for the XRF analysis from each of the pits were prepared for XRF analysis by crushing, grinding, and pulverizing the ore to 600 μm. Coning and quartering sampling method was used to divide the pulverized samples into smaller portions for the analysis to be carried out. The prepared samples from the different pits were mixed to homogeneity and put in a sample bag. Thereafter, an XRF analyser pointer was place on the sample, with the result of the elemental constituents of the samples shown on the display screen of the hand-held XRF analyser. The results of the analysis are presented in Table 1.

Table 1 XRF analysis results of the Run-of-Mine (ROM) sample of Riruwai Cassiterite

Elements	Sn	Nb	Zr	Bi	Pb	Zn	Fe	Mn	Al	Si
% Wt	9.352	0.927	3.998	0.909	1.776	3.628	28.786	2.818	6.232	42.802

Table 2 Particle size analysis for feed to ball mill of Cassiterite

Sieve size range (μm)	Weight retained (g)	Weight % retained	Nominal aperture	Cumulative % retained	Cumulative % passing	% Sn
+500	7.83	3.92	500	3.92	96.08	8.95
−500 + 350	34.83	17.42	350	21.34	78.66	8.72
−350 + 210	32.87	16.44	210	37.78	62.22	10.45
−210 + 180	24.60	12.30	180	50.08	49.92	18.59
−180 + 125	5.69	2.85	125	52.93	47.07	9.13
−125 + 90	23.57	11.79	90	64.72	35.28	6.52
−90 + 63	31.88	15.94	63	80.66	19.34	6.17
Pan (−63)	38.67	19.34		100		2.12
Total	**199.94**	**100**				

Sieves Size and Liberation Size Analysis

A sample of 200 g of ground sample was subjected to sieve analysis to establish the distribution of mineral phases present in the various size fractions. The sieves with size fractions (+500 μm, −500 μm + 350 μm, −350 μm + 210 μm, −210 μm + 180 μm, −180 μm + 125 μm, −125 μm + 90 μm, −90 μm + 63 μm, and −63 μm (pan)) were used for the test, arranged in a stack, with the coarsest sieve on the top and the finest at the bottom. A tight-fitting pan was placed below the bottom sieve to receive the final undersize, and a lid was placed on top of the coarsest sieve to prevent the escape of the sample. The 200 g sample of alluvial cassiterite was charged into a set of sieves on the vibrating sieve screen shaker and shaken for 20 min. The duration of vibration was controlled by the sieve shaker automatic timer. After the required time, the stack was taken apart and the quantity of material retained on each sieve was weighed and analysed with the X-Ray Fluorescence (XRF) Machine. The sieve size analysis result is presented in Table 2, also showing the chemical constituents of tin in each sieve size, and Fig. 1 shows the graph of cumulative percentage (%) Passing and Retained versus Particle Size Analysis for Feed to Ball Mill of Cassiterite.

Gravity Separation Beneficiation of Riruwai Cassiterite

Gravity separation technique of the ore was employed in beneficiating the ore, because cassiterite is usually heavier than all other associated minerals found with

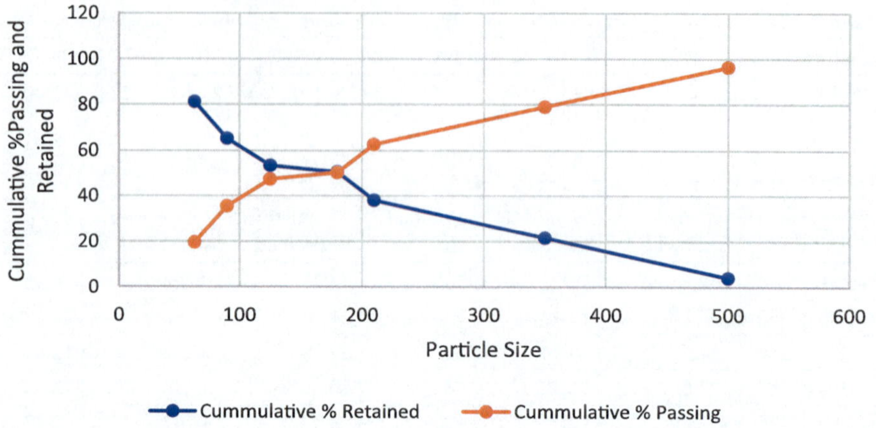

Fig. 1 Cumulative % passing and retained versus particle size analysis for feed to ball mill of Cassiterite

it. The 500 g of the Riruwai cassiterite was ground to the liberation size (−210 + 180 μm) and then fed to the Kip Kelly Air-Float Machine model My-1151 at National Metallurgical Development Centre (NMDC), Jos. The machine separated the Riruwai cassiterite into two constituents, i.e. concentrate and tailing; both constituents were analysed using XRF to determine their chemical composition. The results of the analyses are presented in Table 3. Also, the metallurgical balance sheet for the deposit was determined, showing the percentage assay and percentage recovery in Table 4.

Results and Discussion

The result of the chemical analysis (XRF) of Riruwai cassiterite run-of-mine presented in Table 1 shows that the ore contains 9.352% Sn, 42.802% Si, 2.818% Mn, 28.786% Fe, and other related elements. Based on this, it is evident that the Riruwai deposit is a potential deposit for cassiterite exploitation, as the result meets the typical standard cut-off grade required to mine cassiterite ore deposit, as cassiterite is considered workable if it has a cut-off grade of 0.35% Sn [5, 15].

The mineral distributed in each sieve size was analysed to determine the percent assays. The sieve size that contains the highest percentage assay of cassiterite is the liberation size of Riruwai cassiterite. In Table 2, it can be noted that the cumulative weight percent retained and the cumulative weight percent passing at particle size −200 + 180 μm are, respectively, 50.08 and 49.92 with an assay of 18.59%Sn. Hence, −210 + 180 μm sieve size is the liberation size of the Riruwai cassiterite deposit. At this particle size, beneficiation of the Riruwai cassiterite deposit can be said to be optimum and effective when compared to the other particle sizes because the

Table 3 Chemical analysis (XRF) result of the beneficiated Riruwai Cassiterite using air-float machine

Elements	Concentrates (heavy metals) (wt%)	Tailings (light metals) (wt%)
Sn	26.534	3.657
Nb	1.436	0.837
Zr	4.198	1.450
Bi	1.096	0.043
Pb	3.029	1.534
Zn	5.765	1.093
Fe	30.102	7.964
Mn	3.000	0.546
Si	9.034	66.030
Al	4.672	9.383
Mg	0.945	0.049
V	1.325	0.042
Ag	0.913	0.032
Ti	3.245	1.885
K	3.503	1.224
Ca	0.675	2.038
Y	0.528	0.043
Total	100	100

Table 4 Metallurgical balance sheet for gravity beneficiation of Riruwai Cassiterite using air-float machine

Product	Weight (g)	% Assay	Unit	% Recovery
Feed	500.00	9.352	4676	100
Concentrate	317.875	26.534	8434.495	63.575
Tailings	170.050	3.657	621.873	34.01
Loses	12.075	–	–	2.415

particle sizes at the liberation size are coarse [12]. The interception of cumulative % retained curve and cumulative % passing curve at 180 μm in Fig. 1 also suggests the liberation size of Riruwai cassiterite [16].

Tables 3 and 4 present the chemical analysis (XRF), percent assays, recoveries of tin (Sn), and metallurgical balance sheet of the concentrate and tailing beneficiated using Air-Float Gravity Separator.

Table 3 shows the chemical (XRF) analysis result of concentrate and tailings produced after beneficiation using Air-Float Gravity Separator, while Table 4 shows the metallurgical balance sheet for the Riruwai cassiterite deposit. The chemical analysis shows that the concentrate consists of 26.534% Sn and tailings consist of

3.657% Sn of the gravity beneficiated sample using Air-Float Gravity Separator, the result shows an upgrade in the element of interest (Tin) from 8.952% Sn of the head sample to 26.534% Sn after beneficiation. This shows that the Riruwai cassiterite can be beneficiated by the gravity method using the Air-Float Machine.

The metallurgical balance sheet of processed Riruwai cassiterite (products) also shows that the concentrate produced contained 26.534% Sn with recoveries of 63.575% Sn, while the tailing produced contained 3.657% Sn and 34.01% Sn recovery. The result could be ascribed to the reason that the tin bearing mineral (Cassiterite) was liberated from its associated gangue minerals (silica, annite, microcline, albite, etc.) because tin mineral is heavier than it associated minerals. The ore has been improved to meet the standard concentrate required as charge in the furnace for tin smelting [7, 17].

Conclusion and Recommendation

The chemical analysis of the Riruwai deposit reveals that the sample contains 9.352% Sn, 42.802% Si, 2.818% Mn, 28.786% Fe, and other related minerals. The liberation size of the ore was found to be $-210 + 180$ μm which has the highest percent assay tin of 18.59% Sn. From the physical and chemical characteristics of the deposit, gravity beneficiation used shows an upgrade of the cassiterite from 9.352% Sn to 26.534% Sn. The metallurgical balance sheet indicates the percentage recovery of tin to be 63.575%.

It is recommended that further characterization such as, but not limited to, Atomic Absorption Spectroscopy (AAS), Transmission Electron Microscope (TEM), and Scanning Transmission Electron Microscope (STEM) analysis of the ore should be carried out. Other possible beneficiation techniques such as magnetic separation, froth flotation, and electrostatic separation, or a combination of these separation methods should be considered.

References

1. Sainsbury CL (1969) Tin resources of the world. US Geol Surv Bull 1301:55
2. Olade MA (1980) Geochemical characteristics of tin bearing and tin-barren granites, Northern Nigeria. Econ Geol 75:71–82. https://doi.org/10.2113/gsecongeo.75.1.71
3. Carlin JF (2012) Tin: statistics and information. United States geological survey mineral commodity summaries, pp 170–171
4. International Tin Research Institute Ltd. (ITRI Ltd) (2016) Report on global tin resources & reserves (Security of Long-term Tin Supply) 20
5. Ogunyele AC, Akingboye AS (2018) Environ Earth Sci Res J 5(1):15–23. https://doi.org/10.18280/eesrj.050103
6. Abere DV, Oyatogun GM, Ayodele TJ, Ogar MO, Ohwoekevwo JU, Adejo OH (2018) Upgrading of Tin in Kudedu Cassiterite (Tin Ore) deposit using air floating and 3 D-magnetic beneficiations techniques. Open Sci J Analyt Chem 3(3):28–32

7. Metallic Minerals, October 2, 2010. http://www.onlinenigeria.com/minerals/?blurb=517/. Accessed 24 July 2021
8. http://www.mining.com/markets/commodity/Tin/ Accessed 3 Aug 2021
9. Angadi SI, Sreenivas T, Jeon H-S, Baek S-H, Mishra BK (2015) A review of cassiterite beneficiation fundamentals and plant practices. Miner Eng 70:178–200
10. Leistner T, Embrechts M, Leißner T, Chehreh Chelgani S, Osbahr I, Möckel R, Peuker UA, Rudolph M (2016) A study of the reprocessing of fine and ultrafine cassiterite from gravity tailing residues by using various flotation techniques. Miner Eng 96–97:94–98
11. Feng Q, Zhao W, Wen S, Cao Q (2017) Activation mechanism of lead ions in cassiterite flotation with salicylhydroxamic acid as collector. Sep Purif Technol 178:193–199
12. Sun L, Hu Y, Sun W (2014) Effect and mechanism of octanol in cassiterite flotation using benzohydroxamic acid. Miner Engng 2(4):384
13. Holland-Batt AB (1998) Gravity separation: a revitalized technology. society for mining. Metallurgy and Exploration, Inc., Littlelton, Colorado. SME preprint, 45–98
14. Wills BA (2006) Mineral processing technology, 7th edn. Pergamon Press
15. New Age Exploration Ltd. (NAE Ltd.) (2016) Redmoor resource update
16. Alabi OO, Araoye BO, Bala MB, Igbonwelundu MT, Abere DV, Dalhatu AA (2016) Comparative test for the upgrading of lead in Sabon Layi lead-zinc ore (Alkaleri Local Government Area, Bauchi State), using gravity and froth flotation beneficiation methods. Int J Adv Eng Manag Sci (IJAEMS) 2(6). Infogain Publication (Infogainpublication.com) ISSN: 2454-1311
17. Khanna OP (1999) Material science and metallurgy book. (Dhanpat Rai Publications LTD, India, Rev. Edition), 32.1–32.6

Method for Producing High-Purity $LiOH{\cdot}H_2O$ Using $Ba(OH)_2$

H. Liu and G. Azimi

Abstract In this study, a process for producing high-purity lithium hydroxide monohydrate ($LiOH{\cdot}H_2O$) from synthetic lithium sulfate (Li_2SO_4) solution by using barium hydroxide ($Ba(OH)_2$) is proposed. Various operating conditions are investigated, and results show that 90% of the lithium in the initial Li_2SO_4 solution can be converted to aqueous LiOH in a one-step precipitation process, and the final $LiOH{\cdot}H_2O$ product is determined to have over 99.5% purity. Additional steps including product washing and carbonation are also proposed to improve the lithium conversion and product purity. The by-product barium sulfate ($BaSO_4$) has high market value and can be used in many applications. This process has a simple operating procedure, low capital, and energy cost and can achieve high product yield and battery-grade purity in one batch of reactions without complicated procedures.

Keywords $LiOH{\cdot}H_2O$ · $Ba(OH)_2$ · Precipitation · Purification

Introduction

Lithium hydroxide monohydrate ($LiOH{\cdot}H_2O$) is widely used in many applications including the preparation of lithium-ion battery cathode material, carbon dioxide scrubber, grease, and ceramics [1]. Currently, there are several methods to produce $LiOH{\cdot}H_2O$ such as electrodialysis of brine, roasting of minerals followed by water leaching, and causticization of lithium carbonate (Li_2CO_3) or lithium sulfate (Li_2SO_4) [2–5]. The first two methods are not preferred because of high equipment and maintenance cost, and low purity of product, respectively. The causticization method is more industrially favoured because of its relatively low capital and operating cost, straightforward procedures, and good product quality. Using calcium

H. Liu · G. Azimi (✉)
Department of Chemical Engineering and Applied Chemistry, University of Toronto, 200 College Street, Toronto, ON M5S34, Canada
e-mail: g.azimi@utoronto.ca

G. Azimi
Department of Materials Science and Engineering, University of Toronto, 184 College Street, Toronto, ON M5S34, Canada

T. Ouchi et al. (eds), *Rare Metal Technology 2022*, The Minerals, Metals & Materials Series, https://doi.org/10.1007/978-3-030-92662-5_27

hydroxide ($Ca(OH)_2$) for causticization of Li_2CO_3 results in very low $LiOH{\cdot}H_2O$ yield due to their low water solubilities. Using sodium hydroxide (NaOH) for causticization of Li_2SO_4 results in low product purity because it is difficult to remove the highly soluble sodium salt [6–8]. Thus, there is an urge to develop a method capable of achieving high product yield and purity at the same time.

In this work, barium hydroxide ($Ba(OH)_2$) is used as the causticizing agent to react with Li_2SO_4 to produce $LiOH{\cdot}H_2O$. Both reactants have over 2 mol/L solubility, and the Ba^{2+} and $SO_4{}^{2-}$ are easily precipitated because of the low K_{sp} (10^{-10}) of barium sulfate ($BaSO_4$). Theoretically, this method produces $LiOH{\cdot}H_2O$ with high yield and purity. The effect of operating variables including reagent type (solid or aqueous $Ba(OH)_2$), temperature, and reactant concentration is investigated to determine the optimum operating condition. Additional steps such as solid washing and carbonation are also included in this process to improve the lithium conversion and product purity.

Methodology

Materials

The $Li_2SO_4{\cdot}H_2O$ (99.9%), $Ba(OH)_2{\cdot}8H_2O$ (99.5%), and $BaCl_2{\cdot}2H_2O$ (99.8%) were purchased from VWR. Carbon dioxide gas (CO_2, 99%) was purchased from Messer. Solutions were prepared with deionized water.

Experimental Procedures

In the first step, slightly more (2%) than stoichiometric amount of $Ba(OH)_2$ was mixed with the Li_2SO_4 solution with 400 rpm stirring to ensure complete removal of $SO_4{}^{2-}$, and then the precipitate $BaSO_4$ was filtered and the filtrate was kept for further processing. In this step, tests were performed at different $Ba(OH)_2$ type (aqueous or solid), reaction temperature (25–70 °C), and reactant concentration (0.5–2.0 mol/L). The $Ba(OH)_2$ solution was prepared by dissolving the desired amount of $Ba(OH)_2$ solid in deionized water. As the solubility of $Ba(OH)_2$ is sensitive to temperature (0.5 mol/L at 25 °C and over 3 mol/L at 70 °C), at 25 °C, the $Ba(OH)_2$ solution exists as slurry, and the effective concentration of Ba^{2+} is around 0.5 mol/L no matter the amount of $Ba(OH)_2$ input. However, with the ongoing reaction, the Ba^{2+} is precipitated and the solid $Ba(OH)_2$ can continue to release Ba^{2+} until all dissolved. This will be discussed in detail in the result section. Liquid samples were taken before, during, and after the reaction to monitor the ion concentrations change.

In the second step, the mass of the precipitate $BaSO_4$ was measured before and after drying and the difference was regarded as the moisture content. Then the solid was washed with deionized water with 400 rpm stirring. Because lithium may be coprecipitated with $BaSO_4$, this step aims at recovering the trapped lithium. The metal concentrations in the washing water were measured to evaluate how much lithium can be recovered. The washing solution was cycled back to the filtrate in the first step for future processing.

In the third step, as there existed extra Ba^{2+} in the solution, carbon dioxide (CO_2) gas was fed into the solution to precipitate Ba^{2+} as barium carbonate ($BaCO_3$). Liquid samples were taken during the gas feeding period to monitor the ion concentration change.

In the fourth step, because the LiOH solution can easily react with the CO_2 in air and produce impurity Li_2CO_3, the filtrate solution gathered from steps 1 and 3 was quickly transferred to a sealed container with two outlets allowing argon gas to flow in and out. A water bath was kept at 90 °C to heat the solution until all water was evaporated to obtain the final $LiOH{\cdot}H_2O$ solid.

Characterizations

The concentration of ions was measured by inductively coupled plasma optical emission spectroscopy (ICP-OES) (PerkinElmer Optima 8000). Because ICP-OES cannot detect the polyatomic ions, the S content in the solution is used to represent the SO_4^{2-} content. The viscosities of solutions were measured by a viscometer (BCV-8S, spindle type L2, 60 rpm speed). Solid samples were dried in an oven at 60 °C for 24 h. The lithium conversion percentage is defined as the ratio between the amount of lithium that is converted to $LiOH{\cdot}H_2O$ and the lithium in the initial Li_2SO_4 solution. The purity of product $LiOH{\cdot}H_2O$ is determined by dissolving it in water and measuring the ion distributions. Any ion other than lithium is regarded as an impurity.

Results and Discussion

Effect of $Ba(OH)_2$ Type

To study the impact of the reagent type, slightly more than stoichiometric amount of solution and solid type $Ba(OH)_2$ were added separately into two sets of 2 mol/L Li_2SO_4 solution at 70 °C. As shown in Fig. 1, it was found that in the homogeneous reaction ($Li_2SO_{4(aq)}$ + $Ba(OH)_{2\ (aq)}$), the sulfur content decreased to 0 within 5 min, meaning the SO_4^{2-} was nearly completely removed from the solution. However, in the heterogeneous reaction ($Li_2SO_{4(aq)}$ + $Ba(OH)_{2\ (s)}$), it took around 60 min to precipitate all SO_4^{2-}. There are two possible reasons. First, the solid $Ba(OH)_2$ needs time to dissolve and slows down the reaction. Second, the solution viscosity is significantly increased in the heterogeneous reaction as shown in Table 1. In a viscous solution, the particles move relatively slower and thus the frequency of ion

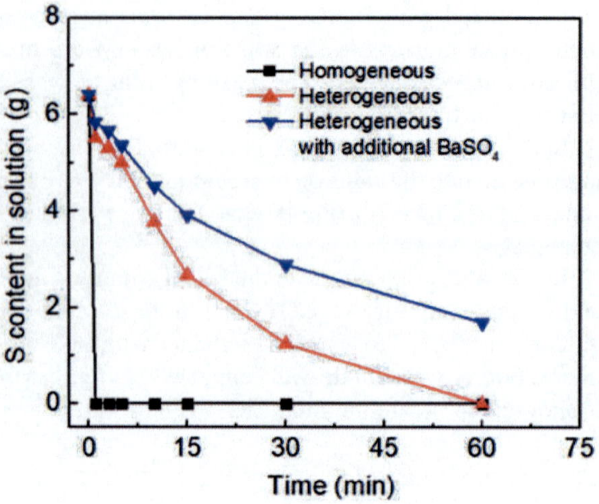

Fig. 1 Sulfur content change as a function of time in the following reactions: **a** homogeneous, **b** heterogeneous, and **c** heterogeneous + additional $BaSO_4$

Table 1 Viscosity of solution at the following conditions: (a) homogeneous, (b) heterogeneous, and (c) heterogeneous + additional $BaSO_4$

	Homogeneous	Heterogeneous	Heterogeneous + additional $BaSO_4$
Viscosity (cP)	67.2 ± 2.4	433.3 ± 17.2	698.5 ± 20.1

collision decreases, resulting in a reduced reaction rate [9]. To prove this hypothesis, additional $BaSO_4$ solid was added to the heterogeneous system and it is found the reaction rate further decreased as shown in Fig. 1. At the same time, the solution viscosity increased as shown in Table 1. Therefore, in the following tests, aqueous type $Ba(OH)_2$ was used.

Effect of Reaction Temperature

To study the effect of temperature, $Ba(OH)_2$ solution was added to 2 mol/L Li_2SO_4 solution at various temperatures (25, 40, 55, and 70 °C). It was found at 25 °C that the SO_4^{2-} was not completely removed even after 60 min while at 70 °C, it took less than a few minutes to remove it. This is possibly because the solubility of $Ba(OH)_2$ at 25 °C is much lower than that at 70 °C as shown in Fig. 2. At 25 °C, over 90% of the $Ba(OH)_2$ remains as solid, and only the dissolved free Ba^{2+} can participate in the reaction. Only after free Ba^{2+} are precipitated, the solid $Ba(OH)_2$ releases Ba^{2+} and continues to react with the SO_4^{2-} ions. This additional step may lead to a decrease in the reaction rate. Additionally, at higher temperatures, the particles move faster and increase the frequency of effective collision. At the same time, as shown in Table 2, the viscosity of the solution mixture decreases with increasing

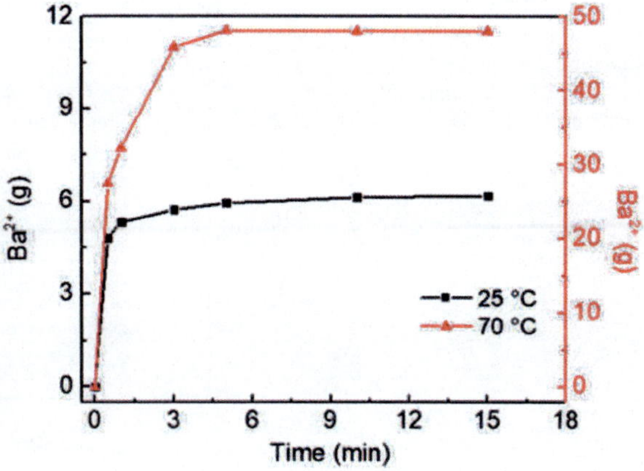

Fig. 2 Dissolution of $Ba(OH)_2$ in 100 mL water as a function of time at 25 and 70 °C

Table 2 Solution viscosity as a function of temperature

	25 °C	40 °C	55 °C	70 °C
Viscosity (cP)	122.1 ± 10.5	104.1 ± 8.9	83.2 ± 8.6	66.8 ± 8.4

temperature. Both reasons may contribute to the increase of reaction rate at higher temperatures. Furthermore, the decreasing viscosity also improves the filtration rate based on Darcy's law [10, 11]. Thus, the following tests were performed at 70 °C.

Effect of Reactant Concentration

To investigate the effect of reactant concentration, the initial concentrations of $Li_2SO_{4(aq)}$ and $Ba(OH)_{2\,(aq)}$ were varied between 0.5 and 2.0 mol/L, and the reaction temperature was kept at 70 °C. As shown in Fig. 3, in each concentration, lithium loss occurs, meaning some lithium is coprecipitated with $BaSO_4$ solid. With the increase of initial reactant concentration, the lithium conversion decreases from around 90% to 80%, meaning more lithium gets trapped. This is possibly because more $BaSO_4$ is produced at a higher initial concentration and traps more liquid that contains soluble lithium ions. As shown in Fig. 3, the moisture content increases with increasing initial concentration. To recover the trapped lithium, solid $BaSO_4$ was washed with pure water with agitation, and nearly 100% of the lithium was released into the washing water, confirming lithium was physically coprecipitated with $BaSO_4$.

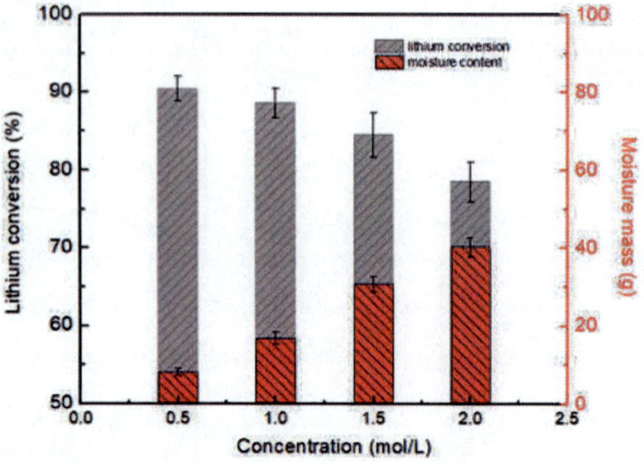

Fig. 3 Lithium conversion (%) and precipitate $BaSO_4$ moisture content as a function of initial reactant concentration

Additional Ba^{2+} Removal by Gaseous CO_2

As mentioned in section "Experimental Procedures", over a stoichiometric amount of Ba^{2+} was introduced to ensure the complete precipitation of SO_4^{2-}, thus there is extra Ba^{2+} in the crude LiOH solution obtained from step 1. Here, CO_2 gas was fed into the solution with a flow rate of 0.5 L/min to precipitate the Ba^{2+} as $BaCO_3$. As shown in Fig. 4, Ba^{2+} is almost completely precipitated at the beginning, while later, the concentration of Li^+ drops because of the formation of a Li_2CO_3 solid. This phenomenon matches well with theory because the K_{sp} of $BaCO_3$ ($10^{-8.3}$) is much lower than that of Li_2CO_3 ($10^{-3.1}$). Then the solution was filtered and purified LiOH solution was obtained. With proper control of flow rate and feeding time, the Ba^{2+} can be removed with little Li^+ loss.

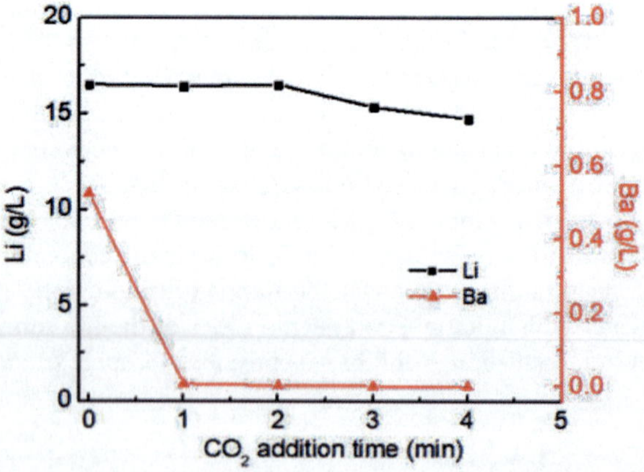

Fig. 4 Concentration of lithium and barium as a function of CO_2 addition time

Table 3 Solution composition

	Li	Ba	S
Concentration (g/L)	16.1 ± 0.3	Not detected	Not detected

Evaporative Crystallization

The purified LiOH solution was then evaporated to obtain solid $LiOH{\cdot}H_2O$ in an argon atmosphere to minimize the contamination by CO_2 in air [12]. The product purity was determined by dissolving 10 g of the solid in 100 mL deionized water, and the solution composition was analyzed by ICP-OES and the result is shown in Table 3. It is found that there was no element other than Li. However, there could still be minor carbonate contamination. To determine the amount of CO_3^{2-} existing in the product, a known amount of barium chloride ($BaCl_2$) solid was added to the solution to precipitate any CO_3^{2-} as $BaCO_3$. Then the solution was filtered and the amount of Ba^{2+} in the filtrate was measured by ICP-OES. The difference between the mole of Ba^{2+} in the $BaCl_2$ solid and the filtrate is the amount of CO_3^{2-} in the $LiOH{\cdot}H_2O$ product because Ba^{2+} and CO_3^{2} have a 1:1 stoichiometric ratio in $BaCO_3$. Based on the calculation, the final product purity is over 99.5% and the only impurity is Li_2CO_3. As the $LiOH_{(aq\ or\ s)}$ can easily absorb CO_2 in air and form Li_2CO_3, if higher product purity is required, the whole process can be carried out in the CO_2-free environment. This can prevent the formation of Li_2CO_3 from the beginning. But commonly, 99.5% purity is enough for most applications including battery cathode material synthesis.

Conclusion

A new method for producing $LiOH{\cdot}H_2O$ from Li_2SO_4 by using $Ba(OH)_2$ is proposed. Experimental results show that aqueous type reagent and 70 °C reaction temperature are preferred as the SO_4^{2-} can be completely removed within a few minutes under these conditions. Increasing initial reactant concentration may lead to more coprecipitation of lithium. However, through washing, almost 100% of the trapped lithium can be recovered. The extra Ba^{2+} remaining in the solution can be removed by feeding CO_2 gas with a controlled flow rate and time. The final $LiOH{\cdot}H_2O$ product is obtained by evaporative crystallization under an inert atmosphere and the purity is determined to be over 99.5%. Compared with traditional methods, this process has the advantage of achieving high $LiOH{\cdot}H_2O$ productivity with battery-grade purity without complicated procedures. Additionally, the valuable by-product $BaSO_4$ can be used in many applications and does not cause environmental issues.

References

1. Graber TA, Morales JW, Robles PA et al (2008) Behavior of $LiOH{\cdot}H_2O$ crystals obtained by evaporation and by drowning out. Cryst Res Technol 43:616–625. https://doi.org/10.1002/crat.200711110
2. Xixiao W, Tang Chunmei LG (1994) Manufacture of lithium hydroxide from lithionite and lime stone by sintering 8
3. Li, Faqiang wang M (2014) Method for preparing lithium hydroxide through electrolysis of salt lake brine
4. Yuanli L (2017) Method for producing battery-grade lithium hydroxide 7
5. Xia G, Yao K, Tu M (2007) Method for preparing battery-stage monohydrate lithium hydroxide 12
6. Kim KJ (2008) Recovery of lithium hydroxide from spent lithium carbonate using crystallizations. Sep Sci Technol 43:420–430. https://doi.org/10.1080/01496390701784088
7. Jiang C, Wang Y, Wang Q et al (2014) Production of lithium hydroxide from lake brines through electro-electrodialysis with bipolar membranes (EEDBM). Ind Eng Chem Res 53:6103–6112. https://doi.org/10.1021/ie404334s
8. Ryabtsev AD, Nemkov NM, Kotsupalo NP, Serikova LA (2004) Preparation of high-purity lithium hydroxide monohydrate from technical-grade lithium carbonate by membrane electrolysis. Russ J Appl Chem 77:1108–1116. https://doi.org/10.1023/B:RJAC.0000044158.61704.93
9. Wang S, Zhang D, Guo J, Guan B (2018) Experiment and analysis of the reaction kinetics of temperature control viscosity acids with limestone. J Pet Sci Eng 165:305–312. https://doi.org/10.1016/j.petrol.2018.02.020
10. Neuman SP (1977) Theoretical derivation of Darcy's law. Acta Mech 25:153–170. https://doi.org/10.1007/BF01376989
11. Liu H, Azimi G (2021) Process analysis and study of factors affecting the lithium carbonate crystallization from sulfate media during lithium extraction. Hydrometallurgy 199:105532.https://doi.org/10.1016/j.hydromet.2020.105532
12. Zilberman P (2015) The CO2 Absorber Based on LiOH. Acta Medica Marisiensis 61:4–6. https://doi.org/10.1515/amma-2015-0023

Part V
Electrochemical Processing and High Temperature Processing

Direct Synthesis of Intermetallic Compounds Through Thermo-Reduction and Electrochemical Deposition

Shuhan Wang, Chao Du, Bao Zhang, Qi Wang, Xin Lu, Osamu Takeda, and Hongmin Zhu

Abstract Intermetallic compounds such as titanium aluminides are currently produced by arc-melting of pure metals. The large difference in the melting point of the elements and the compounds causes the difficulty to make uniform accurate stoichiometric compounds. Intermetallic compounds such as titanium aluminides are currently produced by arc-melting of pure metals. The large difference in the melting point of the elements and the compounds causes the difficulty to make uniform accurate stoichiometric compounds. This paper presents the possibility of direct synthesis of intermetallic compounds with precise components through thermo-reduction from the mixtures of oxide. The possibility of electrochemical deposition for stoichiometric intermetallic compounds was also be discussed.

Keywords Intermetallic compound · Titanium aluminide · Direct synthesis

Introduction

Intermetallic compounds are considered as prospective materials because of their high melting points and large specific strengths at elevated temperatures. Among these intermetallic compounds, titanium aluminides such as Ti_3Al, TiAl, and $TiAl_3$ are attractive as potential materials exhibiting fairly high melting point, low density, good oxidation resistance, and high-temperature strength retention. These materials are applicable for turbine blades in aircraft engines and stationary gas turbines [1–5].

S. Wang · Q. Wang · X. Lu · O. Takeda · H. Zhu (✉)
Department of Materials Science and Engineering, Tohoku University, Sendai 980-8579, Japan
e-mail: hzhu@material.tohoku.ac.jp

C. Du · B. Zhang · H. Zhu
School of Metallurgical and Ecological Engineering, University of Science and Technology Beijing, 30, Xueyuan Rd, Beijing 100083, China

C. Du
School of Materials Science and Engineering, Anhui University of Science and Technology, Huainan 232001, Anhui, China

T. Ouchi et al. (eds), *Rare Metal Technology 2022*, The Minerals, Metals & Materials Series, https://doi.org/10.1007/978-3-030-92662-5_28

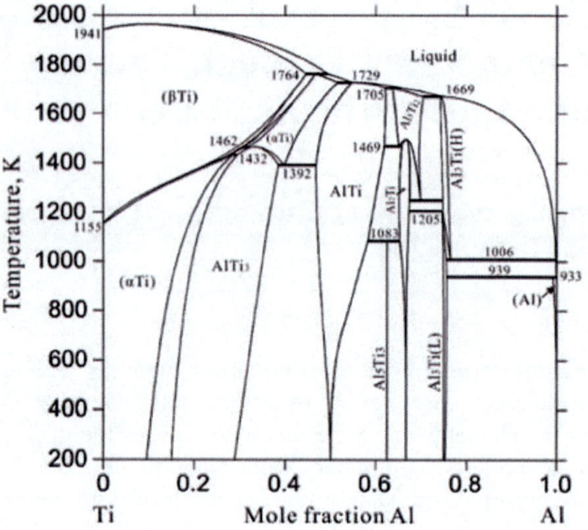

Fig. 1 Phase diagram of Ti–Al binary system

Currently, titanium aluminides are produced by arc-melting from Ti and Al. The large difference in the melting point of the elements and the compounds causes the difficulty to make uniform accurate stoichiometric compounds [6–8]. An efficient way to synthesize titanium aluminides is through chemical reaction or electrochemical reaction. A large number of research has been conducted for electrochemical reduction from TiO_2 and Al_2O_3 to the alloys and intermetallic compounds [9–11]. However, there always remains a question that whether the strict chemical composition of the intermetallic compounds can be controlled through the chemical reaction or electrochemical reaction route.

Herein, we report two synthesis methods for Ti–Al intermetallic compounds (Fig. 1). One is sodiothermic reduction from TiO_2 and Al_2O_3 in molten $CaCl_2$. The other one is electrochemical deposition from the molten salt solution of $AlCl_3$ and $TiCl_3$.

Experimental

Sodiothermic Reduction

Mechanical ball-milling of a certain proportion of TiO_2 (purity > 99.5%) and Al_2O_3 (purity > 99.5%) powders was carried out in a planetary high-energy ball mill.

Raw material of TiO_2/Al_2O_3 composites prepared by mechanical ball-milling together with $CaCl_2$ salt was dehydrated at 400 °C in vacuum. Then the furnace was heated up to the reaction temperature and filled with argon. The Na grains were then fed into the $CaCl_2$ molten salt through a feed pipe for a period of 0.5 h. During the feeding of Na, argon gas was supplied through the same feed pipe for stirring.

After the reaction, the furnace was cooled down to room temperature. The mixture of the product and solidified salt was taken out and leached in water with supersonic vibration and gas bubbling. The powders were obtained by filtering, washed with water, and then dried at 50 °C in a vacuum oven.

Electrochemical Reduction

The electroanalytic measurements were carried out in an equimolar LiCl–KCl melt. All the electrochemical experiments were conducted in a sealed vessel under a dry Ar atmosphere.

The electrolysis and electrochemical experiments were performed using a potentiostat/galvanostat. The reference electrode was the Ag/AgCl electrode, which consisted of a silver wire (1 mm in diameter) in contact with the solution of AgCl (5*wt*%) in a LiCl–KCl molten salt mixture, contained in a Mullite tube. The reference electrode was calibrated against chlorine evolution, and all the potentials were recorded with respect to the potential of Cl_2/Cl^- electrode.

Characterizations

The phase constituents of as-prepared products were determined by X-ray diffraction (XRD, M21X, MAC) with Cu Kα radiation. The morphology and element analysis of products were characterized using a field emission scanning electron microscope (FESEM, ULTRA55, ZEISS) with an energy dispersive spectrometer (EDS, X-MAX50, OXFORD).

Results and Discussion

Sodiothermic Reduction

Calcium metal dissolves into molten $CaCl_2$ with the solubility of 5–15 mol%, forming a homogeneous solution with high reduction potential [12]. Such unique property of high solubility of Ca metal together with the high solubility of CaO in molten $CaCl_2$ [13, 14] enables sufficient calcium thermo-reduction of oxides. Instead of calcium, sodium is also possible to be used as the reductant since Na is even more active than Ca in a chloride melt. Therefore, when sodium metal is fed into $CaCl_2$ melt, calcium will be reduced forming a Ca solution in $CaCl_2$.

$$2Na(l) + \underline{Ca^{2+}} = \underline{Ca} + 2\underline{Na^{+}} \quad (1)$$

The underline marks species dissolved in the solvent molten $CaCl_2$. The dissolved Ca can reduce TiO_2 as well as Al_2O_3 if they are introduced into the solution.

$$2\underline{Ca} + TiO_2(s) = Ti(s) + 2\underline{Ca^{2+}} + 2O^{2-} \quad (2)$$

$$3\underline{Ca} + Al_2O_3(s) = 2Al(l) + 3\underline{Ca^{2+}} + 3\underline{O}^{2-} \quad (3)$$

Reactions (2) and (3) were confirmed in the previous research [15]. Then the question is whether we can synthesize the stoichiometric Ti–Al compounds with strict components as designed from the mixture of TiO_2 and Al_2O_3 through the same route.

Figure 2a, b shows the SEM micrographs of TiO_2 and Al_2O_3 raw materials. The particle size of TiO_2 and Al_2O_3 were in the range of 2–3 μm and 15–40 μm, respectively. Figure 2c, d shows the SEM micrographs, and Al and Ti element mapping of TiO_2/Al_2O_3 composites with the mole ratio of 2:1, milled for 0.5 and 12 h, respectively. As shown in the figure, individual particles in the size of 1–2 μm for TiO_2 and 5–10 μm for Al_2O_3 were observed after 0.5 h ball-milling. The element mapping also presents the ununiformity of Al and Ti element distribution indicating that the TiO_2/Al_2O_3 composite was not uniformly mixed. After milling for 12 h, the TiO_2 $/Al_2O_3$ composite was composed of relatively small particles below 1 μm. Al and Ti elements were distributed uniformly, which indicated that the TiO_2/Al_2O_3 composites were homogeneous in the scale of micrometers. Figure 3 shows the XRD patterns of the reduction products of TiO_2/Al_2O_3 composite milling for 0.5 h and 12 h, reacted at 900 °C. In both tests, the compound of TiAl was designed in the total amount of the raw TiO_2/Al_2O_3 composite.

$$7\underline{Ca} + 2TiO_2(s) + Al_2O_3(s) = 2TiAl(s) + 7\underline{Ca^{2+}} + 7\underline{O}^{2-} \quad (4)$$

All the TiO_2/Al_2O_3 oxides were reduced to Ti–Al intermetallic in both cases. In the product of 0.5 h milled composite, phases of Ti_3Al, and $TiAl_3$ were also found other than TiAl. Obviously, the lack of uniformity in the short-time milled composite resulted in the compounds of Ti–rich (Ti_3Al) and Al-rich ($TiAl_3$) phases. When the TiO_2/Al_2O_3 composite was milled for 12 h, in which TiO_2 and Al_2O_3 were mixed uniformly in the scale of micrometer (Fig. 2d), the reduction product contained only TiAl single phase, as designed.

Figure 4 shows the SEM micrographs of the TiAl intermetallic compound powders obtained with different reaction temperatures. As can be seen from the pictures, there are two types of particles in the powders. One is very small particles with a size of several tens of nanometers, and their agglomerates. The other is relatively bigger particles with the size of some hundreds of nanometers. It is presumed that the fine

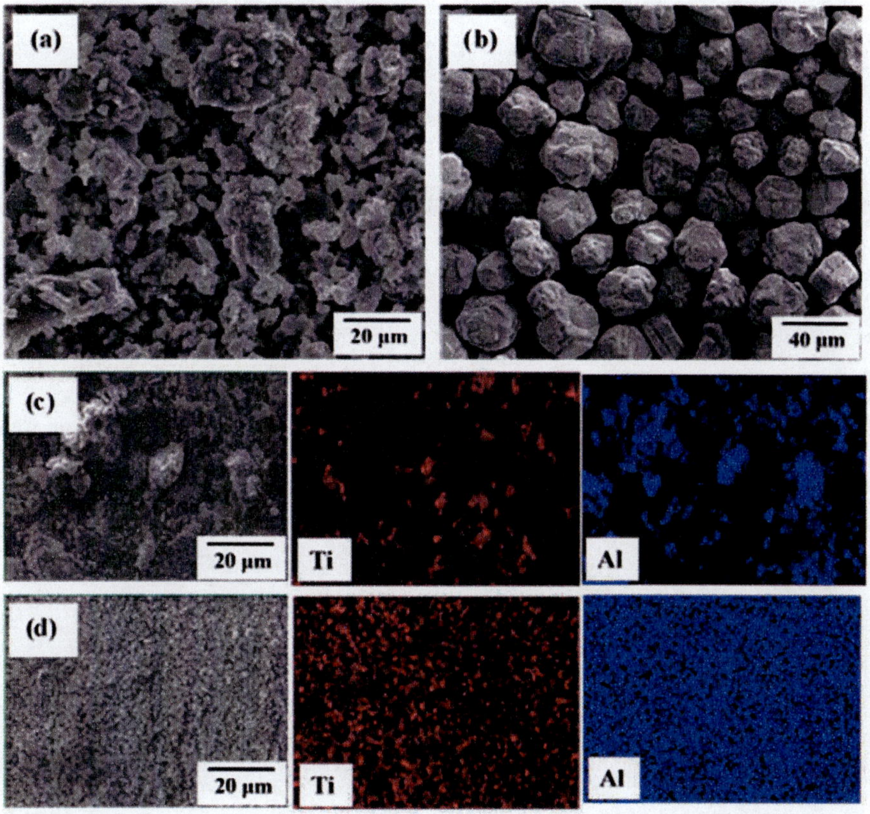

Fig. 2 SEM micrographs of **a** TiO_2 and **b** Al_2O_3; SEM and element mapping micrographs of TiO_2/Al_2O_3 composites with mole ratio of 2: 1 after milling in the ball mill for **c** 0.5 h and d 12 h

small particles were the primary particles of the reaction, while the big particles were from the combination and growth of the fine particles.

Figure 5 shows the XRD patterns of the reduction products with the TiO_2/Al_2O_3 ratio of 6:1 and 2:3. As can be seen from the figure, a single stoichiometric compound of Ti_3Al and $TiAl_3$ was obtained. These results demonstrate the possibility of direct synthesizing powders of Ti–Al intermetallic compounds, with the tunable content.

Electrochemical Reduction

Another approach to synthesize the intermetallic compounds would be electrochemical co-deposition directly from dissolved ions in molten salts. In this case, the feed sources of titanium and aluminum are dissolved ions. However, the difference

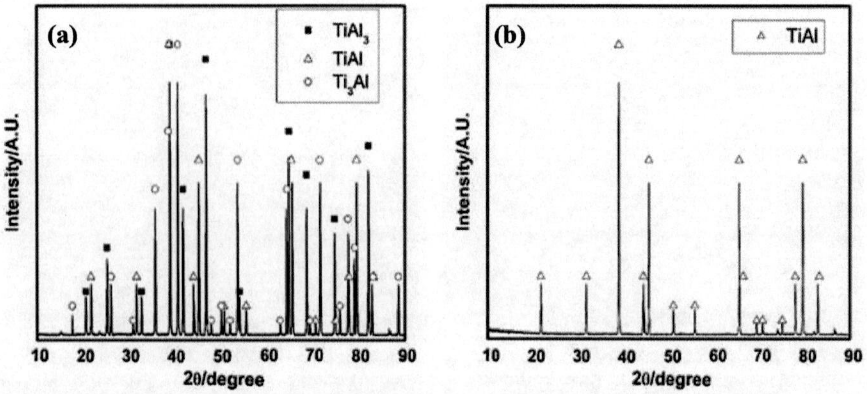

Fig. 3 XRD patterns of the products by sodiothermic reduction of the TiO_2/Al_2O_3 composite with mole ratio of 2:1, milled for **a** 0.5 h and **b** 12 h, 900 °C

Fig. 4 SEM micrographs of the TiAl intermetallic powders obtained at **a** 900 °C and **b** 1100 °C

between the deposition potential of titanium and aluminum will be another limitation for the co-deposition with a stable component of intermetallic compounds.

Figure 6 shows the voltammograms and square wave voltammograms of LiCl–KCl melt with the addition of $AlCl_3$, $TiCl_3$, and $TiCl_3$–$AlCl_3$. As can be seen from the figure, the cathodic reaction of Al^{3+} happens at the electrode potential of −2.32 V versus Cl_2/Cl^-. The cathode reaction of Ti^{3+} consists of steps, and the Ti deposition happens at −2.39 V.

$$Al^{3+} + 3e = Al \quad -2.32\,V\ versus\ Cl_2/Cl \tag{5}$$

$$Ti^{3+} + e = Ti^{2+} \quad -2.09\,V\ vs\ Cl_2/Cl \tag{6}$$

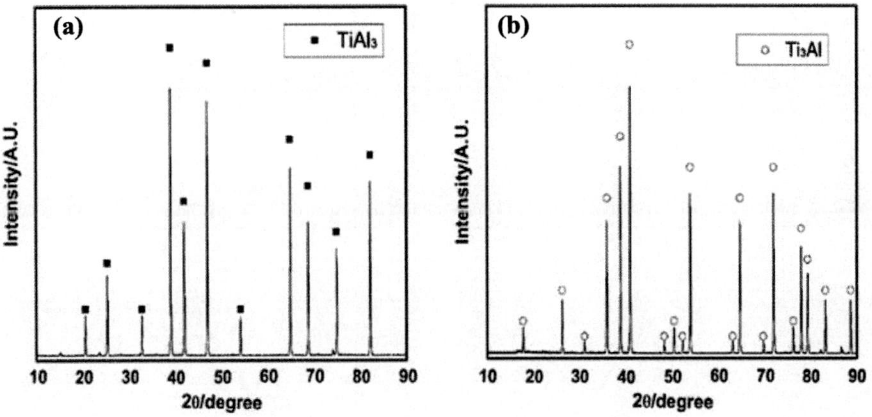

Fig. 5 XRD patterns of the products by sodiothermic reduction of the 12 h milled TiO_2/Al_2O_3 composite with mole ratio of **a** 6:1, **b** 2:3, and 900 °C

$$Ti^{2+} + 2e = Ti \quad -2.39\,V\,vs\,Cl_2/Cl \tag{7}$$

The small difference (70 mV) between the deposition potentials of Al and Ti will enable the co-deposition for the intermetallic compound. The voltammogram of solution contains both Al^{3+} and Ti^{3+}; there is only one metal deposition peak after the peak for $Ti^{3+} + e = Ti^{2+}$. The peak potential is around −2.27 V versus Cl_2/Cl^-, which is slightly higher than both potentials for Ti^{2+}/Ti (−2.39 V), and Al^{3+}/Al (−2.32 V). This potential shift comes from the difference in chemical potential in pure metal and compounds. The exchange electron number for the compound formation electrode reaction was estimated to be 2.7 from the half-wave width of the square wave voltammogram. This exchange electron number corresponds to the formation of $TiAl_3$, which is also reasonable for the potential order between Ti^{2+}/Ti and Al^{3+}/Al.

$$Ti^{2+} + 3Al^{3+} + 11e = TiAl_3 \tag{8}$$

On the basis of the electroanalytic results above, constant current electrolysis was performed in LiCl–KCl melt with 3 mol% $TiCl_3$ and 3 mol%$AlCl_3$, at a current density of 200 mA cm^{-2}. Figure 7 shows the image, XRD pattern, SEM micrograph, and element mapping micrographs of electrodeposited powders. Fine powders with a primary particle size of less than 5 μm were obtained. The XRD pattern and the element mapping micrographs confirm the formation of Ti Al_3 with a single uniform phase.

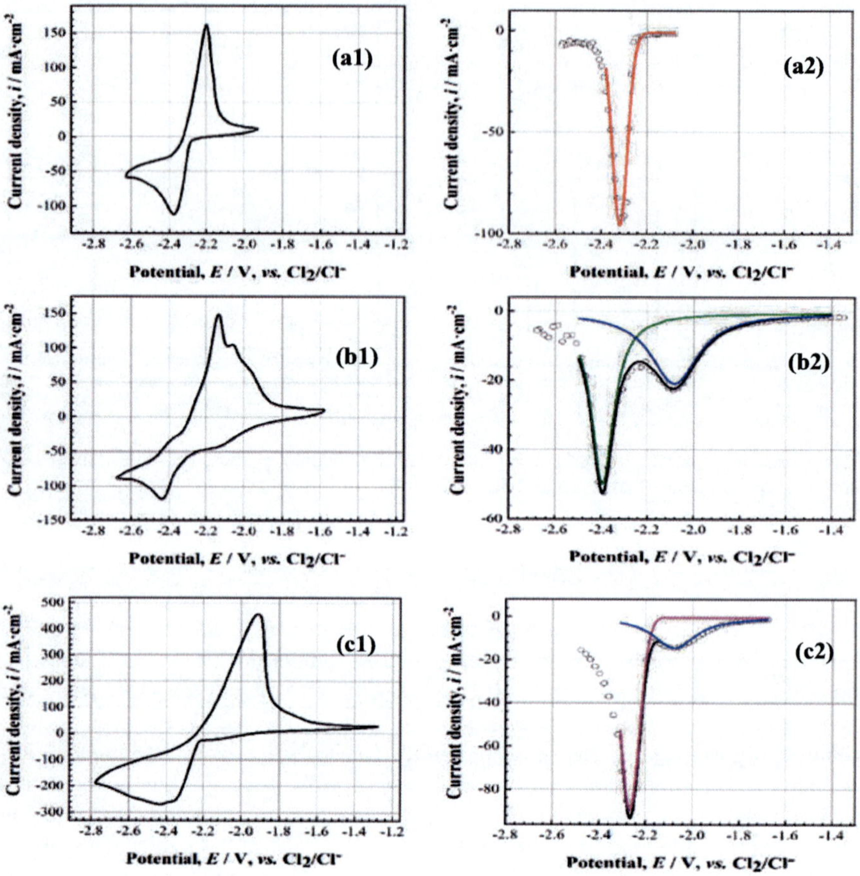

Fig. 6 Voltammograms and square wave voltammograms of LiCl–KCl melt with **a** 0.3 mol%$AlCl_3$, **b** 0.3 mol%$TiCl_3$, **c** 0.3 mol%TiCl3-0.5 mol%AlCl3. 500 °C, v = 200 mV s−1, f = 12.5 Hz

Conclusion

By using uniformly mixed TiO_2/Al_2O_3 composites, all the titanium aluminates (Ti_3Al, AlTi, and $TiAl_3$) can be directly synthesized through sodiothermic reduction in $CaCl_2$. The formation of an accurate stoichiometric intermetallic compound can be controlled by uniformly mixing through ball-milling.

Through electrochemical co-deposition from Al^{3+}, Ti^{3+} dissolved in LiCl–KCl melt, $TiAl_3$ compound powders with a single phase were obtained.

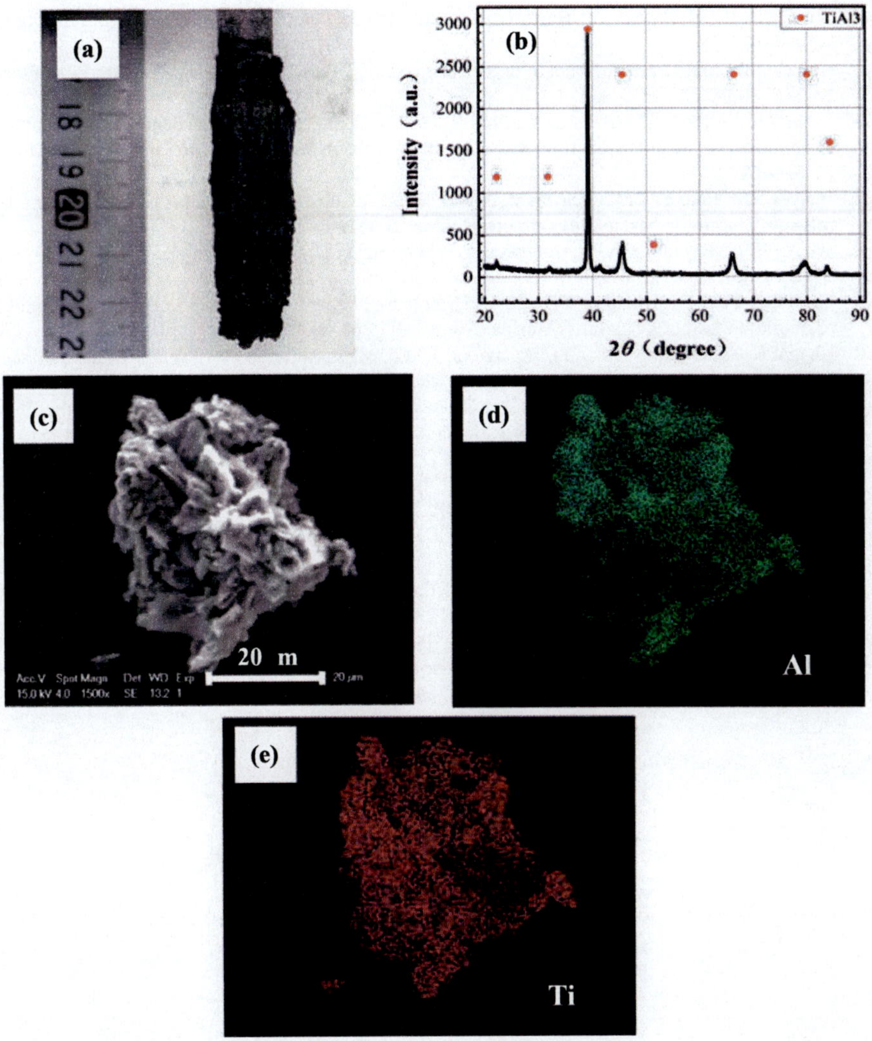

Fig. 7 **a** Image, **b** XRD pattern, **c** SEM micrograph, **d, e** element mapping micrographs of $TiAl_3$ powders obtained by electrolysis of 3 mol%AlCl3-3 mol%TiCl3 solution in LiCl–KCl melt, 200m Acm^{-2}, 500 °C, 3 h

References

1. Kim YW (1995) Gamma titanium aluminides: their status and future. JOM 47:39–42
2. Loria EA, Gamma titanium aluminides as prospective structural materials. Intermetallics 8:13339–1345
3. Dimiduk DM (1999) Gamma titanium aluminide alloys-an assessment within the competition of aerospace structural materials. Mater Sci Eng A 2:281–288

4. Das G, Kestler H, Clemens H, Bartolotta PA (2004) Sheet gamma TiAl: status and opportunities. JOM 56:42–45
5. Gebauer K (2006) Performance, tolerance and cost of TiAl passenger car valves. Intermetallics 14:355–360
6. Sakamoto K, Yoshikawa K, Kusamichi T (1992) Changes in oxygen contents of titanium aluminides by vacuum induction, cold crucible induction and electron beam melting. ISIJ Int 32:616–624
7. Imayev RM, Imayev VM, Oehring M, Appel F (2007) Alloy design concepts for refined gamma titanium aluminide based alloys. Intermetallics 15:451–460
8. Wu X (2006) Review of alloy and process development of TiAl alloys. Intermetallics 14:1114–1122
9. Chen GZ, Fray DJ, Farthing TW (2000) Direct electrochemical reduction of titanium dioxide to titanium in molten calcium chloride. Nature 407:361–364
10. Bhagat R, Jackson M, Inman D, Dashwood RJ (2008) Production of Ti-W alloys from mixed oxide precursors via the FFC cambridge process. J Electrochem Soc 155:63–69
11. Jackson BK, Inman D, Jackson M, Dye D, Dashwood RJ (2010) NiTi production via the FFC cambridge process: refinement of process parameters. J Electrochem Soc 157:36–43
12. Eastman ED, Cubicciotti DD, Thurmond CD (1950) Temperature-composition diagrams of metal-metal halide systems. In Quill LL (Ed), Chemistry and Metallurgy of miscellaneous materials, McGraw-Hill Book Co., New York, pp 6–12
13. Wenz DA, Johnson I, Wolson RD (1969) $CaCl_2$-rich region of the $CaCl_2$–CaF_2–CaO system. J Chem Eng Data 14:250–252
14. Perry GS, MacDonald LG (1985) Role of $CaCl_2$ in the reduction of PuO_2. J Nucl Mater 130:234–241
15. Du C, Wang Z, Hou JG, Jiao SQ, Zhu HM (2014) Production of titanium powder by sodiothermic reduction in $CaCl_2$ molten salts. Metall Mater Trans B 45:1750–1756

Development of Molten Salt Electrolysis of MgO Using a Metal Cathode and Vacuum Distillation to Produce Ultra-High Purity Mg Metal

Jungshin Kang, Tae-Hyuk Lee, Hyeong-Jun Jeoung, Dong-Hee Lee, Young Min Kim, Kyung-Woo Yi, Toru H. Okabe, and Jin-Young Lee

Abstract To resolve the drawbacks of the current Mg production process, direct use of MgO as feedstock via an electrolytic method has been attempted. In this study, molten salt electrolysis of MgO using a metallic cathode such as Sn, Cu, or Ag and a graphite anode followed by vacuum distillation was developed to produce ultra-high purity Mg metal. During the electrolysis of MgO in MgF_2–LiF molten salt at 1053 K, Mg alloy is produced at the bottom of the electrolysis cell owing to the high density of the cathode. Afterward, 99.999% Mg metal is produced through vacuum distillation of the Mg alloy at 1200–1300 K. In addition, the use of secondary resources containing MgO as feedstock, the stability of the structural materials that are used, and scale-up of the process were also investigated in order to establish a cost-effective and environmentally friendly Mg production process.

Keywords Magnesium metal · Magnesium oxide · Molten salt electrolysis · Metal cathode

J. Kang (✉) · T-H. Lee · H-J. Jeoung · D-H. Lee · J-Y. Lee
Korea Institute of Geoscience and Mineral Resources, 124 Gwahak-ro Yuseong-gu, Daejeon 34132, Korea
e-mail: jskang@ust.ac.kr; jskang@kigam.re.kr

J. Kang · Y. M. Kim · J-Y. Lee
University of Science and Technology, 217 Gajeong-ro Yuseong-gu, Daejeon 34113, Korea

H-J. Jeoung · K-W. Yi
Department of Materials Science and Engineering, Seoul National University, 1 Gwanak-ro, Gwanak-gu, Seoul 08826, Korea

Y. M. Kim
Korea Institute of Materials Science, 797 Changwondae-ro, Seongsan-gu, Changwon, Gyeongnam 51508, Korea

T. H. Okabe
Institute of Industrial Science, The University of Tokyo, 4-6-1 Komaba, Meguro-ku, Tokyo 153-8505, Japan

T. Ouchi et al. (eds), *Rare Metal Technology 2022*, The Minerals, Metals & Materials Series, https://doi.org/10.1007/978-3-030-92662-5_29

Introduction

Magnesium (Mg) is the lightest among structural metals and has superior properties such as high specific strength, high stiffness, high thermal conductivity, good damping capacity, and electromagnetic shielding [1]. Owing to its excellent properties, Mg is used in various applications such as electronics and transportation. Transportation is currently the most important market for Mg metal. For example, approximately 37.5% of the world's Mg metal was used for transportation in 2015 [2]. In addition, the demand for Mg metal in this field will increase in the future for increasing fuel efficiency to reduce carbon dioxide (CO_2) gas emissions [3].

The Pidgeon and electrolytic processes are used for commercial Mg metal production. Although two methods are employed, approximately 85% of primary Mg metal is produced by the Pidgeon process in China [4]. In the Pidgeon process, the silicothermic reduction of calcined dolomite using ferrosilicon (Fe·Si) is conducted at 1373–1474 K under vacuum [5–7]. However, high energy consumption, generation of a large amount of harmful gases such as sulfur dioxide (SO_2) gas, high labor input, and low productivity are the disadvantages [5]. In the electrolytic process, Mg metal and chlorine gas (Cl_2) are produced at the cathode and anode, respectively, through the molten salt electrolysis of anhydrous magnesium chloride ($MgCl_2$) at 928–993 K. However, the preparation of anhydrous $MgCl_2$ from several resources, such as brines and magnesite ($MgCO_3$), is energy-intensive, and the generation of toxic gas during electrolysis and high capital cost are also the disadvantages [5].

Several studies have been conducted to develop an environmentally friendly method for resolving the drawbacks of the current Mg production processes. Among these, the single prominent method is the solid oxide membrane (SOM) process [8–10]. In the SOM process, Mg metal and oxygen gas (O_2) are produced through the electrolysis of magnesium oxide (MgO) in magnesium fluoride (MgF_2)–calcium fluoride (CaF_2) molten salt at 1423 K using a yttria-stabilized zirconia (YSZ) tube. MgO is directly used as feedstock; however, a decrease in the partial pressure of Mg in the molten salt, such as argon (Ar) gas bubbling, is required for maintaining high current efficiency.

To produce Mg metal from MgO feedstock with a simple electrolysis cell, the electrolysis of MgO in MgF_2–lithium fluoride (LiF) molten salt using metallic cathodes such as tin (Sn), copper (Cu), and silver (Ag) at 1053 K has been developed by the authors [11–15]. Figure 1 shows a schematic of the molten salt electrolysis of MgO using a metal cathode and vacuum distillation. As shown in Fig. 1, Mg alloys are produced at the bottom of the cell because of the use of high-density metals as cathode. In addition, CO_2 or O_2 gas is generated depending on the anodic material used. Afterward, vacuum distillation of the Mg alloys is conducted at 1200–1300 K, and Mg metal with a purity of 99.999% is obtained. In this study, recent advances such as the use of various MgO resources as feedstock, process scale-up, and stability of the structural materials will be briefly discussed.

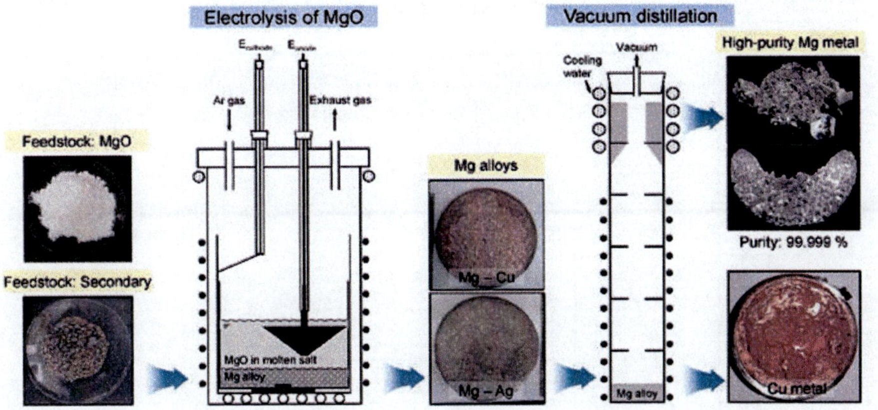

Fig. 1 Schematic of molten salt electrolysis of MgO using metal cathode and vacuum distillation

Principle of the Molten Salt Electrolysis of MgO Using Metal Cathode

When electrolysis of MgO dissolved in 54 mass% MgF_2–46 mass% LiF molten salt using Sn, Cu, or Ag cathode and a carbon (C) anode is conducted at 1053 K, Mg alloys and carbon oxide (CO_x) gas are produced at the cathode and anode, respectively, by the reactions shown in Eqs. (1) and (2) [11–13]. The theoretical decomposition voltage of MgO with C is 1.46–1.49 V at 1053 K, while those of MgF_2 and LiF are 4.87 V and 5.33 V at 1053 K, respectively, under standard state [16]. In addition, O_2 gas is generated when an inert anode such as platinum (Pt) is used.

$$\mathrm{Mg}^{2+}(\text{in molten salt}) + 2\mathrm{e}^- + M(s) = \mathrm{Mg} - M(s,l)(M = \mathrm{Sn, Cu, or\ Ag}) \quad (1)$$

$$\mathrm{C}(s) + x\mathrm{O}^{2-}(\text{in molten salt}) = \mathrm{CO}_x(g) + 2x\mathrm{e}^-(x = 1, 2) \quad (2)$$

Figure 2 shows the binary phase diagrams of the Mg–Sn system with the isobaric line of the vapor pressure of Mg (p_{Mg}) as a function of temperature and Mg concentration [17]. The binary phase diagrams of the Mg–Cu and Mg–Ag systems with the isobaric line of p_{Mg} have been reported in the literature by the authors [11–13, 17]. As shown in Fig. 2, when the concentration of Mg in the Mg–Sn alloy increases at 1053 K, p_{Mg} also increases. The increase in p_{Mg} indicates that the loss of Mg from the alloy increases, which leads to a decrease in the current efficiency. However, active vaporization occurs when the p_{Mg} is larger than 0.01 atm under atmospheric pressure [18]. Consequently, the maximum content of Mg electrodeposited in the Sn cathode is expected to be 30 mass% at 1053 K. As discussed in previous studies, the maximum contents of Mg in Cu and Ag cathodes are expected to be 31 mass%

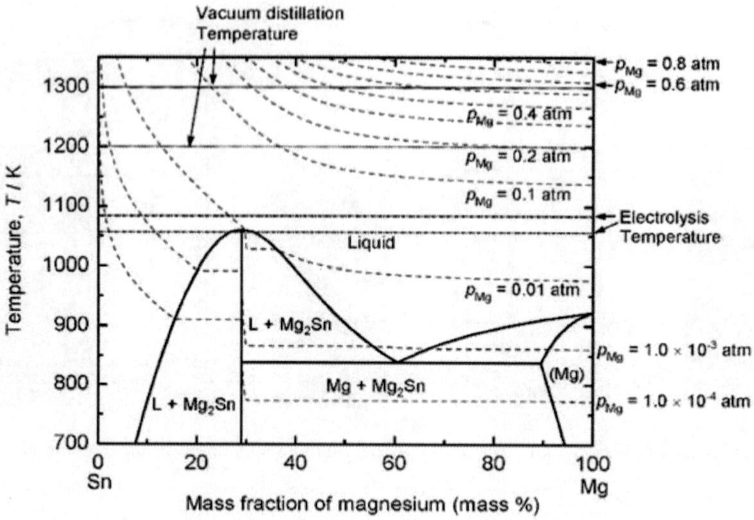

Fig. 2 Binary phase diagrams of the Mg–Sn system with the isobaric line of the Mg vapor pressure (Reprinted with permission from Springer Nature: Springer, *Metallurgical and Materials Transactions B*, Molten Salt Electrolysis of Magnesium Oxide Using a Liquid–Metal Cathode for the Production of Magnesium Metal, Lee et al. [12])

and 27 mass% at 1053 K, respectively [11–13]. In addition, up to these Mg concentrations in Mg alloys, the theoretical densities of Mg alloys are larger than that of 4.8 mass% $MgO–MgF_2–LiF$. As a result, Mg alloys are produced at the bottom of the electrolysis cell until the maximum concentration of Mg electrodeposited on the cathode is reached.

When the pressure of the system is lower than 1 atm, Mg evaporation can occur even though p_{Mg} is in the range of 0.0001–0.001 atm, which is smaller than 0.01 atm. When the high recovery efficiency of Mg is considered, the suitable temperature for the vacuum distillation of Mg–Sn alloy is assessed to be 1200–1300 K, as shown in Fig. 2. In addition, the suitable temperatures for the vacuum distillation of Mg–Cu and Mg–Ag alloys were assessed to be 1200–1300 K and 1300 K, respectively, as discussed in previous studies [11–13].

Recent Advances in Molten Salt Electrolysis Using Metal Cathodes

Molten salt electrolysis using a metal cathode uses MgO as the feedstock. MgO can be produced from primary resources such as magnesite and seawater. For the production of MgO from primary resources, many processing stages such as mining, mineral processing, and chemical treatments are required; these are costly and have environmental concerns. Therefore, the utilization of secondary resources containing MgO as the feedstock for electrolysis is necessary to be investigated.

Among the several secondary resources containing MgO, such as waste MgO–C refractory brick and ferronickel slag, MgO–C refractory brick was used as the feedstock for the experiments. Table 1 shows the compositions of the oxidized bricks at 1073 K before use. When the electrolysis of the feedstock was conducted in MgF_2–LiF–KCl molten salt using a Cu cathode and C anode at 1043 K with an applied average current of 1.44 A for 12.5 h, the average cell potential measured was 2.48 V and 14.3 mass% Mg–Cu alloy was obtained with a current efficiency of 92.3%, as shown in Table 2. In addition, when vacuum distillation of the Mg alloy obtained was conducted at 1300 K for 10 h, Mg metal with a purity of 99.994% was obtained. These results indicate that the use of waste MgO–C refractory brick is possible for molten salt electrolysis using a metal cathode, and the competitiveness of the process will be enhanced.

A scale-up study of the molten salt electrolysis of MgO using a metal cathode was also conducted [13]. Figure 3 shows a schematic of the experimental apparatus used. The electrolysis of MgO using Cu or Ag cathode and C anode in MgF_2–LiF molten salt was conducted at 1053 K for 75–95 h with an applied average current of 8.81–9.59 A. The average cell potential measured was 2.93–2.95 V, and 20.1 mass% Mg–Cu and 18.2 mass% Mg–Ag alloys were produced with a current efficiency of 82.2–88.0%. The actual weight of Mg metal produced was 286.8–312.0 g for these experiments. When the vacuum distillation of Mg–Cu and Mg–Ag alloys was conducted at 1300 K for 10 h, Mg metal with a purity of 99.999% was obtained. Figure 4 shows the

Table 1 Analytical results of various MgO-containing feedstocks used in this study

Feedstock	Concentration of element *i*, C_i (mass%)[a]							
	MgO	Al_2O_3	Fe_2O_3	Na_2O	K_2O	CaO	Cr_2O_3	MnO
MgO–C refractory brick	Bal	8.22	0.64	0.03	0.08	1.03	0.18	0.04

[a] Determined by ICP-OES and ICP-MS analysis; N.D: Not Detected. Below the detection limit (<0.0001–0.0048 mass%)

Table 2 Results of the electrolysis and analytical results of the Mg–Cu alloys obtained

Exp no.[a]	Average potential, E / V	Current efficiency, η (%)[c]	Concentration of element *i*, C_i (mass%)[b]						
			Mg	Si	Li	Al	K	Fe	Cu
210,228	2.48	92.3	14.3	0.03	0.10	0.08	0.02	N.D	Bal

[a] Experimental conditions;
(1) Weight of MgF_2 = 256.0 g, weight of LiF = 219.0 g, and weight of KCl = 25.0 g
(2) Weight of feedstock, w_{feed} = 40.0 g, weight of Cu cathode, $w_{cathode}$ = 44.99 g
(3) Reaction time, t = 12.5 h, applied average current, i = 1.44 A
[b] Determined by ICP-OES analysis; N.D: Not Detected. Below the detection limit of the ICP-OES (<0.0025–0.0060 mass%)
[c] η = (actual weight of Mg obtained × 100)/theoretical weight of Mg obtained

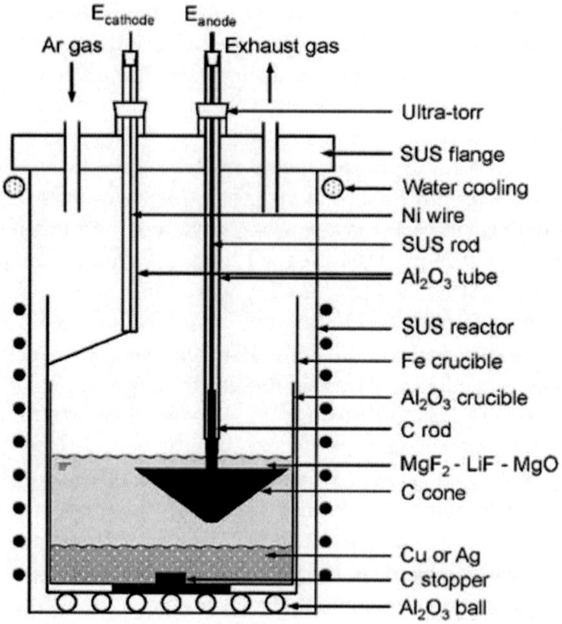

Fig. 3 Schematic of the experimental apparatus used for the scaled-up experiments (Reprinted with permission from Springer Nature: Springer, *Journal of Sustainable Metallurgy*, Scale-Up Study of Molten Salt Electrolysis using Cu or Ag Cathode and Vacuum Distillation for the Production of High-Purity Mg Metal from MgO, Lee et al. [12])

Fig. 4 Microstructure of the cross-section of the Al_2O_3 crucible in contact with the Mg alloy after electrolysis (Reprinted with permission from Springer Nature: Springer, *Journal of Sustainable Metallurgy*, Scale-Up Study of Molten Salt Electrolysis using Cu or Ag Cathode and Vacuum Distillation for the Production of High-Purity Mg Metal from MgO, Lee et al. [12])

microstructure of the cross-section of the alumina (Al_2O_3) crucible contacting the Mg alloy after electrolysis using the Ag cathode at 1053 K for 95 h to produce 18.2 mass% Mg–Ag alloys. As shown in Fig. 4, 220 μm, magnesium aluminum oxide layer with a dense microstructure was generated. The authors have verified the feasibility of the novel electrolytic process and also increased the production of Mg metal per batch from several grams to hundreds of grams to demonstrate the scale-up of the process. However, further scale-up is necessary to demonstrate its practical application (Fig. 3).

Conclusions and Outlook

To develop an efficient and environmentally sound Mg metal production process, the molten salt electrolysis of MgO using a metal cathode at 1053 K and vacuum distillation at 1200–1300 K were investigated. To establish a cost-effective and green technology for Mg production, the application of secondary resources such as MgO–C refractory bricks as feedstock for electrolysis was investigated. In addition, a scale-up of the process using a simple cell was conducted to demonstrate its feasibility for practical applications. To enhance the robustness and extension of the process, the use of various secondary resources, further scale-up, and development of inert electrode will be conducted.

References

1. Kojima Y (2001) Project of platform science and technology for advanced magnesium alloys. Mater Trans 42(7):1154–1159
2. Roskill (2016) Magnesium metal: global industry, markets and outlook to 2020. Roskill Information Services Ltd., London
3. U.S. Department of Energy (2010) Materials technologies: goals, strategies, and top accomplishments. https://www.energy.gov/sites/prod/files/2014/03/f13/materials_tech_goals.pdf. last Accessed 8 Aug 2021
4. Tauber M (2019) 2018 global primary magnesium supply and demand balance and outlook. In: 76th international magnesium association world magnesium conference, Budapest, 15–17 May 2019
5. Friedrich HE, Mordike BL (2006) Magnesium technology-metallurgy, design data, applications. Springer, Berlin
6. Pidgeon LM, Alexander WA (1944) Thermal production of magnesium-pilot plant studies on the retort ferrosilicon process. New York Meeting: reduction and refining of non-ferrous metals. Trans Am Inst Min Met Eng 159:315–352
7. Halmann M, Frei A, Steinfeld A (2008) Magnesium production by the pidgeon process involving dolomite calcination and MgO silicothermic reduction: thermodynamic and environmental analyses. Ind Eng Chem Res 47:2146–2154
8. Krishnan A, Lu XG, Pal UB (2005) Solid oxide membrane process for magnesium production directly from magnesium oxide. Metall Mater Trans B 36:463–473
9. Powell AC IV (2018) Final scientific/technical report of industrial scale-up of low-cost zero-emission magnesium by INFINIUM electrolysis. https://www.osti.gov/biblio/1431302. Last Accessed 11 Jan 2021
10. Guan X, Pal UB, Powell AC (2014) Energy-efficient and environmentally friendly solid oxide membrane electrolysis process for magnesium oxide reduction: experiment and modeling Metall Mater Trans E 1:132–144
11. Lee TH, Okabe TH, Lee JY, Kim YM, Kang J (2020) Molten salt electrolysis of magnesium oxide using a liquid-metal cathode for the production of magnesium metal. Metall Mater Trans B 51:2993–3006
12. Lee TH, Okabe TH, Lee JY, Kim YM, Kang J (2021) Development of a novel electrolytic process for producing high-purity magnesium metal from magnesium oxide using a liquid tin cathode. J Magnes Alloy 9:1644–1655

13. Lee DH, Jeoung HJ, Lee TH, Yi KW, Lee JY, Kim YM, Okabe TH, Kang J (2021) Scale-up study of molten salt electrolysis using Cu or Ag cathode and vacuum distillation for the production of high-purity Mg metal from MgO. J Sustain Met 7:883–897. https://doi.org/10.1007/s40831-021-00367-x
14. Kim Y, Yoo J, Kang J (2020) Applicability of the electrochemical oxygen sensor for in-situ evaluation of MgO solubility in the MgF_2-LiF molten salt electrolysis system. Metals 10:906
15. Kim Y, Kang J (2021) Viscosity of molten MgF_2-LiF-MgO system and structure investigation using classical molecular dynamic simulations. J Non-Cryst Solids 552:120377
16. Barin I (1995) Thermochemical data of pure substances. VCH Verlagsgesellschaft mbH, Weinheim
17. Bale CW, Belisle E, Chartrand P, Decterov SA, Eriksson G, Gheribi AE, Hack K, Jung IH, Kang YB, Melancon J, Pelton AD, Petersen S, Robelin C, Sangster J, Spencer P, Van Ende MA (2016) Factsage thermochemical software and databases, 2010–2016. Calphad 54:35–53
18. Kang J, Okabe TH (2014) Thermodynamic consideration of the removal of iron from titanium ore by selective chlroination. Metall Mater Trans B 45:1260–1271

Electrochemical Cell Design for Emf Measurements of Liquid Nd-Bi Alloys via Coulombic Titration in LiCl–KCl–NdCl$_3$ Electrolyte

Sanghyeok Im, Nathan D. Smith, Stephanie Castro Baldivieso, and Hojong Kim

Abstract Using an electrochemical cell containing a two-phase Nd-Sn alloy ($x_{Nd} = 0.10$) in LiCl–KCl–NdCl$_3$ (1 mol%) as a reference electrode, coulometric titration emf measurements at various cathodic current densities (10–100 mA cm^{-2}) were performed to evaluate the stability of the Nd-Sn electrode and corroborate emf values for Nd (in Bi) at $T = 873$ K. The potential difference between two identical Nd-Sn reference electrodes was measured before/after the electrochemical evaluation for 45 days, and the stability of the cell for the long-term operation was verified. The emf values of Nd (in liquid Bi) titrated at various current densities were consistent at each Nd concentration, demonstrating the reliability of the cell.

Keywords Neodymium · Liquid bismuth · Molten chloride · Coulometric titration · Emf

Introduction

Electrochemical recovery processes for rare-earth elements are necessary for facilitating the removal of fission products accumulated during the treatment of spent nuclear fuel as well as in developing improved recycling cycles for clean energy technologies [1, 2]. Liquid metal electrodes (e.g., Bi and Sn) have been considered for improving the recovery efficiency of alkali, alkaline-earth, and rare-earth metals based on their strong chemical interactions [3, 4]. However, determination

S. Im · N. D. Smith · S. C. Baldivieso · H. Kim (✉)
Materials Science and Engineering, The Pennsylvania State University, 419 Steidle Building, University Park, PA 16802, USA
e-mail: huk29@psu.edu

S. Im
e-mail: svi5106@psu.edu

N. D. Smith
e-mail: nds174@psu.edu

S. C. Baldivieso
e-mail: szc6095@psu.edu

T. Ouchi et al. (eds), *Rare Metal Technology 2022*, The Minerals, Metals & Materials Series, https://doi.org/10.1007/978-3-030-92662-5_30

of reliable thermodynamic properties for rare-earth elements can be difficult due to their high reactivity with certain electrolytes. Electromotive force (emf) measurement is an effective tool for determining the thermodynamic properties of alloys in alkali/alkaline-earth metal chloride, but the high reactivity of Nd in molten chlorides limits electrochemical cell stability and lifetime. For example, the presence of a disproportionation reaction $Nd + 2NdCl_3 \rightarrow 3NdCl_2$ between pure Nd and the electrolyte results in loss of Nd metal to the electrolyte [5, 6].

In a previous study, we determined the thermodynamic properties of the Nd-Bi system through coulometric titration emf measurements using a less reactive two-phase Nd-Sn alloy as a reference electrode [7]. Coulometric titration emf measurement was able to determine thermodynamic properties of the Nd-Bi system including the Bi-rich liquid phase region by fine composition control of the liquid Nd-Bi alloy, and the accuracy of the technique has been verified in relevant studies using liquid metal in molten salts [7–10]. This study describes a cell configuration including a highly stable reference electrode for coulometric titration emf measurement using liquid metals in molten chloride and validates the reliability of the described cell design.

To verify the reliability of the cell, the coulometric titration was repeated with various current densities at 873 K and then the measured open-circuit potential (OCP) values following each titration step were compared. In addition, the stability of the developed RE in our previous study was confirmed by comparing the potential difference between two identical REs before/after the electrochemical evaluation. The measured OCP value was determined as the emf value by correcting the established cell potential between the Nd-Sn alloy ($x_{Nd} = 0.10$) versus pure Nd (s) developed in the previous study [6].

Experimental

LiCl–KCl-$NdCl_3$ electrolyte (eutectic LiCl–KCl + 1 mol% $NdCl_3$) was prepared from appropriate weights of LiCl (Ultra dry, 99.9%, Alfa Aesar), KCl (Ultra dry, 99.95%, Alfa Aesar), and $NdCl_3$ (anhydrous, 99.5%, Alfa Aesar) powders. The powder mixture was loaded into a quartz crucible (Technical Glass Products) for pre-melting in a stainless-steel vacuum chamber. The chamber was loaded into a crucible furnace (Mellen, CC-12), evacuated to less than 10 mtorr, and heated under vacuum at 373 K for 12 h and at 543 K for 12 h to remove residual moisture and oxygen. The chamber was then purged with ultra-high purity Ar three times and heated to 923 K for 3 h under a flowing (50 mL min^{-1}) argon atmosphere. After cooling, the dry and homogeneous electrolyte was ground into a fine powder using a mortar and pestle for use in the electrochemical cell assembly.

The Nd-Sn and Nd-Bi alloys for reference electrodes (RE) and counter electrodes (CE) were fabricated using a laboratory arc-melter (MAM-1, Edmund Bühler GmbH) under an inert argon atmosphere from pure Nd (99.1%, Alfa Aesar, Stock No. 00214), Sn (99.9999%, Alfa Aesar, Stock No. 45481), and Bi (99.999%, Sigma-Aldrich,

Stock No. 556130). The arc-melted Nd-Sn alloy ($x_{Nd} = 0.10$) was re-melted inside a boron nitride (BN, Advalue Technology) crucible (35 mm height, 12 mm outer diameter, 8 mm inner diameter, and 30 mm depth) using an induction heater custom-installed inside a glovebox ($O_2 < 0.5$ ppm). Two 1 mm diameter capillary holes were drilled at 7 mm above from the bottom of the crucible. During induction heating, a tungsten wire was inserted into the RE in the liquid state to establish electrical contact. A liquid Nd-Bi ($x_{Nd} = 0.03$) electrode served as the CE. The CE in each cell was contained in a BN crucible (15 mm height, 25 mm outer diameter, 22 mm inner diameter, and 11 mm depth) with a tungsten wire (1 mm diameter) serving as the electrical lead. The alloy was fabricated by melting Bi using an induction heater and gradually adding Nd pieces into liquid metals until the target composition was achieved. The working electrode (WE) was fabricated by induction melting 2.76 g of pure Bi in a BN crucible (12 mm outer diameter, 8 mm inner diameter, 15 mm height, and 10 mm depth). The nominal surface area of WE was estimated to be 0.5 cm^2. The three-electrode electrochemical cell configuration is shown in Fig. 1.

The final assembly of the electrochemical cell was conducted in an Ar-filled glovebox (O_2 concentration <0.5 ppm). The electrodes were placed in an alumina crucible (Advalue Technology) and electrolyte was poured over the electrodes. The three-electrode cell assembly was loaded in a stainless-steel test chamber. The test chamber was then sealed, removed from the glovebox, loaded into a crucible furnace, and evacuated to ~1 Pa. The chamber was heated to 373 K for 12 h then 543 K for 12 h under vacuum, was purged three times with ultra-high purity Ar, and then heated under flowing argon (~10 mL min^{-1}) to 873 K. The cell temperature was measured using a thermocouple (ASTM type-K) located at the center of the electrolyte, and thermocouple data acquisition system (NI 9211, National Instruments).

Emf measurements were conducted using a potentiostat–galvanostat (Autolab PGSTAT302N) at 873 K. The composition of the liquid Bi WE was controlled by approximately $x_{Nd} = 0.0025$ per each step through coulometric titration by applying

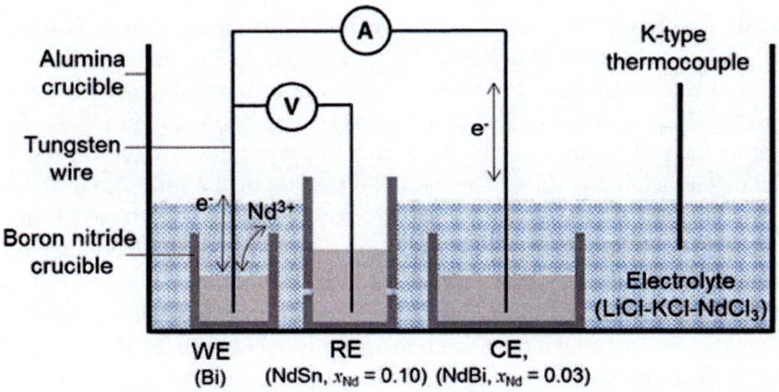

Fig. 1 Three-electrode electrochemical cell configuration used for emf measurement of the Nd-Bi alloys via coulometric titration

constant cathodic current densities of 10–100 mA cm^{-2} for 180–1730s. After each titration, the WE was relaxed under open-circuit condition for 1 h for equilibration of the electrode and then the stabilized OCP was measured to emf. Assuming a perfect coulombic efficiency, the concentration of Nd in liquid Bi was calculated by Faraday's law:

$$x_{Nd} = \left(\frac{m_{WE}}{M_{WE}}\frac{zF}{It}\right)^{-1}$$

where m_{WE} is the initial mass (2.76 g) of the WE, M is the molecular weight of the pure Bi, and I is the constant current applied during the titration time, t. The emf values (E_I) were obtained from the average of steady OCP ($dE/dt = 0.1$ mV/hr, except for the de-alloying process) for the last 10 min of each step and then converted to the emf (E_{eq}) versus pure Nd. The cell potential (E_{II}) of the two-phase Nd-Sn alloy ($x_{Nd} = 0.10$) relative to pure Nd (s) was previously established and it exhibited linear behavior at 724–1075 K [6]:

$$E_{II} = 0.590 + 1.52 \times 10^{-4} T\ [V]\ \text{vs. Nd(s)}$$

In the present work, the cell emf (E_{eq}) of a Nd-Bi alloy relevant to pure Nd (s) was converted by adding the two cell potentials ($E_{eq} = E_I + E_{II}$) at a given temperature 873 K.

Results and Discussion

Reference Electrode Stability

The potential difference (E_{RE2}–E_{RE1}) between the two identical REs (Nd-Sn alloy, $x_{Nd} = 0.10$) before/after electrochemical evaluations at 873 K including coulometric titration emf measurements was observed to monitor the stability of the REs. As shown in Fig. 2a, the potential difference was less than 0.05 mV during the first 2 h of operating the cell, indicating a highly stable RE. After the 45-day experiment, the potential difference between the REs remained at less than 4 mV. The utilization of alloys with two-phase behavior (e.g., Nd-Bi alloys with $x_{Nd} = 0.20$ or Nd-Sn alloys with $x_{Nd} = 0.10$) as REs reduces the reactivity with the electrolyte compared to the reactive pure Nd, leading to greater stability and a significantly longer cell lifetime.

Figure 2a indicates that the lifetime of the RE was improved in the molten chloride electrolyte in comparison with the potential difference between the REs of ± 5.0 mV for about 40 h reported in the previous study using Nd-Bi alloys ($x_{Nd} = 0.20$) as REs [6]. Furthermore, as the Nd-Bi alloy exhibits the peritectic phase transitions [L + $NdBi_2$ = Nd_3Bi_7] and [L + NdBi = $NdBi_2$] at 774 K and 932 K, the emf exhibited

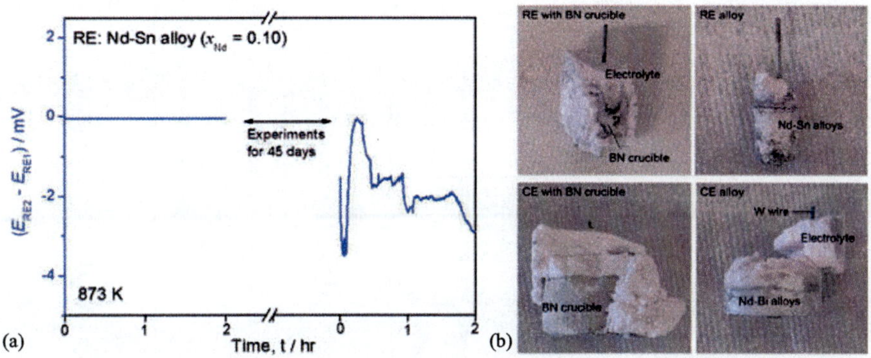

Fig. 2 **a** The potential difference between Nd-Sn alloy ($x_{Nd} = 0.10$) reference electrodes for 2 h at 873 K before/after coulometric titration emf measurement for 45 days. **b** The post-mortem images of the electrodes after electrochemical evaluation for 45 days

discontinuous linearity [6, 7]. On the other hand, the Nd-Sn alloys do not experience any phase transitions in the 724–1075 K range and therefore exhibit linear emf behavior. These observations indicate that the Nd-Sn two-phase RE is a more viable option than the Nd-Bi two-phase RE for use in electrochemical cells.

The post-mortem images of each cooled electrode (RE and CE) are shown in Fig. 2b. The BN crucibles containing electrodes with molten salt electrolyte were structurally intact and the BN surface remained white, unlike in measurements using pure Nd as a RE which were shown to blacken the electrolyte due to loss of pure Nd into the salt. The CE appeared similarly unchanged in structure and color, despite the lengthy measurement time.

Coulometric Titration Emf Measurements

Emf of the Nd-Bi alloys in liquid phase was measured via coulometric titration using various constant cathodic current densities (10–100 mA cm^{-2}) at 873 K against the aforementioned Nd-Sn two-phase RE. The coulometric titration emf measurement of the ± 25 mA cm^{-2} current density is shown in Fig. 3; the electrode potential decreased as Nd was deposited under cathodic constant current and stabilized as the open-circuit potential (OCP) condition was held for 1 h to reach a homogeneous composition of the electrode. As the titration proceeds, the OCP of the Nd-Bi alloy decreases and gradually reaches a constant value due to the phase transition, which is described in detail in our previous study [7].

The measured steady OCP value after titration at various cathodic current densities ($10 - 100$ mA cm^{-2}) including Fig. 3 was converted into the emf (E_{eq}) versus pure Nd and plotted as a function of the mole percent of Nd in Fig. 4. Compared with the emf trajectory, which was separated into a two-phase region (dotted lines in Fig. 4) and

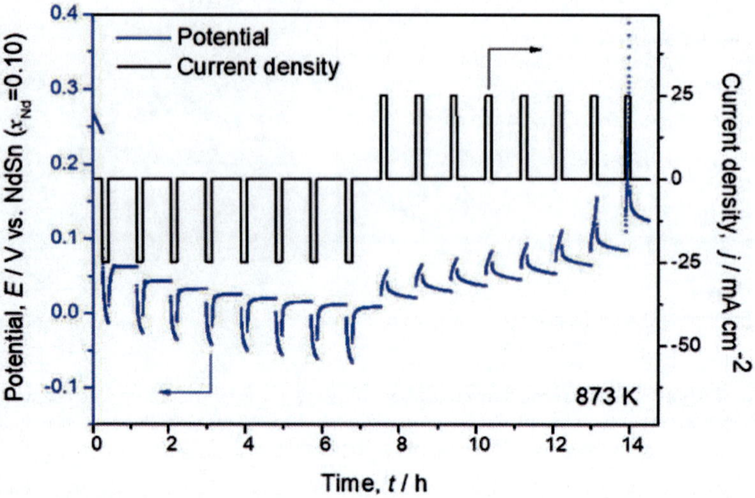

Fig. 3 Representative coulometric titration emf measurement data of liquid Nd-Bi alloy WE (j = ± 25 mA cm^{-2}) as a function of time at T = 873 K

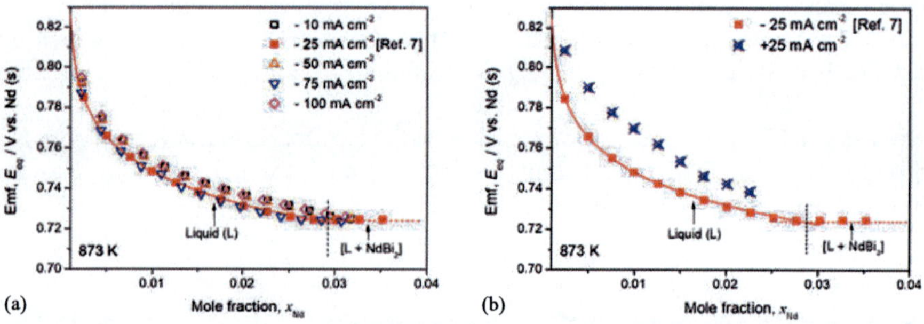

Fig. 4 **a** Emf measured at various cathodic current densities (10–100 mA cm^{-2}) as a function of the mole percent of Nd at 873 K, where solid lines represent logarithmic curve fit in the liquid phase. **b** Comparison of emf values determined during titration (alloying process) and back-titration (de-alloying process) in Fig. 3

the liquid single-phase region (solid line in Fig. 4) to determine solubility in the study on thermodynamic properties of the Nd-Bi system [7], emf values measured in this study show a good agreement within ±5 mV at each Nd concentration. The results in Fig. 4a indicate that the proposed electrochemical cell configuration, including highly stable Nd-Sn alloys as REs, provide reliable emf measurements under various current density conditions and long-term operation at 873 K.

As shown in Fig. 4b, the measured OCP values during the alloying process and de-alloying process under a current density of ±25 mA cm^{-2} at 873 K were inconsistent. The electrode potential quickly stabilized to equilibrium during the alloying process, but did not reach equilibrium within relaxation time during the de-alloying process. The initial flux boundary condition formed at the interface between the liquid metal and the electrolyte can be different according to the deposition/dissolution of Nd by the applied current direction, which is expected to cause a difference in the chemical diffusivity of Nd in liquid Bi. Based on the report of fast charge-transfer properties in the study of electrochemical recovery of reactive elements such as alkali/alkaline-earth using liquid metal in molten salt [8, 10, 11], mass transport property is considered as a dominant factor, and it is thought to be related to the difference in the change of electrode potential during titration and back-titration in Fig. 3.

Conclusion

Coulometric titration emf measurements of liquid Nd-Bi alloys were performed under various current densities using the proposed electrochemical cell containing an Nd-Sn alloy ($x_{Nd} = 0.10$) with two-phase behavior as a stable reference electrode. The reliability of the cell was demonstrated as all emf results measured at various cathodic current densities (10–100 mA cm^{-2}) were consistent within the $\pm$ 5 mV range compared with the equilibrium potential determined in the previous study [7]. In addition, the long-term stability of the cell was validated as the potential difference between the REs was measured within 3.5 mV before/after the electrochemical evaluation for 45 days.

Acknowledgements This work was supported by the US National Science Foundation (grant number: CMMI-1662817) and the US Department of Energy, Office of Nuclear Energy's Nuclear Engineering University Program (Award No. DE-NE0008757).

References

1. Bruno J, Ewing RC (2006) Spent nuclear fuel. Elements 2:343–349. https://doi.org/10.2113/gselements.2.6.343
2. U.S Department of Energy (2011) Critical materials strategy. https://www.energy.gov/sites/prod/files/DOE_CMS2011_FINAL_Full.pdf
3. Kim H, Boysen DA, Newhouse JM, Spatocco BL, Chung B, Burke PJ, Bradwell DJ, Jiang K, Tomaszowska AA, Wang K, Wei W, Ortiz LA, Barriga SA, Poizeau SM, Sadoway DR (2013) Liquid metal batteries: past, present, and future. Chem Rev 113:2075–2099. https://doi.org/10.1021/cr300205k
4. Nigl TP, Lichtenstein T, Kong Y, Kim H (2020) Electrochemical separation of alkaline-earth elements from molten salts using liquid metal electrodes. ACS Sustain Chem Eng 8:14818–14824. https://doi.org/10.1021/acssuschemeng.0c04249
5. Novoselova A, Smolenski V (2013) Electrochemical behavior of neodymium compounds in molten chlorides. Electrochim Acta 87:657–662. https://doi.org/10.1016/j.electacta.2012.09.064

6. Lichtenstein T, Im S, Yu CT, Kim H (2020) Thermodynamic properties of rare-earth alloys by electrochemical emf measurements. J Mater Res 1–9. https://doi.org/10.1557/jmr.2020.218
7. Im S, Shang S, Smith ND, Lichtenstein T, Sun H, Adam M, Bocklund BJ, Liu Z, Kim H (2021) Thermodynamic properties of the Nd-Bi system via emf measurements, DFT calculations, and CALPHAD modeling. Acta Mater Under Rev. https://doi.org/10.2139/ssrn.3894900
8. Kim H, Boysen DA, Ouchi T, Sadoway DR (2013) Calcium-bismuth electrodes for large-scale energy storage (liquid metal batteries). J Power Sources 241:239–248. https://doi.org/10.1016/j.jpowsour.2013.04.052
9. Baldivieso SC, Smith ND, Im S, Kim H (2021) Thermodynamic properties of Gd-Bi alloys determined by emf measurements in LiCl–KCl–$GdCl_3$ electrolyte. J Alloys Compd 886 161229. https://doi.org/10.1016/j.jallcom.2021.161229
10. Ouchi T, Kim H, Ning X, Sadoway DR (2014) Calcium-antimony alloys as electrodes for liquid metal batteries. J Electrochem Soc 161:A1898–A1904. https://doi.org/10.1149/2.0801412jes
11. Kim H, Smith N, Kumar K, Lichtenstein T (2016) Electrochemical separation of barium into liquid bismuth by controlling deposition potentials. Electrochim Acta 220:237–244. https://doi.org/10.1016/j.electacta.2016.10.083

Two-Phase Rare-Earth Alloys as Reference Electrodes in Molten Chlorides for Reliable Electrochemical Measurements

Nathan D. Smith, Stephanie Castro Baldivieso, Timothy Lichtenstein, Sanghyeok Im, and Hojong Kim

Abstract Electromotive force measurements were used to examine pure rare-earth reference electrodes (Nd, Gd) in molten chloride salts (LiCl-KCl-$RECl_3$) over a temperature range of $T = 400$–800 °C to determine the effects of disproportionation reactions (i.e., $Nd + NdCl_3 \rightarrow NdCl_2$) and active metal dissolution on their longevity and stability. Several methods for developing and calibrating a stable two-phase Gd-Bi electrode were investigated including the use of a pure Gd metal reference electrode continuously submerged in the electrolyte, a pure Gd metal reference electrode intermittently dipped into the electrolyte, and a transient technique that deposited pure Gd onto an inert W wire. Electromotive force measurements were also employed to examine the lifetime of the two-phase alloy as the reference electrode in comparison to a pure rare-earth metal as the reference electrode. Pure rare-earth metal did not exhibit stability for more than 15 h, whereas two-phase Gd-Bi electrodes were found to be stable for over 24 days at $T = 500$–700 °C.

Keywords Reference electrodes · Rare-earth elements · Nd · Gd · Two-phase electrodes · Liquid metals

N. D. Smith · S. C. Baldivieso · T. Lichtenstein · S. Im · H. Kim (✉)
Materials Science and Engineering, The Pennsylvania State University, 406 Steidle Building, University Park, PA 16802, USA
e-mail: huk29@psu.edu

N. D. Smith
e-mail: nds174@psu.edu

S. C. Baldivieso
e-mail: szc6095@psu.edu

T. Lichtenstein
e-mail: tvl5297@psu.edu

S. Im
e-mail: svi5106@psu.edu

T. Ouchi et al. (eds), *Rare Metal Technology 2022*, The Minerals, Metals & Materials Series, https://doi.org/10.1007/978-3-030-92662-5_31

Introduction

Thermodynamic properties of rare-earth elements are necessary for developing improved separation and recovery processes as knowledge of activities and phase transition temperatures is essential for constructing efficient pyroprocessing technologies for a sustainable materials cycle. Electromotive force (emf) measurements are a valuable method of obtaining such data, as they directly determine chemical potentials via the Nernst equation [1]. Electrochemical cells have been constructed to measure emf values for various rare-earth elements (e.g., Nd, Gd, La, Dy), producing large quantities of useful thermodynamic data [2–7]. For emf measurements performed in molten chloride salts, the reference electrode is often chosen as the requisite pure rare-earth element (REE) [4–6, 8]. Pure rare-earth metal represents a standard state for rare-earth alloys and can provide a straightforward reference potential for electrochemical measurements. However, in order to qualify as a suitable reference for electrochemical measurements, the reference electrode must be stable, and its potential should be reversible and reproducible throughout the measurement [9].

Recent work by Lichtenstein et al. on the thermodynamics of the Nd-Bi system indicated that the use of pure Nd as a reference in LiCl-KCl-$NdCl_3$ electrolyte at $T =$ 725–1075 K resulted in significant instability (>500 mV) over the ~25 h experiment due to a disproportionation reaction (i.e., $Nd + 2NdCl_3 \rightarrow 3NdCl_2$) [2], which has been reported by several other researchers [10–13]. Emf measurements using pure Nd as a reference electrode in molten chlorides have been performed by Lebedev et al. as well as Kulagina and Bayanov [4, 5]; however, neither work comments on the observed stability of the pure Nd reference. Baldivieso et al. attempted to use a pure Gd reference electrode for emf measurements on the Gd-Bi system at $T = 773$–973 K, as Gd^{3+} is the only stable valence for Gd in molten LiCl-KCl-$GdCl_3$ which removes the possibility for a disproportionation reaction [3]; however, they determined that the dissolution of Gd metal was substantial enough to eliminate pure Gd as a reference electrode choice. Several other authors performed emf measurements using pure Gd as a reference [6, 8], but the only note of reference stability was from Yamana et al. which indicated less than 1 mV of instability in the emf values at 10 min into the measurement [8].

Despite stability issues with using pure REEs as reference electrodes, both Lichtenstein et al. and Baldivieso et al. managed to generate viable thermodynamic data using a two-phase alloy (mole fraction $x_{\text{Nd(in Bi)}} = 0.20$ and $x_{\text{Gd(in Bi)}} = 0.16$, respectively) as a reference electrode in their electrochemical measurements [2, 3]. The use of the two-phase rare-earth element/liquid metal alloys has several advantages:

(1) the strong atomic interactions between rare-earth elements and liquid metals mitigate dissolution of the metal into the electrolyte;
(2) disproportionation side reactions are effectively blocked;
(3) provided the reference electrodes are made at a composition in a two-phase region (e.g., L + Gd-Bi), changes in the composition will not shift the measured potential as activity (i.e., emf) is constant in a two-phase region via the Gibbs phase rule.

This study investigated the long-term stability of pure Nd and Gd elements in molten chloride salts over a ~30 h period at $T = 400–800$ °C, tested several methods of calibrating two-phase Gd-Bi reference electrodes, and compared the long-term stability of the two-phase reference electrodes to the results from pure metals.

Experimental

Two molten salt mixtures, LiCl-KCl-$NdCl_3$ (58.6-40.4-1.0 mol%) and LiCl-KCl-$GdCl_3$ (57.4-39.5-1.0 mol%), were prepared by weighing appropriate amounts of LiCl (Ultra dry, 99.9%, Alfa Aesar), KCl (Ultra dry, 99.95%, Alfa Aesar), and $NdCl_3$ (anhydrous, 99.5%, Alfa Aesar) or $GdCl_3$ (anhydrous, 99.9%, Sigma-Aldrich) to make ~70 g of salt. The salt was loaded into a quartz crucible (Technical Glass Products) for pre-melting in a stainless-steel vacuum chamber, which was placed into a furnace (Mellen CC-12) and evacuated to <10 mtorr. The furnace was then heated to 100 °C for 12 h, 270 °C for 12 h, purged with ultra-high purity Ar, and heated to 700 °C for 3 h under flowing Ar. The electrolytes were then cooled to room temperature and crushed with a mortar and pestle into a fine powder for use in measurements.

Pure rare-earth metal reference electrodes were constructed by cutting a chunk of the metal from an ingot (Nd: 99.1%, Alfa Aesar; Gd: 99.9%, Alfa Aesar) in dimensions of approximately 0.5 cm × 0.5 cm × 2.5 cm and wrapping it tightly at the top with a Mo wire (1 mm diameter, 99.95%, Beantown Chemical).

Two-phase Nd-Bi and Gd-Bi electrodes were prepared by arc-melting (MAM-1, Edmund Bühler Gmbh) pieces of Bi (99.999%, Sigma-Aldrich) and the requisite rare-earth element in compositions of $x_{Nd} = 0.20$ and $x_{Gd} = 0.16$, respectively. The homogeneous two-phase alloys were then re-melted in a BN crucible (12 mm OD, 8 mm ID, 35 mm height, 30 mm depth, AX05 grade, Saint Gobain Advanced Ceramics) using an induction heater (IH15A-2 T, Across International) custom-installed in an Ar glovebox (<0.5 ppm O_2). An inert W wire (1 mm, 99.95%, ThermoShield) was inserted into the molten alloy as an electrical lead.

The electrochemical cells were assembled inside of an Ar glovebox to avoid any reaction between electrode materials and air/moisture. Two pure metal electrodes and two of the two-phase REE-Bi electrodes were placed in an alumina crucible (60 mm diameter, 100 mm height, AdValue Technology) and ~70 g of the crushed electrolyte was poured into the crucible (the Mo lead wrapped around the pure rare-earth elements was kept above the electrolyte level). The electrical leads were placed in insulating alumina tubes, inserted into the stainless-steel chamber, and sealed at the top with epoxy. A K-type thermocouple was also inserted through the setup and into the salt to record the temperature. The stainless-steel chamber was then sealed, removed from the glovebox, and placed into a furnace. The salt was further dried at 100 °C for 12 h and then at 270 °C for 12 h under a vacuum of <10 mtorr to remove any remaining moisture/O_2. The cell was then heated to 400 °C to melt the electrolyte, establish electrical contact, and begin the emf measurements. Electrochemical measurements were performed with a potentiostat (AUTOLAB PGSTAT302N, Metrohm AG) and the temperature was monitored throughout using the K-type thermocouple attached to a data acquisition system (NI 9211, National Instruments). The temperature was cycled between 400 and 700 °C in 25 °C increments with either 1 or 2 h holds at each increment.

Results and Discussion

The emf values of pure Nd were measured versus pure Nd in LiCl-KCl-$NdCl_3$ over $T = 400$–700 °C and the emf values of pure Gd were measured versus pure Gd in LiCl-KCl-$GdCl_3$ over $T = 400$–800 °C as depicted in Fig. 1.

Expected emf values for pure Nd versus pure Nd and pure Gd versus pure Gd are 0 mV based on the Nernst equation. Nd and Gd maintain stability for ~10–15 h; however, the potential then begins to fluctuate for both, deviating by up to 100 mV. This behavior indicates significant instability in the pure REE electrodes after sufficient time in the electrolyte at high temperatures. Following the emf measurements,

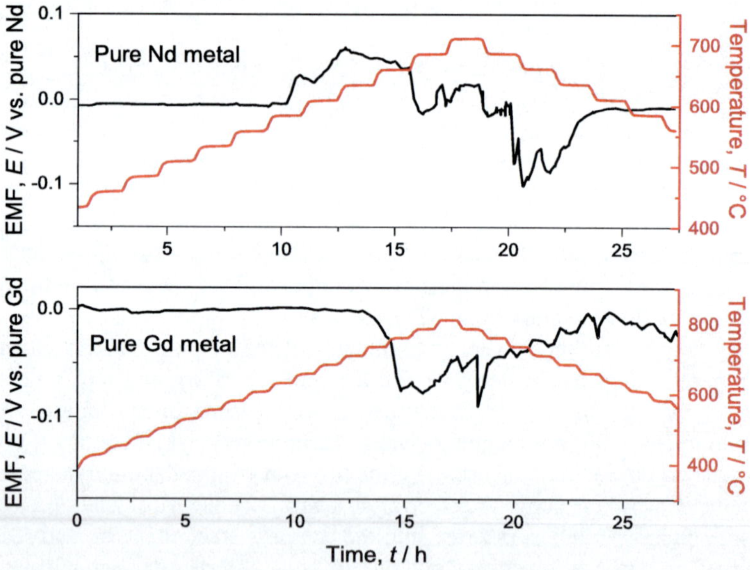

Fig. 1 Electromotive force values of **(a)** pure Nd metal reference electrode versus pure Nd metal working electrode and **(b)** pure Gd metal reference electrode versus pure Gd metal working electrode as a function of time and temperature

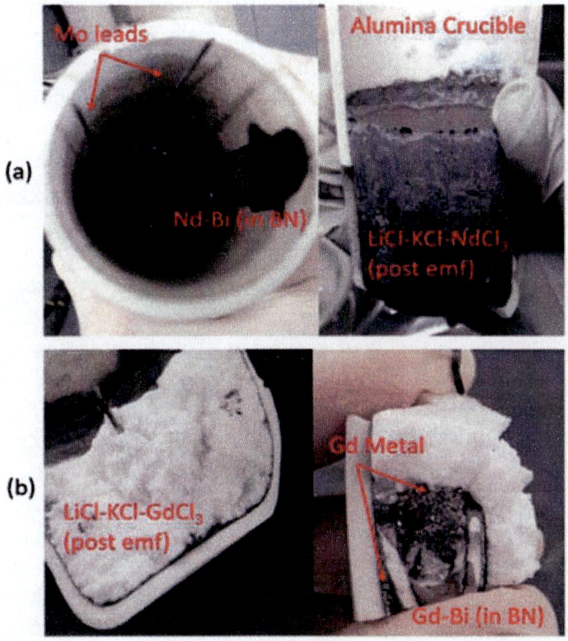

Fig. 2 Photographs of the chloride electrolyte following the emf measurements in Fig. 1 for **(a)** the electrochemical cell using pure Nd as a reference electrode (LiCl-KCl-$NdCl_3$) and **(b)** the electrochemical cell using pure Gd as a reference electrode (LiCl-KCl-$GdCl_3$)

the electrochemical cells were cooled and deconstructed. Images of the salt from the cell with pure Nd (Fig. 2a) and pure Gd (Fig. 2b) are depicted below.

Substantial damage occurred to the LiCl-KCl-$NdCl_3$ electrolyte as evidenced by the discoloration of the electrolyte, which was white in color prior to the measurement. This is likely due to the disproportionation reaction producing large amounts of pure Nd, which accumulated in the electrolyte. This effect is also likely responsible for the failure of the cell after ~10 h of operation. The LiCl-KCl-$GdCl_3$ electrolyte was largely unchanged, retaining its original color; however, some Gd metal was observed around the sides of the cell as well around the surfaces of the Gd-Bi two-phase electrodes with an observable loss of Gd from the pure Gd electrodes. Despite the lack of a disproportionation reaction in the Gd system, the pure Gd electrodes still exhibited comparable instability after ~14 h and metal was observed in the electrolyte.

As the pure REE showed poor stability, efforts were made to calibrate the two-phase REE-Bi electrodes versus the pure REEs so that the two-phase electrodes could be employed as reliable reference electrodes in future measurements. The previous emf measurements indicated that prolonged exposure of the pure REE to the molten chloride electrolyte is detrimental to the cell lifetime, so several methods of measurement were examined and compared. In the first case, the pure Gd electrode was in the electrolyte throughout the entire emf measurement versus the Gd-Bi electrode (continuous submersion). The emf data was then compared with two different techniques utilized by Baldivieso et al. [3]: (1) a method which minimized the exposure of pure Gd to the electrolyte via intermittent dipping and (2) a transient technique in which Gd was deposited onto an inert W wire, which was also successfully used to calibrate an Nd-Bi two-phase electrode by Lichtenstein et al. [2]. In the second case, the pure Gd electrode was lowered into the electrolyte for 180 s of emf measurement versus the Gd-Bi electrode, then raised above the salt until the next temperature increment (intermittent dipping). The purpose of this was to significantly reduce the amount of time during which pure Gd could dissolve into the electrolyte. In the final case (deposition), the pure Gd electrode was omitted entirely and instead Gd^{3+} from the salt was reduced to metal on an inert W wire prior to each emf measurement versus the Gd-Bi alloy. This process effectively created a pure Gd electrode which was used to measure emf against the Gd-Bi electrode. The emf results from each technique are depicted in Fig. 3 as a function of temperature with red points indicating the initial heating cycle and blue indicating the subsequent cooling cycle.

The emf results from the constantly submerged pure Gd electrode were stable during the majority of the heating cycle, but began to diverge quickly at higher temperatures and exhibited poor reproducibility during the cooling cycle (>80 mV). These results align well with the observed behavior of the pure Gd versus pure Gd emf, with promising initial stability lost with increasing time and temperature. Emf values from the intermittently dipped pure Gd electrode versus the Gd-Bi two-phase electrode showed significantly improved stability compared to the "constant" pure Gd electrode; however, the emf does not indicate the expected linear behavior and the ~10 mV difference between heating and cooling is non-ideal. The pure Gd deposited on the inert W wire at each temperature versus the two-phase Gd-Bi electrode showed perfect reproducibility between heating and cooling (<2 mV) as well as linear behavior (adj-$R^2 = 0.997$), indicating high stability.

By applying the linear fit ($E_{cell} = 0.817 - 1.565 \times 10^{-4} * T$) of the measured emf from the deposition technique determined by Baldivieso et al. [3], the two-phase Gd-Bi ($x_{Gd} = 0.16$) electrode can be used in other electrochemical measurements

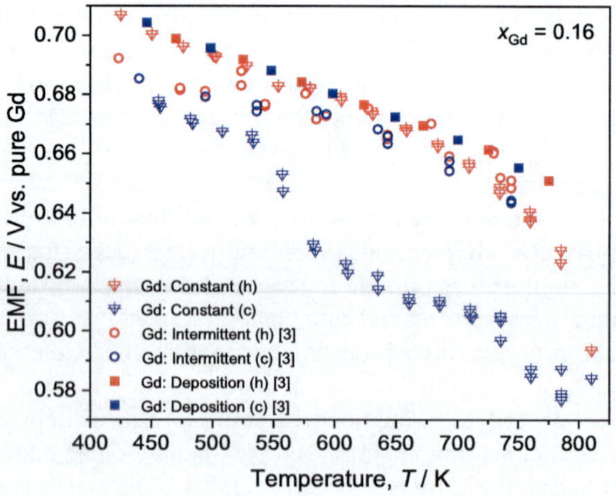

Fig. 3 Comparison of a pure Gd metal reference electrode left in electrolyte, a pure Gd metal reference electrode dipped into electrolyte during emf measurement only [3], and pure Gd deposited on a W wire for emf measurement as a function of temperature relative to $x_{Gd(in\ Bi)} = 0.16$ electrode [3]. This figure was adapted from ref [3]. Copyright Elsevier (2021)

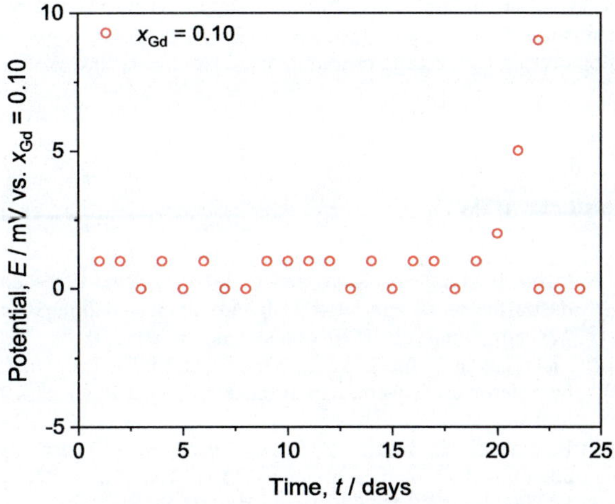

Fig. 4 Emf values of an $x_{Gd} = 0.10$ versus $x_{Gd} = 0.10$ electrode sampled daily over a 24-day period in LiCl-KCl-$GdCl_3$

and results can easily be converted to pure Gd as a reference point. This provides the theoretical advantages of using pure Gd as a reference electrode as well as the practical advantages of the stable Gd-Bi reference electrode. Unlike pure Gd, which was shown to last for only ~14 h, two-phase Gd-Bi electrodes maintained the electrode stability over a 24-day period with <10 mV deviation (Fig. 4). During the 24-day period, the temperature of the cell was held at 500, 600, and 700 °C for ~ 8 days at each temperature. Except for a brief instability up to 9 mV around Day 22, the measured deviation between the Gd-Bi electrodes was less than 2 mV.

Conclusions

This study examined the stability of pure REEs and two-phase REE-Bi electrodes in molten chloride electrolytes over multi-day timescales and $T = 400$–800 °C. Based on the emf data and post-mortem cell analysis, it is not recommended that pure REEs be utilized as reference electrodes due to their rapid degradation after ~10 h of use as a result of either disproportionation reactions (Nd) or active metal dissolution (Gd). A preferable alternative is to use a stable two-phase electrode, which can be effectively calibrated to a pure REE via the use of the transient technique outlined by Baldivieso et al. and Lichtenstein et al. involving the temporary deposition of pure REE onto an inert electrode for emf measurements [2, 3]. Subsequent emf measurements between identical Gd-Bi two-phase electrodes proved that their stability far exceeds that of the pure REEs in molten chloride salts, lasting for more than 24 days compared to less than 15 h.

Acknowledgements This work was supported by the US National Science Foundation (grant number: CMMI-1662817) and the US Department of Energy, Office of Nuclear Energy's Nuclear Engineering University Program (Award No. DE-NE0008757).

References

1. Ipser H, Mikula A, Katayama I (2010) Overview: the emf method as a source of experimental thermodynamic data. Calphad Comput Coupling Phase Diagrams Thermochem 34:271–278. https://doi.org/10.1016/j.calphad.2010.05.001
2. Lichtenstein T, Im S, Yu C, Kim H (2020) Thermodynamic properties of rare-earth alloys by electrochemical emf measurements. J Mater Res 36:70–78. https://doi.org/10.1557/jmr.2020.218
3. Baldivieso SC, Smith ND, Im S, Kim H (2021) Thermodynamic properties of Gd-Bi alloys determined by emf measurements in LiCl-KCl-$GdCl_3$ electrolyte. J Alloys Compd 886:161229. https://doi.org/10.1016/j.jallcom.2021.161229
4. Lebedev VA, Kober VI, Yamshchikov LF (1989) Thermochemistry of rare-earth metals and actinide elements alloys. Metall Chelaybinsk 116–117
5. Kulagina NG, Bayanov AP (1974) Thermodynamic properties of gadolinium tristannide and its solutions in liquid tin. Russ J Phys Chem 48(1):143
6. Sheng J, Yamana H, Moriyama H, January R, May A (2001) Thermodynamics of dilute solution of Gd in Bi. PhysChemComm 4:1–4. https://doi.org/10.1039/b101001h
7. Yamana H, Sheng J, Kawamoto K, Moriyama H (2001) Thermodynamic systematics of the formation of liquid alloys of f-elements with bismuth. J Nucl Mater 294:53–58
8. Yamana H, Sheng J, Souda N, Moriyama H (2001) Thermodynamic properties of lanthanide metals in liquid bismuth. J Nucl Mater 294:232–240
9. Lovering DG, Gale RJ (1987) Molten salt techniques, vol 3. Plenum Press, pp 107–108
10. Liu K, Soc JE (2017) Evaluation of the electroextractions of Ce and Nd from LiCl-KCl molten salt using liquid Ga electrode evaluation of the electroextractions of Ce and Nd from LiCl-KCl. J Electrochem Soc 164:169–178. https://doi.org/10.1149/2.0511704jes
11. Masset P, Konings RJM, Malmbeck R (2005) Thermochemical properties of lanthanides (Ln = La, Nd) and actinides (An = U, Np, Pu, Am) in the molten LiCl–KCl eutectic. J Nucl Mater 344:173–179. https://doi.org/10.1016/j.jnucmat.2005.04.038
12. Novoselova A, Smolenski V (2013) Electrochemical behavior of neodymium compounds in molten chlorides. Electrochim Acta 87:657–662. https://doi.org/10.1016/j.electacta.2012.09.064
13. Hayashi H, Akabori M, Ogawa T, Minato K (2004) Spectrophotometric Study of Nd^{2+} Ions in LiCl-KCl Eutectic Melt. Z Naturforsch 59:705–710

Fundamental Study of a Novel Electrolytic Process Using a Cu Cathode in MgF_2-LiF-KCl Molten Salt for Producing Mg Metal from MgO

Hyeong-Jun Jeoung, Tae-Hyuk Lee, Kyung-Woo Yi, Jin-Young Lee, Young Min Kim, Toru H. Okabe, and Jungshin Kang

Abstract A fundamental study on the molten salt electrolysis of magnesium oxide (MgO) using a copper (Cu) cathode was conducted to develop a novel process for producing magnesium (Mg) metal. The influence of the electrolysis temperature, the concentration of Mg in the Mg-Cu alloy, and cathodic current density on the current efficiency was investigated. In addition, the reaction between the alumina (Al_2O_3) crucible and the Mg-Cu alloy contained in the crucible was analyzed. In the experiments, 1.42–1.63 A was applied to the electrolysis cell using Cu cathode and graphite anode at 1043 K and 1173 K for 9.5–15.0 h in an argon (Ar) gas atmosphere. When the concentration of Mg ranged from 9.25–14.2 mass%, the Mg alloys of $MgCu_2$ and Cu (Mg) mixtures were obtained, with 72.5–91.7% of the current efficiency under certain conditions. Thus, the feasibility of the novel and effective electrolytic method for the production of Mg was demonstrated.

Keywords Magnesium metal · Magnesium oxide · Molten salt electrolysis · Copper cathode

H-J. Jeoung · T-H. Lee · J-Y. Lee · Y. M. Kim · J. Kang (✉)
Korea Institute of Geoscience and Mineral Resources, 124 Gwahak-ro Yuseong-gu, Daejeon 34132, Korea
e-mail: jskang@ust.ac.kr; jskang@kigam.re.kr

H-J. Jeoung · K-W. Yi
Department of Materials Science and Engineering, Seoul National University, 1 Gwanak-ro, Gwanak-gu, Seoul 08826, Korea

J-Y. Lee · J. Kang
University of Science and Technology, 217 Gajeong-ro Yuseong-gu, Daejeon 34113, Korea

Y. M. Kim
Korea Institute of Materials Science, 797 Changwondae-ro, Seongsan-gu, Changwon 51508, Gyeongnam, Korea

T. H. Okabe
Institute of Industrial Science, The University of Tokyo, 4-6-1 Komaba, Meguro-ku, Tokyo 153-8505, Japan

T. Ouchi et al. (eds), *Rare Metal Technology 2022*, The Minerals, Metals & Materials Series, https://doi.org/10.1007/978-3-030-92662-5_32

Introduction

At present, primary magnesium (Mg) metal is commercially produced using the Pidgeon and electrolytic processes. However, several disadvantages are addressed in the current Mg metal production processes. For example, the Pidgeon process requires high energy consumption. It produces Mg metal through the silicothermic reduction of calcined dolomite using ferrosilicon (Fe·Si) under vacuum at 1373–1473 K [1]. In addition, sulfur oxide (SO_x) gas is generated because coal is used as a heat source [2]. In electrolytic processes, the preparation of anhydrous magnesium chloride ($MgCl_2$) feedstock is energy-intensive and produces toxic chlorine (Cl_2) gas during electrolysis at 928–993 K [2].

Several studies have been conducted to develop an environmentally friendly process for producing Mg metal using MgO as a feedstock. The solid oxide membrane (SOM) process is a promising process involving the direct conversion of MgO to Mg and the generation of oxygen (O_2) gas using an yttria-stabilized-zirconia (YSZ) membrane [3]. However, the decrease in current efficiency from 90% to 40–50% and degradation of the YSZ membrane occurred without decreasing the partial pressure of Mg in the flux [4].

Molten salt electrolysis of MgO using high-density metals such as Cu, silver (Ag), or tin (Sn) as a cathode has been developed against this background [5–7]. A schematic of the process is shown in Fig. 1. We investigated the influence of the electrolysis temperature, the concentration of Mg in the Mg alloy, and cathodic current density on the current efficiency of the electrolysis of MgO using a Cu cathode in MgF_2-LiF-KCl molten salt. In this process, a high current efficiency is maintained during electrolysis because the reaction between Mg and reactive gases such as O_2 evolved at the anode is hindered by the production of the high-density Mg alloy at the bottom of the electrolysis cell. In addition, the dissolution of Mg metal into the molten salt is limited by the production of Mg alloy. After electrolysis at 1053–1083 K, Mg metal with a purity of approximately 99.999% was produced through the vacuum distillation of Mg alloys at 1200–1300 K.

Experimental

A schematic of the experimental apparatus used in this study is shown in Fig. 2. The Cu cathode was prepared by placing the Cu pellets in an alumina (Al_2O_3) crucible, which was assembled with a graphite rod shielded by boron nitride (BN) [8]. MgO was charged into the graphite crucible as a feedstock after placing the Al_2O_3 crucible at its bottom. The crucible was physically connected to a nickel (Ni) wire for the current lead. Afterward, the MgF_2-LiF-KCl mixture, which was dried under vacuum at 453 K for more than 72 h and pre-electrolyzed to remove the impurities, was charged into the graphite crucible. The crucible was positioned at the bottom of a vertical gas-tight stainless steel reactor. The interior gas of the steel reactor was

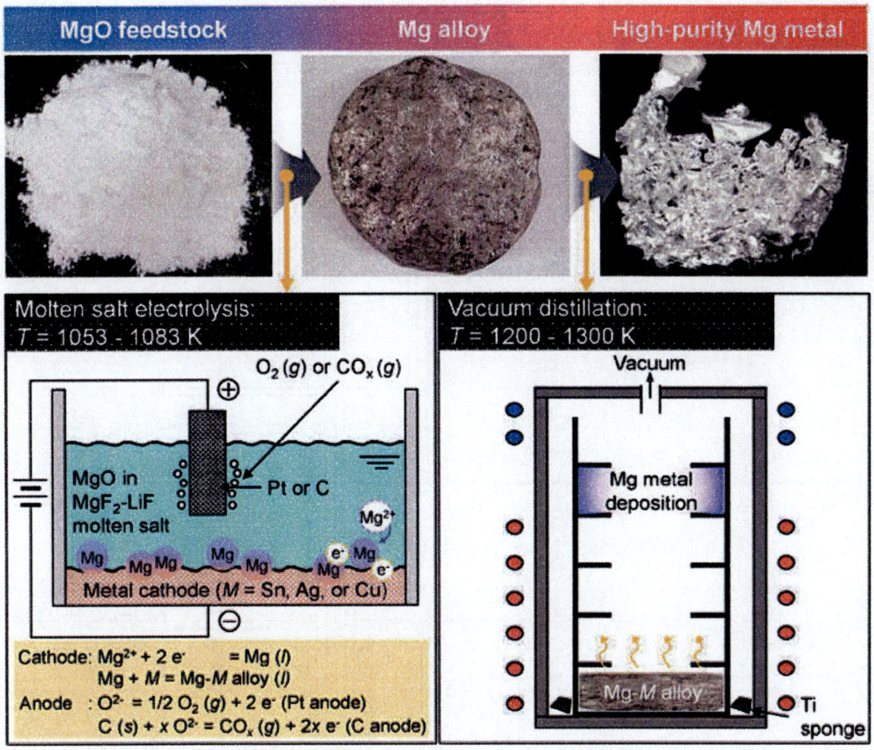

Fig. 1 Schematic of the process for producing high-purity Mg metal from MgO

changed to Ar gas by evacuating followed by filling with the Ar gas. A continuous flow of Ar gas maintained the internal pressure of the reactor at 1 atm during the experiments. Subsequently, the reactor was placed inside a vertical electric furnace. The temperature was increased to 1043 K or 1173 K for the electrolysis. Electrolysis was conducted using the chronopotentiometry method by applying 1.42–1.63 A to the electrolysis cell for 9.5–15.0 h.

After electrolysis, the Mg alloy was recovered from the Al_2O_3 crucible, and the salts deposited on the surface of the Mg alloy were removed. The concentrations of elements in the Mg alloys were analyzed using inductively coupled plasma optical emission spectroscopy (ICP-OES). The crystalline phases of the Mg alloys were analyzed using X-ray diffraction (XRD).

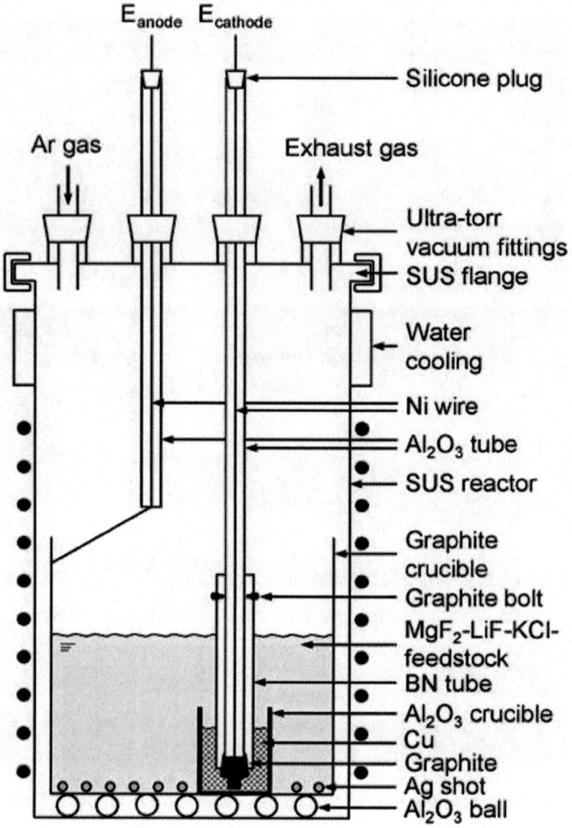

Fig. 2 Schematic of the experimental apparatus for the molten salt electrolysis of MgO

Results and Discussion

Table 1 lists the experimental conditions and results of the molten salt electrolysis of MgO. As shown in Table 1, the average cell potential was 1.95–2.54 V during the experiments. Assuming the generation of carbon dioxide (CO_2) gas, the theoretical decomposition voltages of MgO under graphite are 1.50 V and 1.42 V at 1043 K and 1173 K, respectively [9]. On the other hand, the theoretical decomposition voltage of KCl, which possesses the lowest theoretical decomposition voltage among the components of the electrolyte, is 3.48 V at 1043 K [9]. This indicates that electrolysis occurred by the decomposition of MgO.

Table 1 Experimental conditions and results for the electrolysis of MgO using a Cu cathode in MgF_2-LiF-KCl molten salt

Exp no.[a]	Temp., T/K	Time, t/h	Applied avg. current, i/A	Average potential, E/V	Cathodic current density, $J_{cathodic}$ /A · cm^{-2}	Current efficiency, η (%)[b]	Conc. of element i, C_i (mass%)			
							Mg	Li	Al	Cu
201216	1043	9.50	1.42	2.27	0.217	91.7	9.25	0.02	0.04	Bal
201203	1043	15.0	1.45	2.28	0.222	88.1	13.6	0.06	0.05	Bal
210117	1173	15.0	1.45	1.95	0.222	72.5	11.4	0.05	0.04	Bal
210203	1043	13.5	1.63	2.54	0.326	90.6	14.2	0.02	0.04	Bal

[a] Experimental conditions;
(1) Weight of MgF_2 = 256.0 g, weight of LiF = 219.0 g, weight of KCl = 25.0 g, and weight of MgO = 30.0 g.
(2) Weight of Cu used for the cathode = 55 g.
(3) The purities of MgF_2, LiF, and KCl were 99.9%, 99.9%, and 99%, respectively (Kojundo Chemicals Lab. Co., Ltd.).
(4) The purity of MgO was 99% (Sigma Aldrich).
(5) The purity of the Cu metal pellets (size: 3 mm) was 99.995% (RND KOREA Corp.)
[b] η (%) = (actual weight of Mg obtained × 100)/theoretical weight of Mg obtained

Influence of the Electrolysis Temperature on the Current Efficiency

Table 1 shows that when the electrolysis temperature increased from 1043 to 1173 K with a cathodic current density of 0.2 $A \cdot cm^{-2}$, the current efficiency decreased from 88.1 to 72.5%. The current efficiency can be affected by the electronic current in the flux generated by the dissolution of the Mg metal [4] and the evaporation of Mg metal. However, in the molten salt electrolytic process using a liquid metal cathode, the dissolution of the Mg metal is insignificant owing to the formation of Mg alloy, and high current efficiency was maintained even after long-term electrolysis [7]. Therefore, it was considered that the decreased current efficiency resulted from Mg metal evaporation when the electrolysis temperature was increased. Figure 3 shows the phase diagram of the Mg-Cu binary system with the isobaric line of the vapor pressure of Mg as a function of the temperature and Mg concentration [10]. As shown in Fig. 3, the vapor pressure of Mg in the Mg alloy increases with the increase in the electrolysis temperature at an identical concentration of Mg. The vapor pressures of Mg in Mg alloys obtained in the experiments conducted at 1043 K and 1173 K are 9.21×10^{-4} and 1.18×10^{-2} atm, respectively [10]. It is worth noting that Mg evaporates actively when the vapor pressure of Mg exceeds 0.01 atm [11]. Consequently, current efficiency decreased with increasing electrolysis temperature from 1043 to 1173 K owing to the decrease in the actual weight of Mg obtained by evaporation.

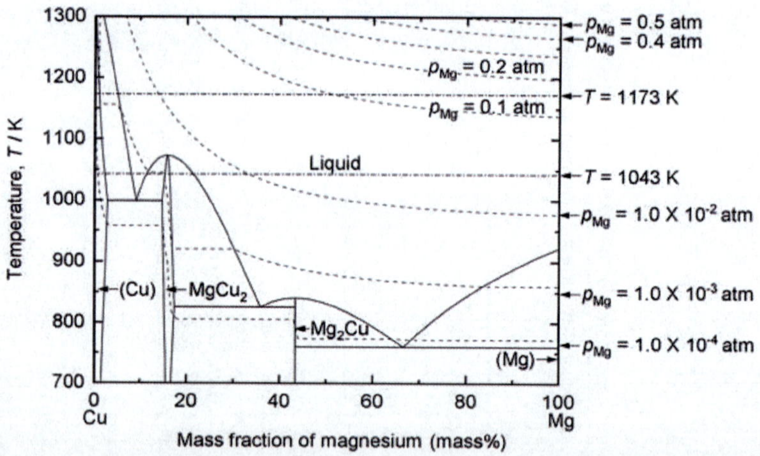

Fig. 3 Phase diagram of the Mg-Cu binary system with the isobaric line of vapor pressure of Mg as a function of the temperature and Mg concentration

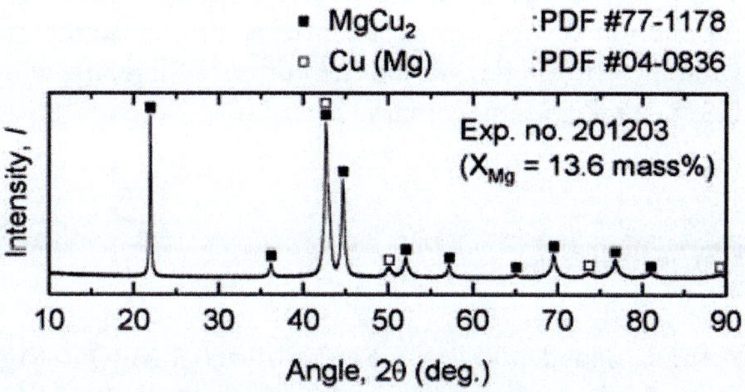

Fig. 4 XRD result of the Mg alloy (Exp. no. 201203) obtained after electrolysis

Influence of the Concentration of Mg on the Current Efficiency

As presented in Table 1, when the MgO was electrolyzed at 1043 K using a cathodic current density of 0.2 $A \cdot cm^{-2}$, the current efficiency slightly decreased from 91.7 to 88.1% as the Mg concentration in the Mg alloy increased from 9.25 mass% to 13.6 mass%. The current efficiencies were relatively high, considering that the current efficiency decreased from 88.1 to 72.5% as the temperature was increased. Figure 3 shows that the vapor pressure of Mg increases as the concentration of Mg in the Mg alloy increases, which leads to a decrease in the current efficiency. However, Fig. 3 also shows that the vapor pressure of Mg is small compared to 0.01 atm, the vapor pressure at which Mg begins to evaporate actively, when its concentration is in the range of 9.25–13.6 mass%. Thus, high current efficiencies were maintained, although the vapor pressure of Mg increased with the concentration of Mg in the alloy.

Figure 4 shows the XRD result of the Mg alloy obtained after electrolysis at 1043 K for 15.0 h. As shown in Fig. 4, the Mg alloy consisted of a mixture of $MgCu_2$ and Cu (Mg) when the alloy had a Mg concentration of 13.6 mass%. This result is consistent with the phase diagram of the Mg-Cu binary system shown in Fig. 3.

Influence of the Cathodic Current Density on the Current Efficiency

Table 1 also shows that the current efficiency was 90.6% when the cathodic current density was 0.3 $A \cdot cm^{-2}$ at 1043 K. Thus, the Mg alloy is obtained with a high current efficiency after electrolysis despite the increase in the current density from 0.2 to 0.3 $A \cdot cm^{-2}$. Although the current density increased from 0.2 to 0.3 $A \cdot cm^{-2}$, the cell potential was in the range of 2.28–2.54 V. The values of the cell potentials indicate that MgO is decomposed without the decomposition of the supporting electrolyte.

The increase in the current density and the resulting current efficiency affected the production of Mg metal. Therefore, further studies regarding the influence of the cathodic current density on the current efficiency, when the current density exceeds $0.3\ A \cdot cm^{-2}$, are necessary.

Conclusions

A fundamental study on the electrolysis of MgO in MgF_2-LiF-KCl molten salt using Cu cathode and graphite anode was conducted. Mg-Cu alloys were obtained by the decomposition of MgO with a cell potential of 1.95–2.54 V when 1.42–1.63 A was applied at 1043–1173 K. The current efficiency was 88.1–91.7% when 9.25–14.2 mass% Mg-Cu alloy was produced by electrolysis at 1043 K with a current density of $0.2\text{–}0.3\ A \cdot cm^{-2}$. However, the current efficiency decreased to 72.5% upon increasing the temperature to 1173 K owing to the evaporation of the electrodeposited Mg. Hence, the novel electrolytic method using a Cu cathode is feasible for the reduction of MgO.

References

1. Pidgeon LM, Alexander WA (1944) Thermal production of magnesium-pilot plant studies on the retort ferrosilicon process. Trans Am Inst Min Mater Eng 159:315–352
2. Friedrich HE, Mordike BL (eds) (2006) Magnesium technology: metallurgy, design data, applications. Springer, Berlin
3. Krishnan A, Lu XG, Pal UB (2005) Solid oxide membrane process for magnesium production directly from magnesium oxide. Metall Mater Trans B 36:463–473
4. Guan X, Pal UB, Powell AC (2014) Energy-efficient and environmentally friendly solid oxide membrane electrolysis process for magnesium oxide reduction: experiment and modeling. Metall Mater Trans E 1:132–144
5. Lee TH, Okabe TH, Lee JY, Kim YM, Kang J (2020) Molten salt electrolysis of magnesium oxide using a liquid-metal cathode for the production of magnesium metal. Metall Mater Trans B 51:2993–3006
6. Lee TH, Okabe TH, Lee JY, Kim YM, Kang J (2021) Development of a novel electrolytic process for producing high-purity magnesium metal from magnesium oxide using a liquid tin cathode. J Magnes Alloy 9:1644–1655
7. Lee DH, Jeoung HJ, Lee TH, Yi KW, Lee JY, Kim YM, Okabe TH, Kang J (2021) Scale-up study of molten salt electrolysis using Cu or Ag cathode and vacuum distillation for the production of high-purity Mg metal from MgO. J Sustain Metall 7:883–897
8. Kang J, Kim HS, Lee TH, Jeoung HJ, Kim YM (2021) Structure for cathode lead used in metal refining method using liquid metal cathode. Korea Patent 10-2284343
9. Barin I (1995) Thermochemical data of pure substances. VCH Verlagsgesellschaft mbH, Weinheim
10. Bale CW, Belisle E, Chartrand P, Decterov SA, Eriksson G, Gheribi AE, Hack K, Jung IH, Kang YB, Melancon J, Pelton AD, Petersen S, Robelin C, Sangster J, Spencer P, Van Ende MA (2016) FactSage thermochemical software and databases, 2010–2016. Calphad 54:35–53
11. Kang J, Okabe TH (2014) Thermodynamic consideration of the removal of iron from titanium ore by selective chlorination. Metall Mater Trans B 45:1260–1271

Electrochemical Reduction of Iron Oxides in Aqueous NaOH Electrolyte Including Iron Residue from Nickel and Zinc Electrowinning Processes

Geir Martin Haarberg, Bo Qin, and Babak Khalaghi

Abstract Experiments were carried out in NaOH-H_2O (50–50 wt%) electrolyte with a suspension of Fe_2O_3 particles at ~100 °C. A rotating disk electrode of graphite or silver was used as the cathode. The current efficiency for iron deposition was consistently higher than 90%. Recent experiments were carried out by using iron-containing residue from industrial electrolysis processes for producing nickel and zinc. The challenge is to purify the residue before electrolysis.

Keywords Iron · Electrolysis · Hematite · CO_2 emissions

Introduction

Iron production and steel production are responsible for a large amount, about 9%, of the world's anthropogenic CO_2 emissions [1]. Different scenarios and low emission pathways have been researched by steelmaking industries together with universities and research institutes to tackle this problem [2]. One of the most promising novel processes to replace the current steelmaking process is the low temperature electrode-composition of iron oxide. This process is currently being developed under H2020-SIDERWIN project in the SPIRE European program. In this process, a suspension of iron oxide particles (20–50 wt%) in a highly alkaline electrolyte of 50 wt% NaOH in water is electrolysed. Hence, the iron oxide is electrochemically reduced to metallic iron at the cathode, while oxygen evolution occurs at the anode. Therefore, this process is capable of producing pure iron with a very low carbon footprint [3–8].

One of the advantages of this process is the fact that in addition to iron oxide (hematite), it is possible to feed this process with other iron-containing raw materials. One alternative raw material which is being studied to be used in this process is bauxite residue (also known as red mud); the waste material from the Bayer process for alumina production. Bauxite residue usually contains 30–45 wt% iron in the form of oxides and hydroxides. Besides, on average, 1–1.5 tons of bauxite residue is

G. M. Haarberg (✉) · B. Qin · B. Khalaghi
Department of Materials Science and Engineering, Norwegian University of Science and Technology (NTNU), NO-7491 Trondheim, Norway
e-mail: geir.martin.haarberg@ntnu.no

T. Ouchi et al. (eds), *Rare Metal Technology 2022*, The Minerals, Metals & Materials Series, https://doi.org/10.1007/978-3-030-92662-5_33

produced per ton of alumina [5]. Therefore, it is a considerable resource of iron. Also, during the purification processes for primary production of nickel and zinc prior to electrolysis in aqueous electrolytes, large amounts of containing waste materials rich in iron are removed during the treatment of the raw materials.

Experimental

Pure iron oxide, Fe_2O_3, and iron-rich waste materials from nickel and zinc purification processes were used as raw materials for electrodeoxidation of Fe_2O_3. The waste materials were treated by washing and calcination before electrolysis.

The electrolysis experiments were carried out in a polytetrafluoroethylene (PTFE) container (Ø 100 mm, 160 mm height). Different types of working electrodes were used in this study. A rotating disc electrode (RDE) was used for galvanostatic electrolysis experiments in the electrolyte of a suspension of particles of Fe_2O_3 or waste material in NaOH 50 wt% aqueous solution. Two grades of graphite G347 (grade 1) and G348 (grade 2) supplied by Tokai Carbon and a pure silver rod, 99.9% Alfa Aesar, were used as cathode materials. For cyclic voltammetry studies, a cavity electrode was used. This electrode was fabricated similar to the electrodes known as "cavity microelectrodes" in electrochemistry literature. Such electrodes are used for electrochemical studies of solid materials insoluble in the electrolyte [6]. The cavity electrode was made of a PTFE rod, Ø 10 and 40 mm long, where a hole of Ø 0.7 mm was drilled into it. A steel rod Ø 4 mm was inserted from the top into the PTFE rod. The powder was crushed in a mortar with a pestle and the cavity electrode was pushed on the powder and the cavity was filled with powder. The powder might be consumed after each scan. Between each sweep, the cavity electrode was removed from the electrochemical cell, cleaned, and reloaded with new powder. The experimental cell and details of the cavity electrode are shown in Fig. 1. The anode was of pure nickel in the form of a mesh surrounding the cathode.

Results and Discussion

Treatment of Iron Residue from Nickel Production

The waste product from Ni production was provided by Glencore Nikkelverk of Norway. Chlorine leaching is used to purify the nickel sulfate concentrate made by smelting. The composition of the main metallic element was analyzed by XRF and presented in Table 1. The material is fine with D50 < 10 μm. In the waste product, the species include iron (III) hydroxide, nickel and cobalt chlorides, and arsenic compound. Among them, the weight percentage of iron hydroxide is 60–100%.

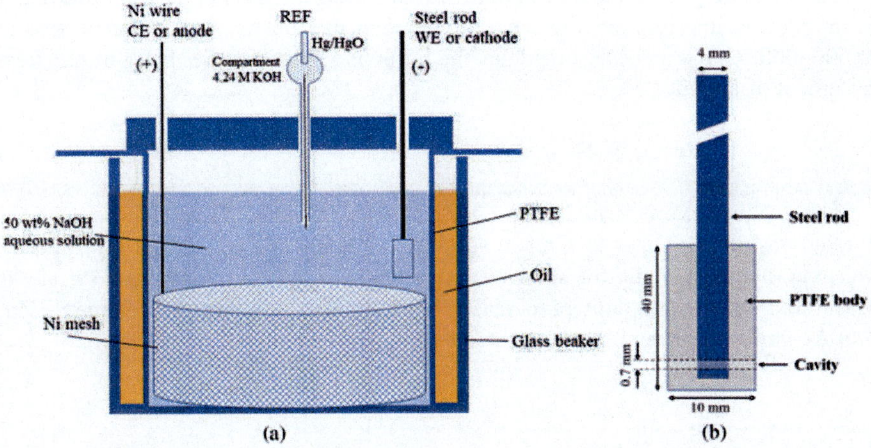

Fig. 1 Schematic diagrams of **a** the experimental setup for electrochemistry and **b** details of the cavity electrode

Table 1 Analysis of iron-rich residue from Glencore Nikkelverk

	Fe	Ni	Co	Cu	As	H_2O
Min	43.4	0.16	0.03	0	0.2	37
Average	52.2	1.4	0.05	0	0.3	41
Max	54.2	8.2	0.07	0	0.7	44

Treatment of Iron Residue from Zn Production

The waste material from purifying the raw material for zinc electrowinning mostly contains the iron-rich compound of jarosite. The material was obtained from the Boliden Zinc plant at Odda, Norway. The results of the analysis of the iron residue by XRF are given in Table 2.

Thermal analysis is beneficial for the study of jarosite. Firstly, thermal decomposition of the jarosite waste was carried out using TG-DSC under nitrogen atmosphere (10 ml min^{-1}). 11.774 mg of the sample was heated to 1000 °C with a heating rate of 10 °C min^{-1}. The decomposition products depend on the treatment temperature of the jarosite. After thermal decomposition, the specific temperature was chosen to treat the jarosite waste, and then the product was washed by two-step processing to enrich hematite.

Table 2 The composition of metallic elements in jarosite from Boliden Odda

Element	Al	As	Ca	Cu	Fe	Mn	Pb	S	Sb	Si	Sn	Zn
Content wt%	1.2	0.18	1	0.5	25	0.55	4.9	12.5	0.08	1.6	0.45	8.6

Thermal analysis of jarosite was carried out by thermogravimetry and differential scanning calorimetry. Hematite can be obtained by thermal decomposition of jarosite at 500–600 °C. It may be assumed that the following reaction takes place during heat treatment of jarosite:

$$ZnFe_2O_4 + Fe_2(SO_4)_3 \rightarrow ZnSO_4 + 2Fe_2O_3 + 2SO_3 \qquad (1)$$

The product after thermal decomposition was washed by two-step processing. Firstly washing in water to remove solid matter such as zinc sulfate, followed by washing in aqueous chloride solution (more than 4 mol/l) to remove lead sulfate. It was found that the iron content increased to ~50% after this treatment. Some Si, Zn, and Al were still present but the contents of other elements were very low.

Electrochemical Studies

Pure Fe_2O_3

Experimental results have been obtained previously and presented in publications [3–8]. High purity iron was deposited on rotating disk electrodes of graphite and silver at current densities from 0.1–0.6 A/cm^2 in aqueous NaOH (50 wt%) at 110 °C. The current efficiency for iron deposition was consistently higher than 90% and independent of rotation rate above 100 rpm. It can be estimated that such a process for producing iron will have an energy consumption of ~3 kWh/kg Fe for the electrolysis process.

Figures 2 and 3 show typical results from an electrolysis experiment using a silver rotating cathode. The current efficiency determined from the weight increase of the cathode was found to be 95.0%. Smooth iron deposits were obtained on silver. Nickel was found to be an excellent anode material for oxygen evolution.

Iron Residue from Ni Production

Results from cyclic voltammetry at 10 mV/s are given in Fig. 4. A graphite rotating disc electrode (RDE) was used as the working electrode, and its rotating speed was 500 rpm. A mechanical stirrer at 100 rpm was placed in the electrolyte. Using a suspension of particles, the content of iron residue was defined by the ratio of solid (iron residue) to liquid (NaOH aqueous solution at 110 °C) [8], the value being equal to 1:90 g. The anodic current peak near −0.9 V (versus the reference electrode) is due to the oxidation of iron.

At higher proportions of solid particles, it was found that some nickel co-deposited with iron. Constant current electrolysis was carried out at 0.6 A/cm^2. The cathode

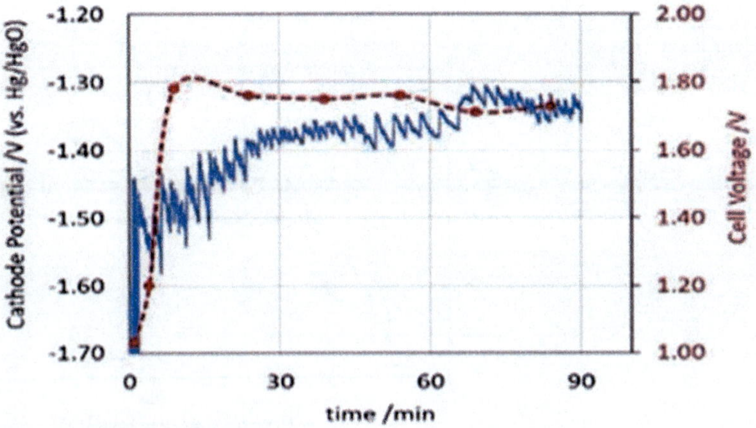

Fig. 2 Cathode potential and cell voltage vs time during electrodeposition of iron on silver cathode at 0.2 A/cm^2 and 100 °C

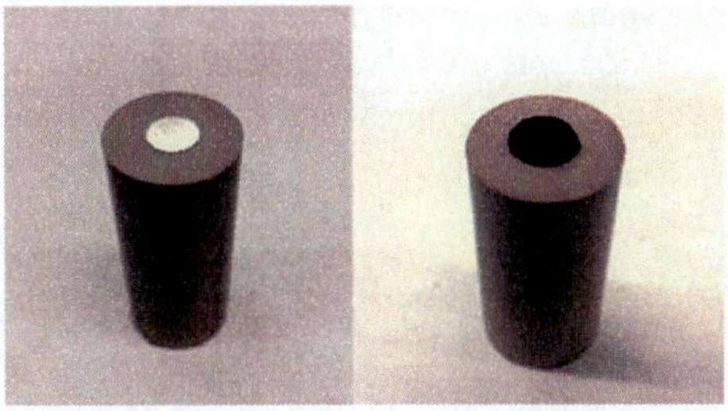

Fig. 3 Pictures of the silver cathode before and after electrolysis

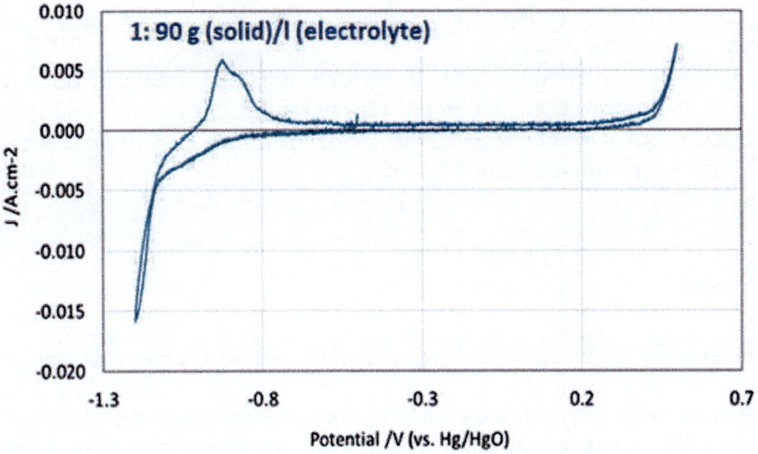

Fig. 4 Cyclic voltammogram of iron residue with graphite RDE at 10 mV/s

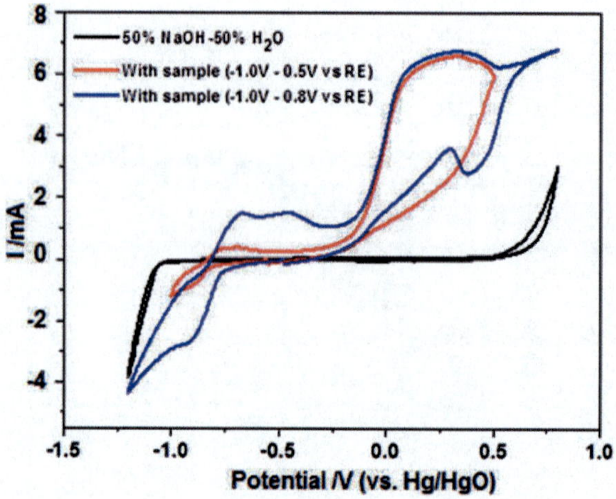

Fig. 5 Cyclic voltammograms of the treated jarosite in a cavity electrode at 100 mV/s

potential was found to be less negative when the content of solid particles increased. The deposit contained a small amount of nickel and no oxide. However, the current efficiency for iron deposition was only 7%. Further treatment to reduce the content of Ni and Co may be beneficial for improving the current efficiency.

Iron Residue from Zn Production

Figure 5 shows voltammetry obtained with a cavity electrode containing jarosite waste. Cathodic peaks are mainly due to iron deposition, but also oxygen reduction may contribute. In the anodic direction, oxidation of deposited iron may take place in two steps by the formation of Fe(II) as an intermediate. Oxidation of produced hydrogen may also take place.

Electrolysis was carried out by applying a constant cell voltage of 1.6 V for 4 h. Some metallic iron was detected in the deposit, but also iron oxide and lead were present. It shows that the efficiency of the iron deposition process depends very much on the purity of the iron oxide raw material.

Conclusions

Iron can be deposited by electrodecomposition of solid Fe_2O_3 particles suspended in a strong alkaline aqueous electrolyte of 50 wt% NaOH at ~100 °C. Pure oxygen is evolved on the anode. The current efficiency is normally above 90%, and the corresponding energy consumption is estimated to be ~3 kWh/kg Fe for the electrolysis process. Iron-rich waste materials from the electrowinning process for nickel and zinc can be purified and used as raw materials for electrodeposition of iron in a similar way. So far the efficiency of these processes is much lower due to the content of impurity elements.

References

1. https://www.worldsteel.org/
2. Abdul Quader M, Ahmed S, Dawal SZ, Nukman Y (2016) Renew Sust Energ Rev 55
3. Yuan B, Kongstein OE, Haarberg GM (2009) Electrowinning of iron in aqueous alkaline solution using a rotating cathode. J Electrochem Soc 156:D64–D69
4. Yuan B, Haarberg GM (2009) Electrodeposition of iron in aqueous alkaline solution: an alternative to carbothermic reduction. ECS Trans 16:31–37
5. Allanore A, Lavelaine H, Valentin G, Birat JP, Lapicque F (2008) Iron metal production by bulk electrolysis of iron ore particles in aqueous media. J Electrochem Soc 155:E125–E129
6. Allanore A, Lavelaine H, Valentin G, Birat JP, Delcroix P (2010) Observation of the reduction of iron oxide particles to metal in alkaline solution by electrolysis. Electrochim Acta 55:4007–4013
7. Tang SH, Haarberg GM (2010) Electrowinning of iron from alkaline solution. ECS Trans 28(6):309–315
8. Haarberg GM, Khalaghi B (2020) Electrodeoxidation of iron oxide in aqueous NaOH electrolyte. ECS Trans 97(7):493–503

Author Index

A
Abdulfattah, Furqan, 279
Adeyemi, Christianah O., 129
Akanji, Fausat T., 129
Alabi, Abdul G. F., 129
Allanore, Antoine, 259
Azimi, Gisele, 147, 289

B
Baba, Alafara A., 129
Baldivieso, Stephanie Castro, 317, 325
Banerjee, Riya, 63
Bhatti, Ijaz Ahmad, 135
Botelho Junior, Amilton Barbosa, 71
Bwala, Daniel Markus, 279

C
Chagnes, Alexandre, 81
Chakladar, Saswati, 63
Chakravarty, Sanchita, 63
Cheema, Humma Akram, 135
Chen, Bowei, 201
Chinwego, Chinenye, 41
Cueva Sola, Ana Belen, 111

D
Dreisinger, David, 3
Du, Chao, 299
Du, Guoshan, 101
Du, Shangchao, 101

E
Echeverry-Vargas, Luver, 17
Espinosa, Denise Crocce Romano, 71

F
Forsberg, Kerstin, 119
Free, Michael L., 51
Fu, Yunfeng, 101

G
Gao, Wei, 179
Ghahreman, Ahmad, 27
Giancola, Emily, 41

H
Haarberg, Geir Martin, 341
Han, Guihong, 93
Han, Haisheng, 211
Hassan, Suleiman Bolaji, 279
Höber, L., 237
Huang, Yanfang, 93

I
Ilyas, Sadia, 135
Im, Sanghyeok, 317, 325
Isakov, Evgeniy, 245

J
Jeon, Jong Hyuk, 111
Jeoung, Hyeong-Jun, 309, 333
Jie, Xiaowu, 179
Jironvil, Jonathan, 41
Jyothi, Rajesh Kumar, 111

K
Kang, Jungshin, 309, 333
Khalaghi, Babak, 341
Kim, Hojong, 317, 325
Kim, Hyunjung, 135

T. Ouchi et al. (eds), *Rare Metal Technology 2022*, The Minerals, Metals & Materials Series, https://doi.org/10.1007/978-3-030-92662-5

Kim, Jonghyun, 169
Kim, Sookyung, 169
Kim, Young Min, 309, 333
Kunanusont, Nattanai, 147
Kurniawan, Kurniawan, 169
Kuznetsova, Nataliya, 245

L
Lee, Dong-Hee, 309
Lee, Jae-chun, 169
Lee, Jin-Young, 111, 309, 333
Lee, Tae-Hyuk, 309, 333
Lesufi, Tsietsi, 189
Lichtenstein, Timothy, 325
Liu, H., 289
Liu, Lei, 101
Liu, Zhaobo, 101
Li, Xiaoyan, 101
Lu, Xin, 299

M
Malatji, Nicholas, 189
Mohapatra, Mamata, 129
Mo, Xiaolan, 201

N
Narita, Hirokazu, 161

O
Ocampo-Carmona, Luz Marina, 17
Okabe, Toru H., 309, 333
Okanigbe, Daniel Ogochukwu, 189
Ostashin, Sergey, 245

P
Peters, Edward Michael, 119
Podder, Prasenjit, 51
Popoola, Abimbola Patricia, 189
Powell, Adam, 41
Pu, Nianwen, 101

Q
Qin, Bo, 341

R
Rafukka, Ibrahim Abdullah, 279
Raji, Mustapha A., 129
Rojas-Reyes, Néstor Ricardo, 17
Ruan, Shufeng, 179

S
Sadri, Farzaneh, 27
Sarswat, Prashant K., 51
Sekgobela, Gift, 189
Shang, He, 201
Shimoyama, Yusuke, 147
Smith, Nathan D., 317, 325
Srivastava, Rajiv R., 135
Steinlechner, S., 237
Stinn, Caspar, 259
Sun, Jianzhi, 201
Sun, Ninglei, 101
Sun, Wei, 211
Su, Shengpeng, 93
Suss, Aleksandr, 245
Svärd, Michael, 119

T
Takeda, Osamu, 299
Tenório, Jorge Alberto Soares, 71
Trinh, Ha Bich, 169

W
Wagner, Hunter, 41
Wang, Chengyan, 179
Wang, Dehua, 101
Wang, Hanyu, 93
Wang, Qi, 299
Wang, Shijie, 225
Wang, Shuhan, 299
Wei, Zhao, 211
Wen, Jiankang, 201

Y
Yi, Kyung-Woo, 309, 333

Z
Zhang, Bao, 299
Zhang, Jiakai, 147
Zhang, Yonglu, 179
Zhang, Zhongyu, 101
Zhu, Hongmin, 299

Subject Index

A
Acidification, 105, 107, 108
Aluminum sulfate, 245, 247, 251, 252, 254, 257
Ammoniacal solution, 169, 170, 175, 176
Anode slimes, 226–228, 235
Antagonistic solvent extraction, 169, 176
Antimony, 179–188
Anti-solvent crystallization, 119–121, 127

B
Ba(OH)2, 289–293, 295
Bauxite residue, 71, 72, 76
Beneficiation, 279, 281, 284–286
Biohydrometallurgy, 201, 204, 206, 207
Bionanominig, 189, 195, 197

C
Calcium sulfate dihydrate (CSD), 27–31, 34, 37
Cassiterite, 279–286
Cementation, 227–234
Cesium, 245, 247, 248, 251, 252, 254, 256, 257
Chemical degradation, 81, 82, 85, 86
Chromium, 102, 104–108
CO2 emissions, 341
Coal ash, 63–70
Cobalt, 169, 170, 174, 176
Collector, 211, 212, 217–221
Complex gold ore, 189–191, 194, 198
Copper, 169–171, 174, 176
Copper cathode, 333
COSMO-vac (vacancy) model, 148
Coulometric titration, 317–323
Crystallization, 29, 31, 32, 34
Crystal size, 119–121, 126, 127
Cyanex 923, 71, 72, 74–76

D
D2EHPA, 72–76
Deep Fox, 4, 5, 8
De-silication, 111, 112, 114–116
Direct synthesis, 299
Distillation, 41–49

E
Electrolysis, 341, 342, 344–346
Emf, 317–323
Energy efficient processing, 189, 194
Extractant, 161–166
Extraction, 3–5, 9, 11–13
Extraction mechanism, 162, 166

F
Flotation reagents, 211, 212, 221
Flotation technology, 211
Fluoride leaching, 248, 251, 252, 254, 257
Foxtrot, 4–9

G
Gd, 325–331
Gold, 179–182, 186, 188
Gold mining, 190, 193, 198
Gold nanoparticles, 195, 196
Gravity Separation, 279–281, 283

T. Ouchi et al. (eds), *Rare Metal Technology 2022*, The Minerals, Metals & Materials Series, https://doi.org/10.1007/978-3-030-92662-5

H
Hematite, 341, 343, 344
Hydrometallurgy, 22, 71, 225–227, 232, 235

I
Indium, 237–239
Industrial Steel, 129–131, 133
Intermetallic compounds, 299, 300, 302–306
Iron, 341–347
Iron precipitate, 238

J
Jarosite, 237–243

L
Lanthanides, 259, 260, 272
Leaching, 63–69, 111–116, 129, 130, 132, 226–229, 232, 234, 235
Lead, 239, 242, 243
LiOH·H2O, 289–291, 295
Liquid bismuth, 317, 319, 320, 323
Liquid metals, 326
Lithium carbonate, 245–248, 250, 251, 255–257

M
Magnesium metal, 310, 312, 333, 334
Magnesium oxide, 310, 312, 333
Metal cathode, 310–313, 315
Metal-organic complex, 211, 213, 221
Mica concentrate, 245, 247–252, 256, 257
Microbial ecology, 201, 205
Mineral, 279–281, 283, 284, 286
Mineral resources, 211, 212, 221
Mining waste, 17, 18
Mixed Rare-Earth Oxide, 5
Modelling, 81, 82
Molten chloride, 318, 320
Molten salt electrolysis, 309–314, 334, 336
Molybdenum, 129–131, 135, 136, 138, 140
Monazite, 17, 18, 20–25
Morphology, 119–123, 125–127

N
N1923, 101–108
Nd, 325–331
NdFeB magnet, 147, 149, 152, 155
Neodymium, 147, 149, 150, 154–156, 317–323
(NH4)3ScF6, 119–123, 125, 127
Nickel, 169, 170, 174, 176, 237, 238, 243
Nickel industry, 237–243
Nigeria, 129, 130

P
Palladium, 161–166
Permanent Magnets, 41
Phosphoric acid, 81–83, 86–88, 90, 91
Physical separation, 259–262, 264, 268–270, 272, 274, 275
Physicochemistry, 81, 82, 91
Potassium sulfate, 245, 247, 251, 254–256
Precipitation, 289, 294
Purification, 294, 295
Pyrometallurgy, 263

R
Rare earth elements, 17, 19, 63, 66, 67, 69, 148, 154–156, 326, 327
Rare earths, 3, 4, 42–45
Rare earth separation, 259, 261
Rare metals, 135, 201, 203–206
Recycling, 41–43, 45
REE, 27–38
Reference electrodes, 325–327, 329, 331
Rhenium, 135–138, 140, 142–144
Rhodium, 161, 162, 164–166
Roasting flue-dust, 136, 137
Rubidium, 245, 247, 248, 251, 252, 254, 256, 257

S
Scandium, 259, 260, 262, 264, 267–269, 274, 275
Sc(III), 71–75
SCR catalyst, 111–116
Selective chlorination, 237–239, 242, 243
Selective precipitation, 227, 232
Selective separation, 93, 94, 99
Silver, 237–239
Slurry electrolysis, 179–182, 184, 186–188
Solution purification, 28
Solvent extraction, 81, 82, 85, 86, 90, 91, 93, 94, 101, 102, 108, 136, 144, 161, 162
Stibnite, 179–181, 184, 186–188
Sulfidation, 259, 261–271, 274

Sulfuric acid, 245, 246, 248, 251, 252, 254, 257
Sulphuric acid solution, 129
Supercritical fluid extraction, 148, 154

T
TBP, 135–139, 141–144
Tellurium, 225–235
Titanium aluminides, 299, 300
Tungsten, 93, 94, 111–116
Two-phase electrodes, 329, 331

U
Uranium, 81–91

V
Vanadium, 93–99, 101, 102, 104–108, 111–116

Z
Zinc, 237–239, 243
Zinc industry, 238–243